MECHANISMS OF LEARNING AND MOTIVATION:
A Memorial Volume to Jerzy Konorski

List of Contributors

M. Sebastian Halliday
University of Sussex
Brighton, England

Eliot Hearst
Indiana University
Bloomington, Indiana

Allan R. Wagner
Yale University
New Haven, Connecticut

Robert A. Rescorla
Yale University
New Haven, Connecticut

John W. Moore
University of Massachusetts
Amherst, Massachusetts

Nicholas J. Mackintosh
University of Sussex
Brighton, England

Michael J. Morgan
University of Durham
Durham, England

Anthony Dickinson
University of Cambridge
Cambridge, England

Michael F. Dearing
University of Sussex
Brighton, England

Robert A. Boakes
University of Sussex
Brighton, England

Kazimierz Zielinski
Nencki Institute
Warsaw, Poland

Jeffrey A. Gray
University of Oxford
Oxford, England

J. N. P. Rawlins
University of Oxford
Oxford, England

J. Feldon
University of Oxford
Oxford, England

Jadwiga Dabrowska
Nencki Institute
Warsaw, Poland

Ronald G. Weisman
Queen's University
Kingston, Canada

Peter W. D. Dodd
Queen's University
Kingston, Canada

Vincent M. LoLordo
Dalhousic University
Halifax, Canada

Sara J. Shettleworth
University of Toronto
Toronto, Canada

William K. Estes,
The Rockefeller University
New York, New York

MECHANISMS OF LEARNING AND MOTIVATION:

A Memorial Volume to Jerzy Konorski

Edited by
Anthony Dickinson
UNIVERSITY OF CAMBRIDGE
Robert A. Boakes
UNIVERSITY OF SUSSEX

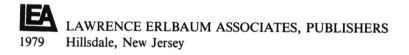 LAWRENCE ERLBAUM ASSOCIATES, PUBLISHERS
1979   Hillsdale, New Jersey

DISTRIBUTED BY THE HALSTED PRESS DIVISION OF
JOHN WILEY & SONS
New York   Toronto   London   Sydney

Lawrence Erlbaum Associates, Inc., Publishers
62 Maria Drive
Hillsdale, New Jersey 07642

Distributed solely by Halsted Press Division
John Wiley & Sons, Inc., New York

Library of Congress Cataloging in Publication Data

Main entry under title:

Mechanisms of learning and motivation.
 Includes bibliographical references and indexes.
 1. Learning in animals.   2. Learning, Psychology of.
3. Motivation (Psychology)   4. Konorski, Jerzy. I. Konorski,
Jerzy. II. Dickinson, Anthony. III. Boakes, R. A.
QL785.M5   156'.3'15   78-11370
ISBN 0-470-26567-1

Printed in the United States of America

Contents

Preface

This volume consists of a series of chapters honoring a Polish psychologist and neurophysiologist who died in 1973. Although his name was familiar to all of the contributors, many had had no personal contact with him and had gained acquaintance with his ideas only through his publications. This unusual venture deserves some explanation.

In the autumn of 1975 the editors had become increasingly interested in Konorski's research and theories, and noted that his name was being cited more frequently in current research journals. We discussed the idea of organizing a third in the series of occasional conferences on animal psychology that have taken place at the University of Sussex, with the aim of providing a forum for discussion of Konorski's work. A list of possible topics, drawn from the range represented in Konorski's *The Integrative Activity of the Brain* (1967), was tentatively sketched. Reactions to this proposal, and to the prospect of delivering a paper that would analyze some aspect of Konorski's work and relate this to other theoretical approaches or to recent research developments, were sought from a number of psychologists. Somewhat to our surprise the reactions were uniformly favorable, frequently very enthusiastic, and of great help in determining the program for the conference. With this evidence of widespread interest, plans went ahead and the event took place between June 30 and July 2, 1977.

The chapters in this volume are revised versions of the papers presented at this conference. The theme ensured a degree of commonality at such events, and a major purpose of many of the revisions was to take account of the exchange of ideas and information that had occurred. Although Konorski's last book (1967) covered a wide spectrum of issues in both animal and human psychology, the overwhelming emphasis in the present volume is on learning and motivation in animals. This reflects both the editors' own interests and the fact that Konorski's major *empirical* contribution was in this field. This choice should not be taken as a judgment that his theoretical ideas do not have wider importance.

In the past decade, major changes have occurred in the study of animal behavior within Western psychology. New developments have taken a variety of forms and, although workers in different research areas have shared the adoption of many basic assumptions and the rejection of others, there has been no widely accepted framework providing a guide to these changes. As Halliday discusses in the first chapter of this book, there appears to have developed a rare situation whereby a number of people have independently found that their present ideas and interests are more closely related to those pursued in Warsaw over the past half-century than to the work of their own intellectual predecessors in the West. It is as if groups from different starting points had entered an apparently new region, started to explore, and then discovered — often unexpectedly — the existence of a sketch, uncertain in parts, but of the region as a whole. The points of departure represented here include concern with the nature of classical and of instrumental conditioning; the analysis of inhibitory effects; attempts to understand the physiological basis of learning; research on autoshaping and its relevance to problems studied in the context of operant conditioning; and problems arising when various kinds of stimuli, responses, reinforcers, and species are compared. Within each area of special interest there naturally exist different points of view and various opinions of the value of specific ideas suggested by Konorski; but overall there seems agreement that Konorski's map provides a better general guide than any other.

There are strong pressures on researchers to take a somewhat parochial view of their field. These, we feel, have contributed to the neglect in the West of Konorski's work. One purpose of the conference was to encourage contributors to present an account of research on some specific topic in a more general context than is usual. Our hope is that this will be considered valuable and that the present book will help to place Konorski's work in the proper perspective.

Permission to reproduce quotations from Konorski's publications was given by the Cambridge University Press, the University of Chicago Press, and Prentice-Hall, Inc. The bibliography of Konorski's work is based on one published in *Acta Neurobiologiae Experimentalis,* with the permission of its editor, and supplemented by additional references kindly supplied by Professor Zielinski.

Our thanks go to the contributors for their willing and prompt cooperation and to our colleagues from the Laboratory of Experimental Psychology at the University of Sussex, especially to Sebastian Halliday and Nick Mackintosh for their valuable advice. The conference was made possible by grants from the Experimental Psychology Society and the European Training Program in Brain and Behavior, and by the help of many undergraduate, graduate, faculty, and staff members of the Laboratory of Experimental Psychology. We are grateful for the additional support toward travel expenses provided by the Society for the Experimental Analysis of Behavior, the Royal Society, and the Polish Academy of Sciences. Finally, we would like to thank our wives for their patient help during the conference and the preparation of this book.

A. DICKINSON
R. A. BOAKES

1 Jerzy Konorski and Western Psychology

M.S. Halliday
University of Sussex

The boundaries between the disciplines of physiology, physiological psychology, and psychology are ill-defined and constantly shifting; Konorski always described himself as a physiologist, but he made major contributions in all these areas during his long and productive life. The focus of this volume is on conditioning, and so this chapter concentrates on this aspect of his work, while recognizing that he himself would certainly have deplored any arbitrary divisions between his various interests. In particular, I would like to consider why 20 years ago Konorski was largely ignored in the West, whereas today he is seen as a figure of great importance.

The mid-1950s were a significant turning point both in Konorski's career and in the development of American learning theory. In an autobiographical sketch Konorski (1974) wrote:

> There are at least two reasons why I can divide my postwar life into two distinct periods, with the demarcation point being 1955. First, 1955 was the year in which, two years after Stalin's death, the "thaw" began, when Kruschev came to power in the USSR and dissociated himself sharply from the Stalinist period. This was immediately reflected in all fields of cultural life in the USSR and even more so in Poland. In my own field the pseudo-Pavlovian indoctrination vanished completely, and I stopped being a revisionist and a servant of capitalism [p. 207]. © 1974. Adapted by permission of Prentice-Hall, Inc., Englewood Cliffs, N.J.

He is here referring to a period, beginning about 1949, during which Pavlovian theory was assimilated to Stalinist orthodoxy, and in which he, as a critic of Pavlov, came under heavy attack; there was even a time during which he was given "leave of absence," though remaining as head of the

1

Department of Neurophysiology at the Nencki Institute (Dr. W. Wyrwicka, cited in Mowrer, 1976). Konorski refused to give way under this pressure, and his integrity was ultimately rewarded by full recognition in his own country. 1955 was also significant because in that year the Nencki Institute moved from its temporary location in Lodz to a new and specially constructed building in Warsaw, replacing the original Institute which had been destroyed by the Nazis. Konorski had played a major part in planning this building and remarked (1974), "I can boast that the Department of Neurophysiology, with all its virtues and defects, is my own child (p. 208)." Thus for the first time since before the war, he had satisfactory facilities for carrying out his research.

Around 1955 American learning theory was also at a transitional point. Up to that time the subject had been dominated by the comprehensive learning theories developed in the 1930s, which attempted to explain all learned behavior using a limited number of relatively simple principles and largely without regard to physiological evidence. By the end of the decade this approach seemed inadequate. Attention was being concentrated on much more limited areas of behavior and on smaller-scale theories; at the same time physiological psychology was making major advances, and Skinnerians were advocating a nontheoretical approach. A representative statement of the old viewpoint, together with some disquiet about its adequacy, can be found in *Modern Learning Theory: A Critical Analysis* (Estes et al., 1954). In this scholarly work the authors attempted a critical survey of the major theories of learning, in the belief that (1954) "the major contributions of the outstanding theorists of our time have been made" and that "their contributions have set the framework for the larger part of the current theoretical literature and have placed their stamp on the vigorous output of our experimental laboratories (p. xi)." Their intention was not merely critical but also constructive. Their interest in the future was clearly expressed in the introduction, where they state that their "primary concern is with learning theory as it may function in the long-range development of a science of behavior (p. xv)." At some time during the meeting at which their book was planned, some of the participants must surely have looked to the future and wondered who would be seen as the significant learning theorists in a generation's time. If that group of psychologists had been asked to draw up a list of likely candidates from those born between about 1900 and 1915, it is almost certain that, however long the list, the name of Konorski would not have appeared. It is not surprising that there is no mention of him in *Modern Learning Theory*, which is primarily a study of five American theorists; but in ignoring him, the authors were confirming the judgment of American psychologists in general. In 1954 there were no references to Konorski in the *Journal of Comparative and Physiological Psychology*, the *Journal of Experimental Psychology*, the *Psychological Bulletin*, or the *Psychological Review*. He was not mentioned in the two major textbooks that had recently appeared, Osgood's *Method and Theory in Experimental Psychology* (1953) or Woodworth and Schlosberg's *Experimental Psychology* (1954), and both editions of Hilgard's *Theories of Learning* (1948 and 1956) are equally silent. In 1954, therefore, there appeared to be universal agreement about the unimportance of Konorski's contribution to psychology.

In the second half of the 1950's and in the 1960s, sporadic references began to appear in the literature, and their frequency has increased up to the present day. The Science Citation Index gives six references to Konorski in 1961; by 1970 this had risen to 35, while in 1976 there are no less than 130 citations. Over the past 20 years Konorski has moved from a position of total obscurity, at least in the West, to that of a major figure in the study of learning. It is interesting to consider why there has been this remarkable reversal in the perceived importance of his work.

There are, of course, straightforward reasons for this change. Scientific contacts with Poland in the 1930s were slight; they were totally disrupted by the war and in the postwar years. Much of Konorski's work had been published in Polish or Russian and was unavailable in England or America. Furthermore, apart from a visit to London in 1946, it was not until the late 1950s that Konorski was free to visit Western countries. As he remarked in his autobiography (1974):

> This first visit of mine to America [in 1957] had very great significance for the further development of our scientific work and its relation to the work going on in the United States. Owing to my connections with American scientists, almost every member of our laboratory was able to visit the United States for one or two years, which gave him (or her) the opportunity to become acquainted directly with the American investigations; this significantly enlarged their scientific horizons. Consequently, our work became much better known by American scientists than it was before [p. 209].

Together with the publication of *Integrative Activity of the Brain* by the University of Chicago Press in 1967, this does much to explain the increasing prominence of Konorski in the 1960s and 1970s. However, it does not entirely account for the neglect he had suffered in previous years. To understand this fully, we need to consider more carefully both the nature of Konorski's contribution and the changing climate in American psychology over the past 20 years.

It is important to realize that much of the work for which Konorski is now celebrated had been published well before 1955, a considerable part of it in the 1930s. This is now seen as pioneering work which went largely unrecognized at the time, but which is directly relevant to modern thinking about animal psychology. But if the work was unrecognized, it should not have been unknown. Konorski and Miller had published two theoretical papers in the *Journal of General Psychology* (1937a, 1937b) in which they described their own work and criticized Skinner's distinction between two types of conditioned reflex. These papers were widely cited for a time and are discussed in both Skinner's *The Behavior of Organisms* (1938) and in Hilgard and Marquis's *Conditioning and Learning* (1940), two books which would have been on the shelves of any animal psychologist in the 1940s and 1950s. Apparently the topics discussed in these papers simply did not seem important at that time. This is borne out by the treatment of the operant–respondent distinction in the chapter on Skinner in Estes et al., *Modern Learning Theory*; only a paragraph is devoted to this topic, and this is dismissive in tone. For example (1954):

> No attempt is made to reduce this differentiation [operant/respondent] experimentally. On the contrary there is an insistence on maintaining it. Although many experiments have shown that behavior which Skinner terms "operant" obeys the same laws as the static laws which define "respondents" when the appropriate experimental conditions are introduced, these are rejected because measures deemed inappropriate to operant behavior are introduced [p. 289].

If the discinction between operant and classical conditioning could be treated in this cavalier fashion, it is hardly surprising that the work of Konorski, whose main contribution up to that time had been to illuminate the nature of that distinction, passed without notice.

A later source for Konorski's ideas which should also have been familiar is *Conditioned Reflexes and Neuron Organization,* (1948). In England this book received favorable reviews in *Mind,* the *British Medical Journal,* and the *Times Literary Supplement,* but no mention in *Nature* or any of the psychological journals. Nevertheless, it was widely read by psychologists and was recommended, for example, to Cambridge undergraduates around 1960. In America, however, the book received little or no attention: Only 399 copies were sold in the United States (Mowrer, 1976) and it was not reviewed by any major psychological journals, although it received some attention in medical publications. This can hardly have been accidental, for it was published by the Cambridge University Press and consequently must have come to the attention of the review editors of the psychological journals; they presumably decided that a review would not elicit much interest. Konorski himself commented (1974):

> My explanation of this fact is that at that time experimental psychology was strongly Skinnerish or Hullian, and physiological explanations of the mechanisms of conditioned reflexes were utterly unpopular. I suspect that either people did not read the book at all, not being attracted by its title... or else if they had it in their hand, they rejected it [p. 204].

Despite this lack of interest in his book, Konorski was gratified to discover on his first visit to the United States in 1957 that he was by no means unknown. The overall picture is therefore one of neglect rather than of ignorance, and the apparent blindness of Western psychologists to the value of Konorski's work was less a consequence of sensory deprivation than of unwillingness to see.

What, then, were the characteristics of his work that made it so unattractive to Western behaviorists at that time? A number of somewhat interrelated factors have combined to produce this effect. First, Konorski was deeply concerned with conditioned reflexes, even describing instrumental learning as a conditioned reflex of Type II. Western behaviorists from the 1920s onward had taken over the primitive ideas of classical conditioning, but had gone little further than this. Konorski himself made this point very clearly (1948):

> As is well known, conditioned reflexes are of particular importance in the American behaviourist school of psychology. But if the manner in which behaviourism exploits the science of conditioned reflexes is analysed more closely, it will be found that it consists chiefly in utilizing the basic and most elementary concepts of this field, i.e. the Pavlovian *nomenclature,* for the denomination of particular phenomena.... How little the achievements of the Pavlov school have been taken into account, even by those investigators who have been most directly concerned with the study of conditioned reflexes, may be judged, for example, in the chapter on the conditioned reflex written by Dr. H. S. Liddell of Cornell University in Fulton's famous monograph, *Physiology of the Nervous System.* Stating that... Pavlov's theory... "is at present of historical interest only," Liddell simply disregards the huge body of experimental evidence which forms the bases of the theory and which must find explanation in some way or another.... Liddell takes the inadmissible course of drawing a veil of silence over most of the facts and deals only with the most elementary phenomena in the field. One regrets to have to say that many American authors behave in exactly the same way [p. 2].

It was precisely this state of affairs that Konorski hoped to remedy by the publication of his book, but the method he chose took little account of the nature of behaviorism. He believed that the main source of resistance to Pavlovian work arose from dissatisfaction with Pavlovian physiological explanations—a dissatisfaction which he fully shared. If, therefore, Pavlovian findings could be brought into line with the neurophysiological tradition of Ramón y Cajal and Sherrington, the way would be cleared for a rapprochement between the American and Russian approaches to the benefit of both. Konorski had first been introduced to Sherringtonian neurophysiology in 1933 on his return from Pavlov's laboratory in Leningrad, by Dr. Liliana Lubinska, who later became his wife and scientific collaborator. Pavlov's view of the activity of the cerebral cortex involved the interplay between waves of cortical excitation and inhibition, and was totally alien to Western neurophysiology, with which Konorski now became familiar. In writing about this period of his life he noted that (1948) "as the years passed I became more and more convinced that Pavlov's theory was not correct, as it could not be reconciled with the evidence of general physiology of the central nervous system (p. xix)." As he wrote later (1974), "One of the two theories should be rejected in toto, and the facts so far explained by the rejected theory should be reinterpreted in the framework of the other theory (p. 198)." His earlier book, which was the fruit of many years thought and study, was a remarkably effective attempt (1974) "to explain the whole bulk of the experimental work collected by Pavlov's school by the Sherringtonian principles of functioning of the central nervous system (p. 198)." Unfortunately the method that he chose might have been deliberately calculated to repel the Western psychologists to whom it was addressed. At that time learning theorists were profoundly unimpressed by physiology and were concerned with explanations of behavior in terms of intervening variables and hypothetical constructs independent of possible physiological mechanisms. This is, I believe, a second reason that Konorski's work was largely ignored, as he himself later recognized. There is a certain irony in the fact that, while he was rejected in the West because of his interest in classical conditioning and physiology, he was under ideological attack in the USSR because of his interest in conditioned reflexes of the second type (instrumental learning) and his rejection of Pavlovian physiology.

Another feature of Konorski's work, which probably contributed to its unpopularity in America, was its concern with processes of inhibition. This was a natural consequence of taking Pavlov's experimental findings seriously; the concept of inhibition is integral to the analysis of classical conditioning, but was anathema to most behaviorists. The objections were, first, that the use of the concept smacked of the worst kind of neurologizing and, secondly, that it was in any case redundant. Skinner spoke for many who would have disagreed with him in other ways when he said (1938): "We do not need the term [inhibition] because we do not need its opposite. Excitation and inhibition refer to what is here seen to be a continuum of degrees of reflex strength and we have no need to designate its two extremes (p. 18)." The general distaste for the concept of inhibition needs no further documentation, but it must have done much to reduce enthusiasm for an author who used the term as freely as a Hullian would talk of reinforcement. Of course, reactive and conditioned inhibition were familiar terms in the context of Hullian theory, but they had always aroused disquiet and were tolerated largely because of Hull's immense prestige; by the early 1950s these concepts were under heavy attack as the weakest part of the Hullian edifice.

Another feature of Konorski's work, which he had borrowed from Pavlov, was his use of single animals in experiments and the total absence of statistics. This, too, was very foreign to the Western approach, which at that time depended almost entirely on group experiments and statistical analysis.

The final aspect that produced withdrawal rather than approach is in some ways the most interesting, because it illustrates most clearly the changes in Western psychology that have occurred in the last 20 years. Throughout his career, Konorski was primarily concerned with the mechanism of association—how particular stimuli, or stimuli and responses, become associated with one another. His early enthusiasm for Pavlov's work and his training in Pavlov's laboratory between 1931 and 1933 ensured that he never made the mistake, so common among Western learning theorists, of assuming that the process of association itself was a relatively simple one. Familiar as he was with Pavlovian concepts of irradiation and concentration of excitation and inhibition, positive and negative induction, and the rest, Konorski was well aware that the study of the mechanism of association in conditioned reflexes of Type I or Type II was no easy matter. In time he became dissatisfied with the Pavlovian account, largely because it did not take the problems of association seriously enough. As a part of a powerful criticism of Pavlovian theory, Konorski wrote (1948):

> The fundamental feature of Pavlov's theory, which sets its mark on it and determines all its statements, is the assumption that not only the process of excitation evoked by the application of an active conditioned stimulus, but also the process of inhibition... is localised in the cortical centre of this stimulus, in the point of the cerebral cortex to which it is "addressed."... This assumption predetermines that all the fundamental processes which occur... when positive and inhibitory conditioned reflexes are in action... take place at the very beginning of the cortical part of the corresponding reflex arcs, and that one or another conditioned reaction, or the absence of it, is only to be regarded as an indicator which gives notice of the "sign" and intensity of the process. Thus the reflex arc as a whole disappears completely from sight, and attention is concentrated on unspecified states of excitation and inhibition... unspecified because it does not matter in the least which executive neurons they are linked to. In our view this assumption is the "original sin" of all Pavlov's theory [p. 38].

Much of his work before 1955 was devoted to providing a more satisfactory model, couched in physiological terms, of the basic conditions required for establishing excitatory and inhibitory associations in conditioned reflexes of Type I and Type II. In this respect Konorski was representative of the Russian psychophysiological tradition, descending, through Sechenov, from nineteenth-century German physiology. Within this tradition the reflex arc, including "reflexes of the brain," was the fundamental subject matter of physiology and psychology. Indeed, insofar as Pavlov can be seen as having a psychological rather than a physiological theory, Konorski was firmly within the Pavlovian tradition; for example, at no point in this book did he question the essentially Pavlovian idea of classical conditioning as stimulus substitution. By contrast, behaviorism was ultimately rooted in Darwinian theory, with its emphasis on adaptation to the environment. Learning was seen as the major mechanism of adaptation, and learning theorists were concerned with the way in which trial-and-error learning adapted an animal to a changing environment. Within this context the important issues appeared to be such things as the order of elimination of errors in a maze, place versus response learning, and, above all, the role of reinforcement in learning. Most Western psychologists either were, like Skinner, largely uninterested in the mechanisms of association, or believed that it was a primitive mechanical process, adequately described by writing a big H between a little s and r—thus leaving them free to get on with the important business of deciding, for example, whether drive reduction was necessary for reinforcement. Thus, Konorski was addressing questions which seemed largely irrelevant to American behaviorists.

It is notoriously difficult, as any reader of Konorski must know, to convert an "aversive" stimulus into an "attractive" one; but over the last 20 years this is exactly what has happened to Konorski's ideas vis-a-vis Western psychology. The change has come about as a result of profound alterations in our approach to animal learning; one could almost say that the appetitive and aversive systems have changed roles. It is common for one generation to react against the preconceptions of the preceding one, but, with respect to the assumptions of 25 years ago, the reversal in thinking in the West has been remarkably complete. Classical conditioning is now of central theoretical interest, the change being conveniently marked by the publication of *Classical Conditioning* (Prokasy, 1965). At the same time, following the appearance of Rescorla and Solomon's classic paper (Rescorla & Solomon, 1967), the interactions between classical and instrumental conditioning provide one of the most vigorous areas of research in psychology. Inhibition has also become a respectable concept in the study of both classical and instrumental learning (e.g., Boakes & Halliday, 1972; Rescorla, Chapter 4; Zielinski, Chapter 10 in this volume). Changes in attitude toward physiology are perhaps less marked, and some prejudices linger on; but there are clear signs of a useful dialogue deveoping between physiological psychologists and students of learning — as is shown, for example, in the physiological contributions to this volume (Moore, Chapter 5; Gray, Chapter 11). Finally, as a result of Kamin's work on blocking (e.g., Kamin, 1968) and Wagner and Rescorla's theoretical and experimental work (Wagner & Rescorla, 1972; Rescorla & Wagner, 1972), the study of the formation of associations has moved to the center of the stage as possibly the most important problem in the study of animal learning. I have mentioned these changes only briefly; they are recent and familiar enough to require little comment. What is remarkable is that one looks in vain to American behaviorism of 25 years ago for the intellectual antecedents of contemporary views; whereas Konorski, who comes from a tradition largely independent of Western psychology, speaks clearly and convincingly, over an interval of more than a generation, on the very issues that are now seen as most important.

Other chapters discuss in detail the significance of Konorski's work for modern approaches to a variety of aspects of learning and motivation. Nevertheless it seems worthwhile to enumerate here some of the ways in which his experimental and theoretical work anticipated contemporary developments, in some cases by as much as 30 years.

In 1928 Konorski and Miller published their first experimental papers, one of which was entitled 'Sur une forme particulière des reflexes condition-nels.'' This paper described experiments carried out with a single dog and makes the first clear distinction between classical and instrumental conditioning (or conditioned reflexes of the first and second type, as they call them). They found that if, during the presentation of a conditioned stimulus, the dog's leg was passively flexed on some of the trials and if food was delivered only on those trials, then the movement would begin to appear spontaneously in response to the conditioned stimulus. If, on the other hand, the reflex was reinforced with an aversive stimulus—such as blowing air into the dog's ear—on those trials when its leg was passively flexed, the dog would increasingly extend its leg until it became so rigid that it was almost possible to pick the dog up by that leg. In discussing these findings, they wrote (1928):

> The phenomena that we have just described have the same general properties as conditioned reflexes: they originate without doubt in the cortex, and they are not innate but are formed and disappear during the life of the individual. It is for this reason that we regard them as conditioned reflexes, but their mechanism is different from the conditioned reflexes of Pavlov. . . . Hence, unable to reduce the reflexes of the second type to the conditioned reflexes of Pavlov, we should consider them as the second fundamental mechanism of the function of the cerebral cortex. . . . Differences between conditioned reflexes of the first and second type are considerable. (1) In the ordinary conditioned reflex, the conditioned stimulus always elicits the same reaction as the reinforcing stimulus, while in conditioned reflexes of the second type these reactions are different. . . . (2) The role of Pavlov's conditioned reflex is limited solely to signalisation while the role of conditioned reflexes of the second type is completely different [according to their relationship to the positive or negative reinforcing stimulus]. (3) In conditioned reflexes of the first type, the reaction is effected by organs innervated through the central or autonomic nervous system, while, in conditioned reflexes of the second type, the effector can probably be only a striate muscle [p. 188; 1969, translation by Skinner].

In their next papers they showed that similar results were obtained when the flexion was induced by passive movement, or by mild electric shock, or when the experimenter merely waited for the response to occur spontaneously. In the 1928 paper, without reporting any findings, they also outlined the procedures for avoidance learning and omission training and illustrated their relationship to instrumental reward and punishment training. This is an impressive achievement for a first publication, and much confusion could have been avoided, both in Russia and the West, if the paper had been widely known and discussed.

It is interesting to know something of the circumstances under which this early experimental work was done. Konorski and Miller were medical students who had become disenchanted with the pedestrian character of their studies and had, quite accidentally, come across a copy of Pavlov's works. As Konorski reported (1974):

> From these works we learned for the first time about conditioned reflexes and we immediately realized that this was exactly the field of science we were looking for. The extent of our excitement brought about by this discovery is difficult to describe. We became entirely involved in studying Pavlov, and only by some miracle, not quite clear to me yet, did we succeed in being graduated in medicine [p. 186].

At that time there was little psychology in Poland, and certainly no work on conditioned reflexes; however, they eventually persuaded a professor at the Free Polish University to let them use a small third-floor room in an apartment block, in which part of his department was housed. Konorski recalled (1974), "After this the first thing we had to do was buy a dog. For this purpose we went to the market place, found the area where people sold dogs, and after long deliberation, chose a young and nice bulldog, which cost us ten zlotys (about one dollar). We called him Bobek. He immediately became friendly with us and we brought him to our 'laboratory.' The housekeeper agreed to let him stay in her apartment." The laboratory arrangements were somewhat primitive: "putting together two square stools we made a 'Pavlovian stand' and used cardboard for a screen. . . . Pieces of food were thrown from the small aperture in the screen by the experimenter." For kymograph recording they used toilet paper, "which was both cheap and convenient, provided that it was relatively smooth and did not have transversal perforations. You can imagine the comical picture presented by two serious young men going to a paper store and asking to be shown all the possible varieties of toilet paper, scrutinizing them thoroughly, and choosing the one which fulfilled both conditions (pp. 187–188)." This period of his life could be seen as an almost perfect illustration of the rather hackneyed point that good research arises from good ideas, not from good equipment.

Konorski and Miller were certainly the first to publish experiments using the omission procedure. These experiments follow essentially the same pattern as those already described, except that passive flexion of the leg signaled nonoccurrence of food. Strong extension reflexes developed which were sufficient, on most trials, to keep the dog's paw on the floor against the calibrated pull of the apparatus. In these experiments they were recording salivation as well, and found that it was elicited by the stimulus if the dog was successful in keeping its foot on the floor, but was immediately inhibited if the leg was flexed, signaling the absence of food. In order to demonstrate the importance of the instrumental contingency, they also showed that there was no extension of the leg to a Type I classically conditioned stimulus that signaled food regardless of the dog's behavior. These early papers provided almost the only example of omission training for some 30 years, and it was not until the publication of Sheffield's paper on omission training in salivary conditioning (1965) that the potentialities of this technique were realized in the West. (For more recent applications, see Hearst, Chapter 2; Mackintosh and Dickinson, Chapter 6; Zielinski, Chapter 10.)

The analysis of avoidance conditioning has, from about 1940, provided both a major problem for learning theorists and an important spur to theoretical development. Understanding of avoidance learning was slow to develop, largely because of confusion about the distinction between instrumental and classical conditioning. Yet in his 1948 book Konorski provided a very clear analysis of avoidance which is in many ways similar to Mowrer's (1960) theory and which is based in experiments carried out in the 1930s or before. The first experiments were, in fact, carried out in 1928 on the admirable Bobek. Mowrer has recently written, "If I had known of Konorski's *Neuron Organization and the Conditioned Reflexes,* I would have had special reasons for citing it (which I did not do) in *Learning Theory and Personality Dynamics*" (Mowrer, 1976). Konorski's analysis of avoidance is based on the following type of experimental procedure (1948):

> The external stimulus (say the metronome) is reinforced by a negative unconditioned stimulus, e.g. the introduction of acid into the dog's mouth. The conditioned reaction consists of the secretion of saliva.... When this reflex is formed, from time to time we provoke a passive flexion of the limb in association with the metronome, and on these trials the acid is not introduced. After a comparatively short time a reflex [of the second type] is formed in which the dog itself raises the limb to the metronome—a movement saving it from the introduction of acid. This reflex is soon established so firmly that for weeks and months the dog goes on raising its paw to each application of the metronome, even though the reflex is never reinforced [p. 228].

He also observed that, once established, the avoidance response would be made spontaneously to other conditioned stimuli that had independently been associated with acid in the mouth, or indeed with other unconditioned stimuli, such as a puff of air in the ear. This type of experiment is in many ways superior to the shuttle box, so familiar to Western psychologists: There is a clear separation of the classically conditioned response and the instrumental avoidance response; the leg flexion is a "pure" avoidance response, uncontaminated by any escape contingency; control of the stimulus situation is far better, since the animal cannot move around in the apparatus; and the development of the classically conditioned response can be monitored during all stages of avoidance learning. Thus, for example, Konorski was able to observe that the first spontaneous avoidance response occurred on the trial immediately following the first trial, during which passive movement of the leg had produced a marked reduction in salivation.

Only when the feedback from leg flexion had become a conditioned inhibitor did the avoidance response emerge; this was in accord with his theoretical position — which was that the reinforcement of the avoidance response was provided by the conditioned inhibition generated by kinaesthetic feedback. He also suggested that the termination of the warning signal might itself act as a conditioned inhibitor of the same sort. He was, however, puzzled at the persistence of avoidance behavior and commented (1948):

> So why is it that the character of this [conditioned] stimulus as a conditioned defensive one is somehow "preserved" when it is accompanied by a movement constituting a conditioned inhibitor? One has the impression that this very movement somehow protects the stimulus against extinction.... But we are unable to explain the causes of this phenomenon [p. 231].

We should recognize that Konorski's 1948 explanation of avoidance was not entirely satisfactory; in particular, as he noted later (1967), "it gave precise rules determining whether and when a given exteroceptive stimulus would elicit a given instrumental response, but it did not answer the question why it would do so (p. 393)." But it is a formulation which avoided misunderstandings and confusions and clearly identified the important questions.

This brief description of Konorski's early work on avoidance illustrates another area in which he anticipated modern experiments: the concurrent measurement of classical and instrumental responses. Such measurements were, of course, a natural consequence of his standard experimental procedures. For example, he showed that when passive movements of the leg were followed by presentation of food, spontaneous leg movements appeared soon after the dog began to salivate to the passive flexion, and that in extinction salivation and leg flexion declined together and disappeared at much the same time. However, he also concluded that it was not possible to make any general rules about the temporal relationship between the classical and instrumental components; sometimes salivation preceded the instrumental response, sometimes it followed it, and on occasion it was absent altogether. He therefore turned to the more powerful superimposition technique, which has become so popular in the West in the last 10 years. A typical example of one of these experiments, which was originally reported in 1936, appears in *Integrative Activity of the Brain* (1967):

> A classical alimentary CR was established in a dog to a bell and a metronome set at 120 beats per minute. A metronome set at 60 beats per minute was differentiated by not reinforcing it by food.... Then... the dog was trained to lift his hindleg in the experimental situation, each movement being reinforced by food. [In the test stage] ... for a few seconds food was not presented and consequently the dog performed the trained movement with maximum frequency. Then either a positive type I CS (bell or M120) or the inhibitory one (M60) was presented.... It was found that in response to

the positive CS the dog immediately stopped performing the trained movement, looked at the feeder and salivated copiously. On the contrary, in response to the differentiated stimulus, salivation was much reduced but the movements of the leg continued to appear with maximal frequency [pp. 369–371].

This is but one of a large number of similar experiments and is intended merely to illustrate the way Konorski was making good use of the superimposition technique in the early 1930s and the interesting results he was obtaining. (For further discussion of these procedures, see Boakes, Chapter 9; Morgan, Chapter 7.)

Konorski was also the first to understand the special character of conditioned inhibition. Pavlov had regarded conditioned inhibition as but one among the varieties of internal inhibition and certainly did not accord it any special status; he claimed that (1927) "all the experimental evidence ... establishes conclusively that the nervous processes on which conditioned inhibition depends are identical in character with those of extinctive inhibition (p. 86)." The only difference that Pavlov acknowledged was in the procedure for producing inhibition in the two cases. Konorski (1948) was dissatisfied with this argument and, after subjecting the Pavlovian account of the mechanism of conditioned inhibition to a devastating criticism, he proposed his own theory. This is couched in physiological terms, but can be interpreted in purely behavioral ones. In outline he assumes that the process of extinction involves the formation of inhibitory connections between the center of the CS and that of the US, but not the destruction of the existing excitatory connections. Thus after extinction the CS has both excitatory and inhibitory properties, and these are roughly in balance. In the case of conditioned inhibition, however, because there are no excitatory connections between the center for the conditioned inhibitor and the center for the US, all the connections which are formed are inhibitory (1948).

So in the case of conditioned inhibition we deal with an interesting situation, and one very suitable for analysis, in which the excitatory and inhibitory connexions have their starting point in two different cortical centres the centre [for the conditioned stimulus] is the starting point for almost exclusively excitatory connections ... while the centre [for the conditioned inhibitor] is almost a pure source of inhibitory connections [p. 155].

Not merely did he identify conditioned inhibition as being in a class of its own because it has almost purely inhibitory effects, but he also recognized that this property makes it a uniquely valuable analytical tool. (See Rescorla, Chapter 4.)

It would be possible to extend this catalogue of experimental and theoretical contributions considerably, but to do so might prove an exercise in overkill; it should already be clear why Konorski's work seems particularly relevant to contemporary animal psychologists. I would like, however, to draw attention to one other finding that has an especially modern ring. Konorski pointed out that, in what would now be called a free operant situation, the instrumental responses are under the control of ill-specified background stimuli; under these circumstances the background stimuli alone do not predict food, whereas these stimuli plus the movement of the leg do. Therefore, by analogy, if the passive training procedure is to be used with a discriminative stimulus which is to have the same relationship to the response as the background stimuli had in the free operant situation, it is essential that two sorts of trial are given: those where the CS occurs, the leg is flexed and food is presented, and others where the CS occurs without flexion and no reinforcement is given; these latter trials being equivalent to the periods in the free operant situation when the background stimuli were present but the dog was not lifting its leg and was not getting food. This was the training procedure which Miller and Konorski used in their original experiment and which resulted in rapid acquisition of the instrumental response. When they went to work in Leningrad in 1931, Pavlov suggested that this was not a necessary condition for the development of Type II conditioned reflexes and that all that was needed was mere contiguity between the external stimulus and the response. Miller and Konorski therefore did a further experiment in which the metronome was paired with mechanical flexion of the leg and reinforced with food on every trial. Konorski (1948) noted:

> The result we get then is as follows: first, an alimentary conditioned reflex of the first type is formed to the metronome, the dog displaying the secretory and the motor alimentary reaction (turning the head, . . . licking the lips, etc.). Secondly, the dog begins to raise the leg actively with greater or lesser frequency during the intervals between the conditioned stimuli. . . . But the active movements are not performed to the application of the metronome; in fact, rather the reverse is observed. . . . There is not even a hint of any "connection" between the metronome and the raising of the leg. . . . From the foregoing data it follows that in order to form a conditioned reflex of the second type—metronome——→ raising the leg—it is not sufficient to have a simple concurrence of these two phenomena and to reinforce them by food. In order to form such a reflex, a situation has to be created in which the movement is an *indispensable* condition of the dog's obtaining food [pp. 219–220].

This analysis, with its recognition of the importance of background stimuli and the distinction between continuity and contingency dates back to the early 1930s; yet it is entirely contemporary in conception. (See Mackintosh & Dickinson, Chapter 6.)

One is naturally led to ask how Konorski saw so clearly things to which most other psychologists of the time appeared blind. Such questions can never be satisfactorily answered, but it is possible to suggest some reasons. Much must be attributable to Konorski's exceptional personal qualities; on this topic I shall mention only one or two points derived from memoirs by his students and collaborators. His dedication to his work was total; once having found the subject of his choice, after early disappointment with mathematics and medicine, his commitment was lifelong and complete. More than one of his students saw in him the embodiment of Pavlov's maxim that "science requires one's whole life, and even if one had another life to offer it would still not be enough." Two anecdotes from Dr. Fonberg's memoir of him may serve to illustrate his enthusiasm and energy. The first dates from 1958 when an international physiological convention was being held in Poland. A Sunday morning had been set aside for boating on a nearby lake. It was a beautiful day and everyone was setting off in different directions when the word was passed around that Konorski was holding a seminar. When they turned back to investigate, the boatmen discovered that he was holding a floating seminar from his boat in a reedy bay; and so the morning, although spent on the water, was devoted to physiology rather than recreation. The other story concerns Konorski's return from his first trip to America; as Dr. Fonberg recalled the occasion (1974):

> He landed in Warsaw on a Saturday afternoon; most of the staff at the Institute was already preparing to leave for home. Konorski, however, went straight from the airport to his laboratory; he had a cup of tea in his study, washed his hands and then proceeded to organize a meeting. He wanted to tell everyone about the interesting research being carried out in the United States and went on lecturing until 9 o'clock in the evening. After the strain of a trans-Atlantic flight, the change in climate and the time lag, Konorski was still radiating with energy, so fervently did he want to share with us his scientific "adventures" in the USA [pp. 662–663].

In adition to his phenomenal energy, he possessed more than his fair share of determination and self-confidence. This is perhaps seen most clearly in his relationship with Pavlov. At the time he was in Leningrad, Pavlov was the absolute ruler in his laboratory. Konorski (1974) wrote:

> He assigned the problems to be worked out to each of his co-workers and he controlled all stages of their research. Only in exceptional circumstances did a student follow his own lines of research. . . . In the absence of Pavlov we usually talked about him, quoted what he said, how he behaved, etc. People boasted when Pavlov spoke to them at some length, and . . . the attitude of Pavlov toward an individual was the main factor determining the hierarchy within the group. . . . The atmosphere reigning in the group reminded one of that usually encountered in royal courts, with Pavlov being the indisputable king [p. 193].

Add this to Pavlov's notorious aggressiveness in argument, and it can easily be appreciated that it would be a brave man who would oppose him. According to Konorski, Pavlov strongly disagreed with the view that there were two kinds of conditioning and failed to see any difference between them. However, throughout his two years in Leningrad Konorski carried on his own line of research and braved Pavlov's disapproval, which must have required considerable strength of will for a young and inexperienced scientist. There is, in fact, reason to believe that Pavlov respected Konorski for his willingness to oppose him in argument—even if he disapproved of his scientific views so strongly that when Konorski and Miller published their work in his journal they did not dare use their own terminology of Type I and Type II conditioned reflexes. Konorski's integrity and strength of purpose were once again put to the test in 1949 when he was denounced as a physiological deviationist, and once again he refused to give way. However, his firmness seems never to have degenerated into dogmatism; to the end of his life he was prepared to entertain new ideas, and indeed there was a very marked change in his theoretical outlook between 1948 and 1967. In the closing passage of his autobiography, he wrote that "the full cycle of renewal of my scientific ideas requires about two decades" and went on to regret that there would not be time for another cycle.

Konorski must also have been an exceptionally gifted experimenter. His original experiments with Bobek in the ramshackle laboratory in the Warsaw apartment block have stood the test of time far better than hundreds of experiments carried out in well-equipped laboratories elsewhere. He clearly had an exceptional flair for the right technique and a sensitivity to the relevant experimental details.

Finally, Konorski had the considerable advantage of not being directly involved in either the Pavlovian or the behaviorist tradition. Had he been a normal student of Pavlov's, he might never have been allowed to start on his unconventional line of research; or had there been a behaviorist tradition in Poland, he might have ignored the merits of Pavlov's work. As it was, he was able to develop his experimental techniques and interpret the results untrammelled by conventions and preconceptions; and this led him in directions which seem, at present, to have been remarkably farsighted.

This discussion of Konorski's work and its reception in the West has dealt almost entirely with the period before 1955; this is because, paradoxically, his earlier work is closer to the concerns of modern Western animal psychology than his more recent thinking. *Integrative Activity of the Brain* may be seen as a development and revision of his earlier views, but it is much more loosely structured and ambitious than his earlier book. It is an attempt to provide a model—couched in terms of a "conceptual nervous system"—not merely for learning but also for perception and motivation. In his book Konorski introduced a range of new concepts; gnostic units, preparatory and consummatory conditioned reflexes, and the idea of mutually inhibitory preservative and protective (appetitive and aversive) motivational systems. Unlike his previous book, *Integrative Activity of the*

Brain was widely noticed in the psychological journals; and the reviews, although varying in enthusiasm, were all respectful and regarded the book as an important contribution. However, the general tenor of these reviews was that the book contained intriguing insights and interesting experimental results, rather than a satisfactory overall framework for understanding the brain and behavior. In general it seems that the reviewers' assessment has been confirmed over the last 10 years, but of course it was some 20 years before the importance of Konorski's earlier work was fully recognized. Konorski's (1974) own reaction is both interesting and appropriate:

> Contrary to my expectations the reaction to the book was rather poor. I had a feeling that many of my friends and colleagues simply disliked it, or had not read it. . . .
> I am very curious to know what will be the final fate of the book; will it eventually win general recognition, which I think it deserves in spite of its shortcomings, or will it have no important impact on the further development of behavioural science? [p. 214].

In the light of his previous book, it would be a rash man who would firmly predict the latter outcome.

ACKNOWLEDGMENTS

I am grateful to R.A. Boakes, A. Dickinson, N.J. Mackintosh, and S. Soltysik who read the draft of this paper and made a number of very helpful suggestions.

REFERENCES

Boakes, R.A., & Halliday, M. S. (Eds.), *Inhibition and Learning.* London: Academic Press, 1972.

Estes, W. K., Koch, S., MacCorquodale, K., Meehl, P. E., Mueller, C. G., Jr., Schoenfeld, W. N., & Verplanck, W. S. *Modern learning theory: A critical analysis.* New York: Appleton-Century-Crofts, 1954.

Fonberg, E. Professor Jerzy Konorski. *Acta Neurobiologiae Experimentalis,* 1974, *34,* 655–664.

Hilgard, E. R. *Theories of learning.* New York: Appleton-Century-Crofts, 1948.

Hilgard, E. R. *Theories of learning* (2nd ed.) New York: Appleton-Century-Crofts, 1956.

Hilgard, E. R., & Marquis, D. G. *Conditioning and learning.* New York: Appleton-Century-Crofts, 1940.

Kamin, L. J. "Attention-like" processes in classical conditioning. In M. R. Jones (Ed.), *Miami symposium on the prediction of behavior: Aversive stimulation.* Miami: University of Miami Press, 1968.

Konorski, J. *Conditioned reflexes and neuron organization.* Cambridge: Cambridge University Press, 1948.

Konorski, J. *Integrative activity of the brain: An interdisciplinary approach.* Chicago: University of Chicago Press, 1967.

Konorski, J. Jerzy Konorski. In G. Lindzey (Ed.), *A history of psychology in autobiography* (Vol. 6). New Jersey: Prentice-Hall, 1974.

Konorski J., & Miller, S. On two types of conditioned reflexes. *Journal of General Psychology,* 1937, *16,* 264–272. (a)

Konorski, J., & Miller, S. Further remarks on two types of conditioned reflexes. *Journal of General Psychology,* 1937, *17,* 405–407. (b)

Miller, S., & Konorski, J. Sur une forme particuliere des reflexes conditionnels. *Les Comptes Rendus des Seances de la Société de Biologie,* Société Polonaise de Biologie 1928, *99,* 1155–1157.

Miller, S., & Konorski, J. On a particular form of the conditioned reflex. (Originally published 1928. English translation by B. F. Skinner.) *Journal of the Experimental Analysis of Behavior,* 1969, *12,* 189–189.

Mowrer, O. H. *Learning theory and behavior.* New York: Wiley, 1960.

Mowrer, O. H. How does the mind work? *American Psychologist,* 1976, *29,* 843–857.

Osgood, C. E. *Method and theory in experimental psychology.* New York: Oxford University Press, 1953.

Pavlov, I. P. *Conditioned reflexes.* Oxford: Oxford University Press, 1927.

Prokasy, W. F. (Ed.), *Classical conditioning: A symposium.* New York: Appleton-Century-Crofts, 1965.

Rescorla, R. A., & Solomon, R. L. Two-process learning theory: Relationships between Pavlovian conditioning and instrumental learning. *Psychological Review,* 1967, *74,* 151–182.

Rescorla, R. A. & Wagner, A. R. A theory of Pavlovian conditioning: Variations in the effectiveness of reinforcement and non-reinforcement. In A. H. Black & W. F. Prokasy (Eds.), *Classical conditioning II: Current research and theory.* New York: Appleton-Century-Crofts, 1972.

Sheffield, F. D. Relation between classical and instrumental learning. In W. F. Prokasy (Ed.), *Classical conditioning: A symposium.* New York: Appleton-Century-Crofts, 1965.

Skinner, B. F. *The behavior of organisms.* New York: Appleton-Century-Crofts, 1938.

Wagner, A. R., & Rescorla, R. A. Inhibition in Pavlovian conditioning: Application of a theory. In R. A. Boakes & M. S. Halliday (Eds.), *Inhibition and learning.* London: Academic Press, 1972.

Woodworth, R. S., & Schlosberg, H. *Experimental psychology.* New York: Holt, 1954.

2

Classical Conditioning as the Formation of Interstimulus Associations: Stimulus Substitution, Parasitic Reinforcement, and Autoshaping

Eliot Hearst
Indiana University

I. INTRODUCTION

Jerzy Konorski is best known to Western experimental psychologists for his early articles proposing and analyzing the distinction between classical (Type I) and instrumental (Type II) conditioning (Miller & Konorski, 1928; Konorski & Miller, 1937a, 1937b). His contributions to this topic influenced the expression of Skinner's specific views on respondent and operant conditioning (Skinner, 1937, 1938), and in the long run probably also played a significant role in the development of the more intricate "two-process learning theories" that were subsequently advanced and became extremely popular. (See Mowrer, 1976, and Rescorla & Solomon, 1967, for summaries and comments.) However, Konorski's general theoretical system, the ways in which he modified it over 20 years, and much of the research that he performed in conjunction with his colleagues at the Nencki Institute received relatively little attention in the West during his lifetime. This neglect may be partially due to the comparative inaccessibility of most of the journals in which he presented his ideas and findings, but is more likely attributable to the neurophysiological emphasis and speculative nature of his writings — features that would inevitably have prejudiced a large number of behaviorists against careful examination of his work. Halliday (Chapter 1) develops some of these themes in attempting to explain why Konorski's contributions were largely disregarded for many years.

Konorski died at a time when many Western psychologists were becoming disenchanted with extremely behavioristic, response-centered approaches to the analysis of learning. He probably would have been pleased with the various recent attempts to discuss learning and memory in terms that go well beyond simple relationships among observables. Nowadays workers in the field of animal and human learning seem less reluctant to speculate about potential intervening processes or stages that may help to explain how an objective, environmental stimulus produces some observable response. Expectancies, spatial maps, reinforcer representations, short- and long-term memories, internal clocks, retrieval processes, channel capacities, and perceptions of causal relations have in the last few years been seriously proposed as mechanisms underlying overt behavior, even in species as unpretentious as the rat and pigeon (e.g., Hulse, Fowler, & Honig, 1978). Many workers find that the postulation of such processes or mechanisms has considerable heuristic value, although they recognize the dangers of any move in a mentalistic direction. Therefore, the Zeitgeist is now more receptive to approaches like Konorski's, and increased attention to various aspects of his work (e.g., the conceptualization and measurement of inhibition: see Boakes & Halliday, 1972; Gray, Chapter 11; Rescorla, Chapter 4; appetitive–aversive interactions: see Dickinson & Dearing, Chapter 8; constraints on learning: see LoLordo, Chapter 14; Shettleworth, Chapter 15) was already discernible in the West at the time of his death.

In this paper I will try to briefly summarize Konorski's views on classical conditioning, compare them with other significant alternatives, and evaluate them in the context of some contemporary research on the topic. His views on mechanisms of reinforcement and the basic connections formed during classical conditioning are stressed here, as well as possible systematic and empirical implications of his distinction between preparatory and consummatory reflexes. Consideration of these issues leads us to reexamine Pavlov's concept of stimulus substitution and Konorski's notion of "parasitic" reinforcement, particularly as applied to recent research concerning the classical conditioning of directed movements (Hearst & Jenkins, 1974; Schwartz & Gamzu, 1977). The findings challenge the clarity and validity of the distinction between classical and instrumental conditioning, just as some of Konorski's own final writings (e.g., 1969, 1973) seriously questioned the sharpness of this distinction, which had formed the basis of his first major scientific contribution.

II.   KONORSKI ON CLASSICAL CONDITIONING

Like Hebb (1949), Konorski distinguished between two general kinds of connections that can be formed within the nervous system. One type involves the integration of sensory input and underlies what other writers have called *perceptual learning;* in Konorski's framework such learning entails actualization of potential linkages between units of lower-level (receptive) analyzers and units within the higher level (perceptive or gnostic) areas of the brain. Gnostic units developed in this manner represent stimulus objects with which the subject has become acquainted; for example, as a new acquaintance's face becomes familiar, it is recognized as a unitary percept. The second type of connection provides the basis for *associative learning* and involves the development of linkages between different gnostic units as a result of their synchronous activation. Thus the sight of a familiar person's face may often evoke an image of that person's voice, because the appropriate visual and auditory gnostic units have been simultaneously activated in the past.

According to Konorski, classical or Type I conditioning provides an especially useful arrangement for studying the formation and maintenance of this second type of connection—that is, an association between perceptive units. The laws of classical conditioning do not differ basically from those governing the acquisition of other interperceptive connections, but the Type I procedure possesses great experimental and analytical advantages because the US reliably evokes a definite, observable response. As a result of contiguity between transmittent stimuli (CSs) and the recipient stimulus (the US), the former eventually come to elicit an overt response that (barring certain practical complications, to be discussed below with respect to "stimulus substitution") is the same as the response consistently elicited by the latter, biologically important stimulus. Measurement of this response to the CS permits us to study the acquisition and persistence of the S–S association in an objective way; the CR acts as a "tracer" enabling detection of the status of the connection.

Drive and reinforcement are terms used frequently by Konorski, but he relied much more on drive as an explanatory concept. The facilitating influence of a certain drive or combination of drives is necessary for associative learning to proceed, but, as we will see, Konorski noted differences between the actions and effects of appetitive and aversive drives that have significant implications for his theory. Concerning the concept of reinforcement, Konorski stated that he would use the term only in an operational, nonphysiological sense, to refer to a stimulus consequence that establishes and sustains a CR to the conditioned stimulus.

In 1948 Konorski discussed the behavioral effects of classical conditioning in terms of the acquisition and maintenance of what he later referred to as consummatory responses. The final version of his complete theory (1967) treats classical conditioning as involving the formation of and interaction between two types of CRs, preparatory and consummatory, which are not necessarily conditioned to the same CSs. Preparatory reflexes tend to direct subjects toward attractive stimuli and away from aversive stimuli, and presumably depend on central drive states like hunger or fear. External or humoral stimuli may activate such drives, whose centers are said to be located in the hypothalamus and limbic system; such activation produces a relatively nonspecific excitation of neural units. The behavioral effects of a drive include the facilitation of attention to impinging stimuli and the excitation of general motor activity.

Consider a hungry dog fed for the first time in an experimental room. Because of the simultaneous activation of the gnostic units corresponding to the experimental context and the gnostic units corresponding to the drive state, the room itself rapidly comes to evoke what Konorski called the hunger CR, which reflects a conditioned motivational state and is indexed by motor restlessness and searching behavior, as well as by heightened sensitivity to a variety of stimuli, particularly those appropriate to the prevailing drive—gustatory or olfactory stimuli in this case. Drive (preparatory) CRs comprise relatively diffuse behaviors, often not easily measurable; these responses are typically tonic or prolonged and are most likely to be observed in some general context or in the presence of comparatively long-lasting CSs. In human beings hunger CRs are mainly determined by the amount of time since the last meal, and not by very specific stimuli like the sight of food on the dinner table.

Through its arousal of afferent and efferent mechanisms, the conditioning of a drive CR to the experimental setting serves as a necessary precursor to the conditioning of specific consummatory CRs to CS. These latter CRs are the ones that experimental psychologists have normally focused upon and measured during an experiment in classical conditioning: salivation, leg flexions, eye blinks. Such responses are relatively discrete and phasic; and, besides being originally evoked by quite specific stimuli (USs), they are themselves most easily attached to intermittent, brief stimuli. Thus human beings begin to salivate as their favorite food is placed in front of them, but do not constantly salivate while they are on their way to the restaurant where it is to be served. According to Konorski, consummatory CRs take longer to condition than preparatory CRs and are less resistant to extinction. The classical conditioning of consummatory responses is presumed to be mediated by centers in the thalamocortical system, and involves connections between gnostic units representing the explicit CS and the effective (proximal) stimuli constituting the US—for example, gustatory for food, somatic pin-prick for shock. In contrast to the restlessness and activity that characterize the hunger CR and initially occur also during the CS, the motoric effects produced by a well-established brief CS for food are said by Konorski to involve cessation of movement and directed attention toward the US delivery site.

Interactions between drive and consummatory reflexes, both conditioned and unconditioned, are important for Konorski's general analysis of classical conditioning. In this regard there are basic differences between appetitive and aversive procedures: Although appetitive- and aversive-drive CRs facilitate their related consummatory reflexes (e.g., hunger excites salivary CRs, and fear generally augments leg-flexion responses to shock), in the case of appetitive conditioning the occurrence of the consummatory reflex temporarily inhibits the relevant drive, whereas in the case of aversive conditioning the delivery of the US has no such inhibitory effect. Thus food in the mouth inhibits hunger, as evidenced by the temporary cessation of stomach contractions and by decreased general motor activity, whereas the delivery of shock does not decrease fear.

According to Konorski, the drive inhibition that occurs during consumption of an appetitive US eventually becomes conditioned to CSs immediately preceding US. This process accounts for the reduced activity and "calmness" that are often observed during CSs for food, and also plays a major role in the gradual impairment of alimentary CR performance that Pavlov (1927) frequently noticed, referred to as "extinction or inhibition with reinforcement (pp. 234ff.)" and regarded as dependent on some inevitable functional exhaustion of active cortical elements, which reinforcement delayed but did not prevent. Attribution of these effects to conditioned drive inhibition helped Konorski handle the fact that introduction of a new set of CSs, or interspersed presentation of CS − s that are similar to CS +, often decreases or prevents such performance decrements during CS +. The less certain the subject is with respect to exactly when or whether the US will be delivered, the more strongly is the relevant drive maintained; unexpected nonreinforcement is presumed to evoke the relevant drive and facilitate performance on later trials (cf. Wagner, Chapter 3). Konorski seemed to predict that in many instances partial reinforcement should maintain preparatory and consummatory CRs better than continuous reinforcement, particularly in appetitive situations—a prediction not completely borne out by experimental findings (see Mackintosh, 1974).

Thus Konorski viewed classical conditioning in a connectionistic manner, as a process requiring the synchronous action of two stimuli and the presence of an appropriate drive. An important corollary is that associations are formed not only between the US and contextual or intermittent external stimuli, as stressed earlier and in standard analyses of classical conditioning, but also between the contextual and intermittent stimuli themselves and between drives and the external stimuli. Of course it is the central representa tions of these events (gnostic units) that are actually connected and mediate the development of CRs; Konorski (1967, p.267) presented several arguments against the possibility that classical conditioning involves establishment of direct connections between CS representations and efferent centers. For him, the CS produces either an image or hallucination of the US, which then brings about the CR. Sensory preconditioning is also treated as a form of S–S learning, governed by the same basic mechanisms as classical conditioning; and Konorski speculated that one drive underlying the formation of such associations between neutral stimuli may involve arousal produced by the stimuli themselves ("curiosity drive").

Although Konorski assumes a definite S–S position concerning classical conditioning, he goes beyond most earlier approaches of this kind by offering rather extensive and yet flexible predictions about the types of behavior that should appear and interact in the classical-conditioning situation. (He cannot easily be accused of leaving his animals "buried in thought," a well-known taunt directed at one S–S theorist of the past.) Konorski (1967) described specific preparatory and consummatory reflexes evoked by various drives and USs, and listed several factors (e.g., CS duration, intertrial interval, US magnitude) that should favor domination of one or the other general type of reflex in the presence of situational or intermittently presented cues. For example, because long CS–US intervals are used, Konorski believed that the CER paradigm (perhaps the most common procedure employed by Westerners who study Pavlovian conditioning) involves mainly preparatory (fear) CRs manifested by freezing, crouching, defecation, and so on, rather than consummatory (shock) CRs like limb flexion. It is certainly true that during a CER stimulus lasting one or two minutes, subjects do not display much flinching or jumping; rather, they ordinarily show the more diffuse or tonic forms of behavior that Konorski characterized as preparatory. Along these lines, Konorski obviously expected the optimal interstimulus interval (ISI) for establishment of Type I CRs to differ depending upon the response measure—an expectation that is solidly confirmed by experimental findings (Mackintosh, 1974).

Because, by Konorski's definition, classical conditioning involves the transfer to CS of only those behavioral effects originally elicited by US, it is clear that he conceived classical conditioning in a way very similar to the view implied by Pavlov's concept of stimulus substitution. However, Konorski (1967, pp. 268–270) suggested several reasons why CRs may occasionally not match URs. One type of exception to this rule is explained by the idea of "parasitic" instrumental reinforcement. For example, food that is delivered immediately after the subject has looked at or moved toward the CS could serve to instrumentally condition such movements; these responses may become quite strong because of their accidental coincidence with subsequent food, even though they are not originally responses to food and are, in fact, antagonistic to the motor behavior directed toward the location of food. Konorski's concept of parasitic reinforcement is, for all practical purposes, identical with the notion of "superstitious" operant conditioning employed by Skinnerians, and later in this chapter I will attempt to discuss the relevance of such concepts for our understanding of some recent findings on the Pavlovian conditioning of directed movements.

The variables that Konorski stressed in his analysis of classical conditioning are not very different from the factors that most other theorists would emphasize: CS intensity, US magnitude, drive level, spatial contiguity of CS and US, and CS–US asynchrony. Besides these standard parameters of classical conditioning, he also discussed the facilitating effects on CR performance of CS intermittency and "roughness". Nonmonotonous CSs (e.g., of irregularly changing intensity) produce stronger CRs. This is an interesting type of manipulation, which has been neglected in Western work on Pavlovian and instrumental conditioning (but see Pavlov, 1927, Chapter 14). Konorski related such effects to physiological adaptation; their possible relevance to the effects of stereotypy of stimulation on reflex habituation and sensitization, and to the operant–respondent distinction, has been stressed by others (see Razran, 1971).

Konorski presented an extensive discussion of theory and data concerning the transformation of excitatory CRs into "inhibitory" CRs and of appetitive CRs into aversive CRs. In such cases, the old connections are presumed to remain intact and to influence the formation, maintenance, and memory of the new connections. Therefore, if a CS+ is converted into a CS−, the old excitatory linkage between the CS and US centers survives and exerts an antagonistic influence on learning and performance appropriate to the new (excitatory) linkage between CS and no-US centers. Transformations of appetitive CS+s into aversive CS+s, and vice-versa, generally prove very difficult to achieve because of the strong mutual antagonism between protective and preservative drives. However, conversion of an appetitive CS− into an aversive CS+, or an aversive CS− into an appetitive CS+, is relatively easy because antidrives in one category are not antagonistic to drives in the other category. (See Dickinson & Dearing, Chapter 8, for an extensive discussion of appetitive–aversive interactions.)

To sum up, Konorski offered an interpretation of classical conditioning that relied heavily on S–S contiguity and drive. His approach was not really response-centered, but because he included both preparatory and consummatory reflexes within his framework, it was possible for him to make relatively flexible predictions about the forms of overt behavior that should initially appear and subsequently persist or decrease during various kinds of Type I conditioning. In his writings he anticipated many issues that are of great contemporary interest for students of classical conditioning: the control exerted by contextual cues, the measurement and significance of inhibitory learning, the effects of certainty and of unexpected events, the complications created by possible parasitic or "superstitious" reinforcement of responses, the rebound motivational effects that may follow stimulus offsets, the antagonism existing between certain drive states, the experimental operations that may alter already-established central representations of external stimuli, and the value of treating the "CR" as an interacting set of functionally related behaviors rather than as a single reflex. These are topics that many contributors to the present volume address.

III. ALTERNATIVE THEORETICAL VIEWS

Konorski's treatment of classical conditioning differs from several influential Western approaches to the topic. One general difference involves his view that such conditioning is merely a special case of the learning of interperceptive (S–S) associations and that the overt behaviors that develop during classical conditioning reflect the associative process rather than constitute basic elements of it. This perception-centered interpretation contrasts with response-centered treatments following the Thorndikian tradition, in which associations between the CS and particular *responses* are said to develop during classical conditioning, either because of contiguity alone or as a result of contiguity plus drive reduction (reinforcement). In the 1930s and 1940s theorists took much more definite or extreme positions than today concerning the relative importance of S–S versus S–R associations in conditioning. Nevertheless, similar issues, sometimes rephrased or disguised, are still with us in the 1970s as a source of debate and experimentation. (See Hearst, 1978; Rescorla, 1978.)

In addition to his belief that the basic elements of Type I associative learning involve only stimulus representations, Konorski also differs from other influential writers in terms of the conditions required for establishment of a Type I CR. For him, not only contiguity between CS and US is essential, but also some background drive. However, other workers (e.g., Rescorla, 1967) would conceive "contiguity" as acting in a somewhat more complex manner than Konorski explicitly described; and still other authors would either scrupulously avoid a separate drive concept or would argue that classical conditioning can proceed in a motivationally neutral setting. With respect to the concept of "reinforcement," Konorski (1967) asserted that he employed the term in a nontheoretical sense; but of course various other theorists would insist that some strengthening event, corresponding to Thorndike's "satisfying state of affairs," must follow a CR in order for it to be acquired and maintained. Certain of these latter theorists believe that mere conjunctions of the CR and reinforcer are sufficient for learning, whereas other writers presume that the CR gains strength because it actually affects the value of the reinforcer. And along different lines, which echo one aspect of the classic exchange between Guthrie and Pavlov in the 1930s (Guthrie, 1930, 1934; Pavlov, 1932), many Western learning theorists would strongly object to Konorski's emphasis on physiological speculation and reductionism, regardless of the behavioral research and psychological theory he proposed. A persistent East–West difference in the treatment of learning and behavior concerns the view, much more prevalent in the East, that behavioral research must go hand in hand with neurophysiological research if work on conditioned reflexes is to be truly valuable.

This section of the present chapter briefly examines several differences between Konorski's views and those of some other writers, expecially with respect to basic associative elements, contiguity, and reinforcement. More complete summaries and empirical evaluations of related issues can be found in Mackintosh (1974).

A. Interperceptive Versus Stimulus-Response Associations

Konorski followed the Pavlovian tradition in his stress on classical conditioning as a process that basically involves the formation of connections between neural centers representing the CS and US. This view is not very different in spirit from the interpretation offered by so-called S–S theorists like Tolman (1932), Bolles (1972), and Bindra (1972), nor from the cognitive interpretation of Pavlovian conditioning recently outlined by Rescorla (1978). All these writers make a distinction between learning and performance and would presumably view the specific CRs that appear during classical conditioning as indices of the establishment of connections between stimulus events — for example, learning the "causal texture of the environment" (Tolman & Brunswik, 1935) — rather than as fundamental or primary units in the associative process. In other words, we are fortunate that certain stimuli happen to evoke various unconditioned responses, for this outcome permits us to study CS–US associations in an objective and reliable way; in classical conditioning it is not the responses themselves that are being learned, for they are wired into the organism from the very outset.

Alternative points of view take issue with such a deemphasis of the response and are generally much more peripheralistic. For example, Guthrie (1935) argued that classical conditioning occurs merely because the procedure ensures that the UR is elicited shortly after the CS is presented. Such close contiguity of a stimulus and a response results in the formation of an association between them so that future presentations of the CS will evoke a response closely resembling the UR — provided that the CS is reproduced fairly precisely on subsequent trials and that no other strongly interfering behaviors are evoked during the CS or US. Although Hull (1943) agreed with Guthrie that the basic associations in classical conditioning are S–R in nature and that contiguity of CS and UR was necessary for conditioning, he believed that the US must, in addition, involve reduction of some drive in order for successful conditioning to occur. Skinner (1938, 1953) cannot be categorized as an S–R theorist in the traditional sense of the label, but his discussions of classical conditioning are certainly response-centered. He accepts the validity of the concept of stimulus substitution, but he maintains that reflex responses constitute only a small part of the total behavior of an organism. Thus Pavlovian conditioning, according to Skinner (1953), can add "new controlling stimuli, but not new responses (p. 56)." This opinion provides one of the reasons for the overwhelming emphasis on operant conditioning in his writings and experimental research; he has been particularly interested in using reinforcement to "shape" complex behavior patterns from available but simpler responses.

A variety of studies have attempted to pit S–S and S–R explanations of Pavlovian conditioning against each other, but without definite resolution of the issue. Blockage of the UR during CS–US pairings by means of surgical procedures or the injection of drugs does not prevent conditioning; when the blocked response is again physically possible, it is evoked by the CS without further training. Sensory-preconditioning experiments are often successful, even though they involve the pairing of two stimuli neither of which evokes a definite overt response. And an appeal to the conditioning of specific peripheral responses cannot explain the effects of Type I CSs on instrumental responding (see Rescorla & Solomon, 1967), because for example, a shock signal will reduce VI food-rewarded behavior and generally increase Sidman avoidance behavior.

Even though these results seem to support an S–S approach over conventional S–R interpretations, which are very peripheralistic, the findings might be handled by less extreme forms of S–R theory. The definition of a "response" could be broadened to include central nervous system activity or motivational states, or it might be argued that subtle responses were conditioned that the experimenter failed to record or was technically unable to measure. Consequently, many workers feel that the S–S versus S–R controversy boils down to a semantic one related to the definition of a response. Although this conclusion has some force, it provides no excuse for the experimental neglect of such phenomena as latent learning, sensory preconditioning, place learning, and so on, which were closely linked historically with the S–S versus S–R battle. And in this volume Rescorla (Chapter 4) describes recent research that attempts to analyze aspects of associative learning in terms of response-centered versus "representational" explanations; his findings are important whether or not they are cast in the framework of S–S versus S–R interpretations.

B. Contiguity Versus Informativeness

According to Konorski, associative learning occurs when two gnostic units linked by a potential connection are synchronously activated; optimal conditioning requires that a US closely follow the onset of CS. There is virtually no one, regardless of his theoretical persuasion, who would disagree with the general notion that temporal contiguity promotes classical conditioning, so long as the CS precedes the US.' However, recent empirical and theoretical analyses have suggested that conceptions of classical conditioning that consider only pairings of CS and US are misleading or incomplete. In addition to being paired with the US, the CS apparently must also be informative with regard to deliveries of the US; pairings of CS and US are necessary but not sufficient for excitatory learning. The use of concepts like *predictability* (Kamin, 1969), *contingency* (Rescorla, 1967), and *validity* (Wagner, 1969) represents the attempts of different workers to encompass the fact that an analysis in terms of mere conjunctions of CS and US is insufficient to explain a variety of recent findings in Type I conditioning.

Experiments on "blocking" (e.g., Kamin, 1969) provide one illustration of the need for a theory of classical conditioning that goes beyond simple contiguity. For example, after a noise (CS) has been paired several times with shock (US), a light may be presented simultaneously with the noise while the noise–shock pairings are continued. Subsequent tests with only the light usually show that it has developed little, if any, power as a CS even though it had been paired with shock often enough for powerful CER conditioning to have appeared if the prior noise-alone trials had not been given. One popular interpretation of this finding is that the light, being redundant, supplied no new information to the subject and therefore conditioning did not occur.

Rescorla (e.g., 1972) performed a series of studies that also illustrate the deficiency of theoretical accounts of classical conditioning that rely excessively on contiguity of CS and US. Even though CS and US were frequently paired in Rescorla's research, the delivery of ("extra") USs in the absence of CS, at the same rate as during CS, resulted in little evidence of conditioning to the CS. These experiments demonstrated that the relative frequencies of the US in the presence versus absence of CS, rather than the sheer number of CS–US pairings, determine the strength of conditioning to a CS. Once again, if a formerly neutral stimulus supplies no information about occurrences of the US, the stimulus will apparently acquire little excitatory power even though it occurs frequently in conjunction with the US.

In his initial treatment of such results, Rescorla (1967) offered a molar interpretation based on the concept of "correlation" or "contingency"; subjects were presumed to possess the sophistication needed to calculate over-all US probability and compare it with US probability during CS. A few years later he and Wagner (Rescorla & Wagner, 1972) proposed a theory that handled a wide variety of these "informational" effects, but on the basis of a more molecular, trial by-trial model that clearly asserted the importance of temporal contiguity. Their theory is so well known that there is no point in describing it further in this chapter. However, with respect to Konorski's approach, the Rescorla–Wagner model is not so much inconsistent with Konorski's views as it is a much more detailed and precise exposition of how contiguity acts to determine the amount of associative strength accruing to different stimulus elements — including contextual or background cues — during classical conditioning. Actually, when talking about situational cues, unexpected events, and excitatory–inhibitory interactions in conditioning, Konorski anticipated in an informal way some of the important features of the new model; moreover, Halliday (Chapter 1) comments on Konorski's apparent realization of the distinction between contiguity and contingency. Incidentally, Konorski's earlier (1948) approach to the analysis of classical excitatory and inhibitory conditioning seems to come closer to making specific predictions of the kind offered by Rescorla and Wagner than does the less formal, more qualitative approach proposed by Konorski in 1967. (Rescorla, Chapter 4, discusses related issues in more detail than is possible here.)

C. Mechanisms of Reinforcement

Konorski (e.g., 1967, p. 266) considered the term "reinforcement" imprecise and misleading, partly because it was used differently by Pavlov and Hull. However, for purposes of convenience and readability, he decided to employ it when the presentation of a stimulus, in either the classical or instrumental procedure, leads to the establishment of a conditioned response to the CS and sustains this response after it has been established. He stated that he would use the word reinforcement in a purely operational sense, implying nothing about the physiological mechanisms of associative learning.

One can debate whether Konorski adhered to a strictly operational use of the term throughout his 1967 book. However, a more important question involves his views concerning the function of the US in classical and instrumental conditioning and concerning the relationship between the US and the particular responses that emerge from these procedures. For instrumental conditioning he viewed the US as supplying the drive-reduction mechanism that accounts for learning and performance of the successful instrumental movement, which closely resembles a Law of Effect interpretation (although his concomitant use of the principle of retroactive inhibition has Guthrian implications; see, e.g., his summary on pp. 513–514). In classical conditioning, on the other hand, his approach seems to follow Pavlov's ideas on stimulus substitution quite closely. According to Pavlov's interpretation, pairings of CS and US establish the CS as a substitute for US; evidence for this interpretation is taken to be the CS's acquired ability to evoke the behavioral effects elicited by the US. This approach does not imply that the CR is actually strengthened by its consequences; the US does not act to "reinforce" prior responses, but to produce a set of responses (URs) that can then be transferred to its substitute, the CS.

I will return to the topic of stimulus substitution in my discussion of some contemporary research on autoshaping in the next section of this chapter. However, to foreshadow these later comments, it seems worth mentioning here the basic aspects of alternative conceptions of classical conditioning that view the US as serving the function of a response strengthener. These approaches (e.g., Hull, 1943) have some difficulty handling the close resemblance between CR and UR that is said to characterize classical conditioning, but we will see shortly that identity of

CR and UR may not be so very common in Type I conditioning anyway. In one version of this general response-strengthening interpretation—a version that I will call the "response–reinforcer contiguity" explanation — any response occurring during the CS and closely followed by a drive-reducing event like food delivery or pain termination will be strengthened. Because it evokes specific URs, the US increases the likelihood that responses identical with or resembling URs will occur in the general situation and will become the behaviors strengthened in this manner; however, in theory at least, virtually any response that happens to appear during CS would also increase in probability due to its mere conjunction with US. This interpretation seems very similar to Konorski's description of the establishment of parasitic instrumental responses and to Skinner's (1948) analysis of superstitious reinforcement, for in both cases responses are strengthened even though there is no contingency of any kind between behavior and US delivery; the behaviors are not presumed to actually affect the delivery or value of the US in any way.

In the other version of this response-strengthening view of US function, which I will call the "response–reinforcer contingency" explanation, the CR is definitely assumed to modify the effectiveness of the US (e.g., Perkins, 1968; see Gormezano & Kehoe, 1975, for a critical discussion). Salivation may increase the palatability of food, making it more attractive; and freezing or flinching may reduce the pain produced by a shock, making it less aversive. Although in many cases the exact nature of the contingency between the CR and subsequent changes in US value is not easily specified by a theorist or assessed by an experimenter, the CR–US relationship presumably operates as it would during standard instrumental conditioning: Those responses are strengthened that bring about a more favorable consequence than would occur in their absence. So far as I can tell, Konorski did not specifically assess this "instrumental" explanation of classical conditioning, but the possibility has arisen frequently in discussions of the merit of the classical–instrumental distinction (see Hearst, 1975a).

Of course it is quite conceivable and very likely that these various mechanisms — stimulus substitution, response-reinforcer contiguity effects, and response-reinforcer contingency effects—are all operating in a given example of so-called Type I conditioning. Furthermore, one of the factors may be relatively more important in the initial stages of training (e.g., in producing the first few CRs in the presence of the CS), whereas the other factors may serve mostly to maintain and strengthen the CR once it has been established (cf. Champion, 1969).

We turn now to some recent findings in the field of classical conditioning that bear on these and related issues. The arrangements that I stress involve the measurement of directed actions of the whole organism (autoshaping and sign tracking). A brief review of the most important findings concerning this topic precedes a discussion of their relevance to Konorski's theory and to other interpretations of classical conditioning.

IV.   DIRECTED MOVEMENTS AND S–S ASSOCIATIONS

A. The Phenomena of Autoshaping and Sign Tracking

The discovery of autoshaping by Brown and Jenkins in 1968 served as the stimulus for a large number of North American experiments investigating the effects of Pavlovian conditioning procedures on directed actions of the whole organism. (See reviews and evaluations of this work by Hearst & Jenkins, 1974, and Schwartz & Gamzu, 1977.) The general topic has attracted considerable attention for several reasons. First of all, evidence demonstrating the classical conditioning of complex skeletal responses—for example, orientation movements, approach–withdrawal behavior, and signal-contact responses—surprised many psychologists who viewed classical conditioning as limited to autonomic responses or to relatively isolated, undirected "local" movements like eye blinks, leg flexions, or knee jerks. Therefore, the new findings had clear implications for the general validity of the classical-instrumental distinction, based as it often has been on presumed differences in the conditionability of autonomic and skeletal responses. Furthermore, results of research on autoshaping are relevant for such current issues as (1) the "arbitrariness" of conventional operant responses like the pigeon's key peck and the rat's lever press, because these responses also appear during exposure to appropriate Type I procedures, and (2) the relative merits of response-centered versus perception-centered theories of simple learning, because approach tendencies toward a CS for food develop and persist after mere "observation" of CS–US contingencies and despite actual loss of reinforcement for approach behavior toward CS. The following abbreviated summary of research on autoshaping should provide a background for discussion of its various theoretical implications, particularly with regard to Konorski's views.

   In contrast to conventional (instrumental) procedures that also use pigeons, response keys, and grain reinforcement, Brown and Jenkins (1968) made 4-sec grain delivery (US) contingent only on a prior stimulus (CS = 8-sec key illumination). Such a CS–US contingency was instituted after brief preliminary training in which pigeons learned to eat from a grain dispenser that was illuminated only during the 4-sec periods of US access. As pairings of CS and US continued, birds first exhibited increases in activity during CS and oriented toward the lighted key; then they began to approach it; finally they pecked at it. Brown and Jenkins named this phenomenon autoshaping and intended the term to cover the approach and contact behavior that develops toward a localized signal for an appetitive US. The procedure was *auto*matic and the pigeon could be said to have shaped it*self* to perform a response that in previous research had required the experimenter's assistance—that is, the method of successive approximations or "manual shaping."

Subsequent experiments have demonstrated that the acquisition and maintenance of key pecking in this type of situation depend mainly on a positive contingency between key illumination and grain delivery; birds rarely peck keys that are illuminated randomly with regard to grain presentations (e.g., Wasserman, Franklin, & Hearst, 1974).[1] Furthermore, autoshaping has been observed in experiments involving a variety of different species, responses, reinforcers, and general situations. For example, rats approach, contact, and often depress a lever whose insertion into the chamber signals food or electrical stimulation of the lateral hypothalamus (Peterson, Ackil, Frommer, & Hearst, 1972); fish approach and touch a lighted target that precedes delivery of food (Woodard & Bitterman, 1974); and baby chicks approach and peck a key that signals the onset of a period of heat in a cold chamber (Wasserman, 1973; Wasserman, Hunter, Gutowski, & Bader, 1975). Finally, autoshaping is usually long-lasting; most subjects that are tested for thousands of trials continue to approach and contact the CS.

There is also a negative counterpart to these standard findings. Not only do pigeons approach and contact a localized visual stimulus that signals an appetitive reinforcer, but they also position themselves relatively far from the same stimulus when it signals that the reinforcer is *not* going to appear (Hearst & Franklin, 1977; Wasserman et al., 1974). The greater the frequency of food in the absence of the stimulus, the stronger is such withdrawal behavior. In these studies the spatial position of the pigeons was monitored on every trial by means of switches beneath the floor of the chamber.

---

[1]Several recent papers (e.g., Davol, Steinhauer, & Lee, 1977) have argued that the establishment and maintenance of autoshaped behavior may be explained in terms of (1) pecking generalized from the lighted magazine to the lighted key, plus (2) the peck-food conjunctions that occur after the first (generalized) key peck. Although magazine-key generaliza- tion effects certainly may promote autoshaping in various situations, this overall explanation of the phenomenon is inadequate. As Hearst and Jenkins (1974) pointed out, the first key peck is controlled by the degree of correlation between CS and US; birds are much less likely to peck the key, or they fail to peck it at all, if the CS and US are randomly or negatively related rather than positively related. Furthermore, various experiments (e.g., Peterson et al., 1972; Wasserman, 1973; Woodruff & Williams, 1976) have demonstrated autoshaping in situations in which no magazine was used and no directed US behavior was involved (the US entailed brain stimulation, or heating of a chamber, or injection of water directly into the pigeon's oral cavity). The proponent of the generalization explanation also has difficulty handling the fact that the first autoshaped key peck often takes 40 to 100 trials to appear and depends dramatically on the duration of the intertrial interval. Moreover, the "omission" and "observation" results to be described shortly also cannot easily be incorporated into an explanation that stresses key peck–food conjunctions in the acquisition and maintenance of autoshaped behavior.

A few additional facts about autoshaping, reviewed in more detail by Hearst and Jenkins (1974), should be mentioned here. First of all, autoshaping does not emerge or is relatively weak when other stimuli in the situation (auditory or visual) predict the arrival of a reinforcer as well as does the CS. Thus, autoshaping develops to the extent that the CS is a nonredundant, "informative" predictor of the reinforcer. Second, the type of behavior directed toward the CS depends on the kind of US with which it is correlated. A pigeon's key pecks at a food-predictive signal are evenly spaced, brief, and forceful, whereas its pecks at a water-predictive signal are irregularly spaced, sustained, and relatively weak; birds seem to be "eating" the former signal and "sipping" or "drinking" the latter. Analogously, rats gnaw or lick a CS (insertion of a lighted lever into the chamber) that predicts food, but sniff or explore the same CS when it signals electrical stimulation of the lateral hypothalamus, a US that itself produces sniffing and exploring. Thus the type of behavior directed toward the CS often resembles or is somehow appropriate to the type of US that it predicts—a point to which I return shortly.

The term *sign tracking* was coined by Hearst and Jenkins (1974) to cover these and several other effects in animal and human learning. Sign tracking refers to behavior (e.g., eye movements, bodily orientation, approach, signal contact) that is directed toward or away from a feature of the environment (a sign) as a result of the relation between that environmental feature and another (the reinforcer, in a typical experiment). Studies of sign tracking provide useful arrangements for assessing the relative behavioral control exerted by S–S ("classical") versus R–S ("instrumental") contingencies in simple learning.

B. Autoshaping and Type I Conditioning

On the surface, at least, autoshaping as a procedure and a phenomenon seems to conform to Konorski's definition of classical conditioning. It involves the paired presentation of two stimuli, one of which gives rise to a definite overt response from the outset of the experiment. As a result of the pairing procedure, the initially neutral stimulus—a lighted key, in the case of the pigeon—acquires the capacity to elicit the same response (approach and pecking) as does the biologically important stimulus (grain presentation). An association between a CS and a US is presumably formed.

However, this view of autoshaping is perhaps too simplistic. Several important and troublesome questions come to mind:

1. Is the response to the CS really the "same" as the response to the US? One could argue that the response to the US involves approaching and pecking in the grain dispenser, whereas these behaviors are directed toward

another object and location, often quite some distance away (Peden, Browne, & Hearst, 1977), during CS presentations.

2. Can the response of approaching and eating from the grain dispenser be legitimately considered an unconditioned response? After all, the bird must learn where the food is and when it is available (lighted dispenser), and must go to the dispenser in order to obtain it. These behaviors seem to involve "instrumental" contingencies; do they preclude interpretation of the results in terms of straightforward classical conditioning?

3. Are the approach and contact behaviors toward the CS superstitiously reinforced? Because key illumination is a salient event and should evoke Konorski's "targeting reflex," subjects may be looking at the key or approaching it when the grain dispenser is operated. Perhaps accidental conjunctions of these responses and the US play an influential role in the development and maintenance of sign-tracking behavior, through the mechanism Konorski described as the conditioning of "parasitic instrumental responses." To what degree is autoshaped behavior dependent on CS–reinforcer associations as opposed to response–reinforcer associations?

4. How do the behaviors that appear during exposure to autoshaping procedures relate, if at all, to Konorski's distinction between preparatory and consummatory responses? Does autoshaping involve basically different kinds of responses from those that have been analyzed in most earlier research on classical conditioning?

These questions are not really independent of each other, but insofar as possible I attempt to discuss each of them separately. Another issue involves the significance of research on autoshaping for the classical–instrumental distinction, but this question is so closely related to points already covered at length in Hearst (1975a; 1975b) that I refer readers to the earlier articles and merely comment briefly on the apparent changes in Konorski's views on the topic after publication of his 1967 book.

1. Stimulus Substitution

Konorski (1967) remarked that the whole procedure of classical conditioning is based on the experimenter's expectation that CS–US pairings will cause the effects of the US to appear in response to the CS; by definition, "the true classical CR comprises the same elements as the UR on which it is based (p. 270)." In a similar vein, Gormezano and Kehoe (1975) define classical conditioning as a paradigm in which (1) presentation or omission of the US is independent of CR occurrence and (2) the CR is "restricted to the selection of a target response from among those effector systems elicited as URs by the US (p. 149)."

General discussions of the empirical validity and logical status of the principle of stimulus substitution can be found in Mackintosh (1974, pp. 98–109) and Hearst (1975a, pp. 204–206). Our concern here is with the applicability of the principle to research performed on autoshaping; as outlined by Mackintosh (1974), the principle requires that "the CS alone comes to elicit the set of responses normally elicited by the US (p. 98)". In connection with approach and contact behavior in the autoshaping situation, predictions based on the concept of stimulus substitution are ambiguous; there seem to be two possible general outcomes during CS that could be taken to support the principle (see Jenkins & Moore, 1973). First, the subject might make the same directed response as is evoked by the US — approach toward and contact with the reinforcement dispenser — or, second, the subject might approach and contact the CS, which comes to serve as a surrogate for the US; Boakes (1977) calls these two types of effect *goal tracking* and *sign tracking,* respectively. The latter outcome is typical of autoshaping experiments with pigeons, and it is noteworthy that the ambiguity in the concept of stimulus substitution arises because the target behavior involves skeletal movements toward environmental objects rather than directionless reflexes, such as salivation or chewing, which are usually measured in appetitive classical conditioning experiments.

Because Pavlov provided several vivid descriptions of dogs that licked or moved toward lamps whose illumination signaled food, his conception of stimulus substitution certainly included and perhaps would predict the second of the outcomes just mentioned. ("The CS actually stands for the animal in place of food": Pavlov, 1934, p. 187; see Hearst & Jenkins, 1974, for further discussion of Pavlov's views.) Other authors (e.g., Staddon & Simmelhag, 1971; Williams & Williams, 1969; Zener, 1937; but cf. Bindra, 1972) apparently believed that the appearance of movements directed toward CS was inconsistent with a substitution theory of conditioning, because the set of responses occurring during CS was not directed toward the same place as during the US.

I have been unable to discover any experimental observations mentioned in Konorski's major writings that could be taken as clear examples of autoshaping. (But see Stepien, 1974, for an illustration of strong CS-directed behavior — the so-called magnet reaction — which occurs after prefrontal ablations in dogs, cats, and monkeys.) On the contrary, Konorski frequently described the behavior of his dogs during a firmly established CS as involving an initial short-lasting targeting reflex toward the source of the signal, followed by persistent gazing at the feeder and a state of complete immobility. Unfortunately the degree of restraint of the subject, and the type of CS and its proximity to the US site and to the subject, are not clearly specified in most of his descriptions. These factors could be crucial for establishing and detecting sign-tracking tendencies (Hearst, 1975b).

One way of analyzing the basis for conditioned directed responses toward CS, such as those observed in autoshaping experiments, is to employ a US that itself does not require approach and contact behavior for its receipt.

Will subjects approach and contact a signal of US if the US itself does not evoke approach and contact behavior? Several recent experimental results bear on this question. Using 3-day-old chicks, Wasserman (1973) and Wasserman et al. (1975) paired 8-sec illuminations of a standard key light with 4-sec activation of a heat lamp located in the ceiling of a cold chamber (4–15° C), a US that presumably requires no approach, contact, or overt consummatory behavior for its reception. Peterson et al. (1972) followed insertion of an illuminated lever (CS) with rewarding electrical stimulation of the lateral hypothalamus (US) in rats. Woodruff and Williams (1976) paired key-light illumination with delivery of water through a cannula implanted in the beak of thirsty pigeons. In every one of these experiments, the subjects approached and contacted the CS—a localized, positive predictor of the US — even though the US itself did not entail approach and contact movements.

These results violate the stimulus-substitution principle in its strictest sense; certain responses elicited by the CS do not belong to the set of responses produced by the US. However, a deeper analysis of the response patterns evoked during CS and US in those experiments yields some observations that could be taken as partial support for the principle. Peterson et al. found that the behavior directed toward the signal of forthcoming brain-stimulation US included sniffing and exploring movements, which also occurred in response to the US itself. Among other water-related responses, Woodruff and Williams observed the occurrence of swallowing and "mumbling" behaviors during the CS for water delivery—responses always evoked by actual water-in-the-mandibles. On the other hand, Wasserman reported a definite lack of resemblance between CRs and URs in his experiments. Although presentation of the heat US produced a cessation of the chick's activity, extension of its wings, twittering sounds, lowering of the body, and rubbing the floor with its chest—the behaviors directed at the key light CS were quite different. Initially key pecking predominated, but subsequently "snuggling"—which involved less directed pecking, along with head shaking and beak sliding across the wall near the key— became the most common form of behavior.

In a comment on Wasserman's results, Hogan (1974) suggested that although behavior toward the key light does not resemble behavior in the actual presence of heat, the approach–pecking–snuggling pattern is part of the normal heat-*seeking* response of young chicks; chicks frequently rub and peck the feathers on the hen's underside and thereby induce the hen to sit and brood. In a sense, therefore, the key light may act not as a substitute heat stimulus but as a substitute for the warmth-supplying mother hen. Konorski might describe the responses evoked by the CS in this case as similar to the preparatory reflexes that are likely to be exhibited by baby chicks in cold environments where heat USs are occasionally delivered.

Despite these arguments, it seems clear that simple application of the principle of stimulus substitution will not encompass the variety of behavioral patterns that occur during a CS that predicts a particular US (see also Timberlake & Grant, 1975). Konorski himself (1967, pp. 268–270) gave several examples of instances in which CS-elicited responses are not observed during the US, and of instances in which effects occurring during the US do not appear during the CS. Depending on the particular example, he attributed the discrepancies to competition between consummatory and preparatory CRs, the dominance of flight or freezing as a fear response, the development of parasitic instrumental CRs, or the evocation of natural in-strumental responses required for receipt of the US (see the next section of this chapter). In spite of these complexities, he concluded by reiterating his assumption that the true classical CR involves the same elements as the UR on which it is based. However, in my opinion, this outcome may require severely refined and artificial experimental conditions with very precise con-trol of CS and US application, as well as extreme restraint of the subject's movements and measurement of a single target response — the type of classical-conditioning preparation recommended by Gormezano and Kehoe (1975).

The experimental analysis of classical conditioning would probably suffer if we were forced to limit ourselves to such settings. One of the reasons for the renewed interest in classical conditioning over the past 10 years has been the variety of procedures that have been accepted as valid methods for studying the effects of CS–US contingencies: for example, CER, taste-aversion learning, and autoshaping. In my opinion, restrictive, response-centered definitions of classical conditioning hinder more than aid research in this area. If one views such conditioning as involving the formation of an association between two stimuli — the interpretation Konorski himself gave to the phenomenon — then any response measure that provides a reasonable and reliable assay of presumed changes in the strength of an S–S association seems justified, so long as we continue to explore the possibility that other behavioral indices may not necessarily yield the same conclusions. Further-more, this general point of view encourages the examination of a variety of CRs in a given situation, as well as the study of their intercorrelations; such research should in the long run provide a solid empirical basis for establishing distinct categories of behavior. This approach may eventually enable predictions about which types of responses would be likely to transfer, relatively unaltered, from a particular US to a particular CS and which responses would probably be greatly altered or absent.

Ironically, Konorski opened the door to an approach to classical condi-tioning that is flexible with regard to response measures. He offered the view that Type I conditioning is a typical, although experimentally priv-ileged case of interperceptive association; and he described in detail the variety of reflexes, preparatory and consummatory, that may appear during a classical-conditioning experiment and may interact with "instrumentally" reinforced responses. His whole conception of behavior and conditioning seems so flexible as to be irreconcilable with a simplistic application of the principle of stimulus substitution; yet Konorski continued to insist that identity of CR and UR was very much the rule rather than the exception in classical conditioning. Perhaps for very specific, selected stimuli and responses in highly restrained subjects the rule may hold, but if URs typically involve a system of varied but functionally related behaviors (Lorenz, 1969, p. 47), then the outcomes are likely to be considerably more complicated (cf. Woodruff & Williams', 1976, "learned release" view of autoshaped "preparatory" responses).

2. Specification of the Unconditioned Response

In the standard autoshaping situation, it may be inaccurate to state that approach and pecking directed toward grain in the dispenser are URs to food. These responses actually precede rather than follow delivery of grain to the oral cavity of the pigeon, and they are not elicited by activation of the grain dispenser unless specific training is given. The CS is followed by the visual and auditory cues associated with grain presentations, not by grain-in-the-beak, and the bird must learn the location of food and the cues that indicate its availability. Furthermore, a definite response is required in order for the subject to obtain the available US.

Konorski (1967) has pointed out several complications that may arise from analogous aspects of more conventional classical alimentary conditioning, and Gormezano and Kehoe (1975) have discussed related issues at even greater length. Konorski (1967, pp. 269–270) noted that the CS in most Pavlovian salivary-conditioning experiments is not followed directly by the US (food in the mouth) but by the sight of a bowl with food. This visual stimulus evokes a "natural instrumental CR", which consists of transferring the food from the bowl to the mouth and which may interfere with the appearance of other CRs (e.g., masticatory) in response to the CS. Konorski argued that in experiments involving a CS that is immediately followed by the "proper US" — such as acid delivered directly in the mouth — vigorous mouthing movements "identical with those occurring in response to the US" do occur. Furthermore, he described the experiments of Soltysik, Kierylowicz, and Divac, who compared the CRs established when the US (milk or water) was presented either in a bowl or via direct introduction into the mouth through a hole in the cheek. In the former case the dog's CR consisted mainly of the posture of expectation, whereas in the latter case vigorous mouthing and swallowing movements occurred to the CS. Konorski remarked that these latter CRs were so similar to the UR that mere observation of the dog could not reveal the exact moment at which the fluid was placed in the mouth.

Along the same lines, Gormezano and Kehoe (1975, pp.151–152) criticized many so-called examples of classical conditioning because they violate the stricture that the US be administered independently of the subject's behavior. Gormezano and Kehoe commented that this requirement is usually met in the case of noxious USs, because the aversive stimulus (e.g., electric shock) is normally delivered directly to the appropriate receptor surface. However, with appetitive USs involving delivery of food in a bowl or dispenser rather than direct placement of food in the mouth, the stricture is violated because subjects must approach and seize the food before it can reach the receptor surfaces of the mouth. Gormezano and Kehoe argued that it is a mistake to consider as a UR the instrumental response necessary for receipt of food, and they supplied examples of several well-known experiments in "classical conditioning" (with general activity or licking as CRs and URs) that presumably fall victim to this criticism. It will be recalled that, according to Gormezano and Kehoe's definition of classical conditioning, the CR must be restricted to the selection of a target response from among those effector systems elicited as URs by the US. Because licking, changes in general activity, and seizing the reinforcer are not responses evoked by actual delivery of the US but are involved in its receipt, such behaviors are said to be instrumental responses and not true classical CRs. Furthermore, in terms of precise experimental control, Gormezano and Kehoe pointed out that classical-conditioning studies employing these kinds of responses do not allow rigorous specification of the CS–US interval.

Nevertheless, at this stage of our knowledge it seems unwarranted to demand adherence to such restrictive definitions. For one thing, who can state precisely when a particular "US" begins its most effective action, so that "URs" can be measured from that point? Konorski (e.g., 1967, p. 276) himself wondered about such temporal factors. Furthermore, and as noted earlier, if the emphasis is taken off particular responses and if classical conditioning is viewed mainly as the learning of associations or relationships between stimuli, then no specific response index is necessarily the correct or ideal one. As Rescorla (1978) has remarked, if we are interested in the organism's knowledge of interevent relations, then any behaviors that change differentially as a result of exposure to a particular relationship are potentially of interest. This general point of view and its corollary—that a learning–performance distinction is valuable even within the framework of classical conditioning — will probably be very objectionable only to those who maintain a strong S–R, response-centered, or "reflex" orientation toward Type I conditioning. However, if the basic principles of associative learning that govern, for example, autoshaping, CER, and salivary or eyelid conditioning all prove to be quite similar, then strong support will be garnered for this flexible point of view toward the definition of the CR and

UR in Pavlovian conditioning. Perhaps we should even (temporarily?) abandon the terms CR and UR and talk about the interactions of various behavioral measures as a result of presenting stimuli in certain relationships to each other.

3. Stimulus-US and Response-US Relationships: Parasitic Reinforcement?

Konorski noted that movements toward the source of the CS during a classical-conditioning experiment may happen to be followed by food, an outcome that could provide a basis for "parasitic" instrumental conditioning and subsequent maintenance of such motor responses—according to the "response-reinforcer contiguity" (superstitious conditioning) interpretation I outlined earlier. It seems quite likely that Konorski would attribute some of the skeletalmotor effects observed in autoshaping experiments to such a process. The question of whether stimulus–US or response–US relationships play a predominant role in autoshaped behavior has been analyzed extensively elsewhere (see Hearst, 1975a, 1975b; Hearst & Jenkins, 1974; Schwartz & Gamzu, 1977), and only a brief summary of the relevant techniques and findings, supplemented by an indication of their possible relevance to Konorski's theory, is given here.

Of course, the first approach to and contact with the CS in a standard autoshaping experiment cannot be attributed to prior occasions on which these responses were followed by US delivery. However, as soon as such directed movements toward the CS begin to occur with regularity, they are inevitably associated with receipt of the US. Thus the eventual strength and persistence of CS-directed behavior may depend as much on close temporal conjunctions between these behaviors and the US as on the explicitly programed positive contingency between signal and reinforcer. Two general methods have been developed to assess the separate effects of these "classical" and "instrumental" factors on the maintenance of certain target behaviors. Although neither method, taken by itself, can provide totally persuasive evidence regarding the relative importance of stimulus–reinforcer versus response–reinforcer correlations, parallel findings from both sources can prove very convincing (cf. Jenkins, 1977).

One of these methods involves procedures that pit a response–US relationship against a stimulus–US relationship. The *omission* procedure, introduced into behavioral research by Konorski himself and first applied to the study of autoshaping by Williams and Williams (1969), illustrates this general method. A scheduled US is omitted if the target response occurs during the CS. Thus, while maintaining a positive CS–US correlation (all USs occur immediately after CS), the experimenter makes certain that the US never occurs in conjunction with the specified movement (a negative response–US correlation). If the target response remains strong even when its occurrence cancels the US, then a primary role for the CS-US correlation would be indicated and the Law of Effect would be challenged. Stated another way, persistent motor behavior that prevents food US seems to violate one of Konorski's (1967, p. 389) principles of instrumental conditioning: If an external stimulus is followed by an appetitive US, whereas the stimulus accompanied by a particular movement is not, then the animal learns to perform an antagonistic movement during the stimulus.

The second general method for evaluating the relative contributions of CS-US and response-US relationships in autoshaping entails arrangements in which the subject is physically prevented from approaching or contacting the two stimuli while they are presented in some relationship to each other. If actual occurrences of approach, contact, and consummatory responses are unnecessary either for the learning of a CS-US association or for the development of tendencies to approach and contact the CS, then mere *exposure* of the subjects to some positive contingency between inaccessible CSs and USs should produce the immediate or very rapid appearance of approach and contact behavior toward CS in a subsequent test phase with the CS and US both available.

In the initial autoshaping study that employed the first of the aforementioned two methods, Williams and Williams (1969) paired key illumination and grain delivery in a standard pigeon apparatus but included the provision that any peck occurring during CS would cancel the grain scheduled on that trial. Despite this negative consequence, pigeons continued to peck the illuminated key on many trials and therefore lost a large number of the reinforcers that they would have received had they not pecked. Using an elongated box, Jenkins (1973) subsequently reported that birds would approach and peck a predictive key light 3 ft away from the grain dispenser, even though this behavior prevented the subject from returning to the dispenser in time to consume most of the available grain (access to grain lasted only 4 sec on each trial). When an omission contingency was specifically established for key approach rather than key pecking in another type of elongated box in which the key light CS could appear at either end, Peden, Browne, and Hearst (1977) found that birds continued to "approach" (i.e., moved to within 20 cm of the lighted key) on approximately 60 to 90% of the trials—the overall approach percentage depending on the spatial proximity of the CS and the grain-delivery location—even though each instance of this response canceled the reinforcer that was scheduled after CS. Thus the positive relationship between *stimulus* and US apparently plays the major role in controlling and directing the subject's approach and contact behavior in these situations, because conjunctions of such *responses* and US are never permitted to occur. Parasitic or superstitious reinforcement cannot account for the high level of sign tracking obtained in these experiments, although the fact that approach and contact behavior was usually weaker with than without the omission procedure could be taken to indicate that the response–US relationship does have a definite effect. (For comparison data from various control groups and a discussion of the species and situational generality of findings on this topic, see Peden, Browne, & Hearst, 1977.)

Results obtained with the second general method for assessing the relative roles of CS–US and response–US relationships in autoshaping also support the conclusion that the perception of a relationship between CS and US is the major factor controlling sign-tracking behavior. For example, Browne (1976; see also Hearst & Jenkins, 1974, pp. 24–26) presented pairings of key lights and grain to "observing" birds that were physically prevented, by a transparent barrier, from reaching the source of the grain and the signal, or (in another experiment) from reaching only the grain. Compared with groups of birds that observed a zero or negative correlation between key lights and grain, such pretraining with a positive CS–US correlation resulted in a much stronger tendency for the subjects to approach and contact the signal as soon as the barrier was removed and regular autoshaping was initiated. Overt directed responses did not appear during the "observation" phase, and therefore such behavior apparently does not have to occur in order for learning of a CS–US relationship to take place. In further support of this general conclusion, Deeds and Frieman (1977) found that satiated birds, exposed to a positive CS–US correlation with key light and grain accessible, autoshaped significantly faster in subsequent sessions conducted at 75% of normal body weight than did birds previously satiated while exposed to a zero or negative CS–US correlation. During the satiation phase, the various treatment groups had shown no differences in eating, pecking, or approach behavior; they all ate and pecked the key on fewer than 2% of the trials.

Furthermore, Boakes and Ismail (1971) reported evidence of stimulus–stimulus learning in subjects that had not received any magazine training (i.e., had never even been placed in the experimental chamber) before introduction of either Correlated (100% pairings of key light and an *empty* but illuminated food hopper) or Random presentations of these two stimuli. During subsequent standard autoshaping, birds that had been exposed to the Correlated condition pecked significantly sooner than the Random birds. Evidently, learning of a relationship between key illumination and grain-magazine cues can occur even when subjects have never received food in the situation (a type of sensory-preconditioning experiment).

All these experiments support the notion that subjects can learn about a stimulus sequence or relationship during a period of mere exposure to it. Konorski (e.g., 1967, pp. 286–287) would surely have agreed that interperceptive associations of this kind are important and pervasive, that they can occur in the absence of overt behavioral change, and that they are instances of the same general category of associative learning as that to which classical conditioning belongs. However, his theory certainly does not predict or explain several of the behavioral effects obtained in the foregoing experiments; for example, his concept of parasitic reinforcement cannot handle the great persistence of directed motor behavior that prevents food delivery. He would probably have argued that several response tendencies are competing with each other in such experiments. In any event, our knowledge is obviously limited with respect to many factors that select and control the specific CRs that will be acquired during exposure to various conditioning arrangements.

Despite the inability of Konorski's theory to handle some important specific findings with respect to autoshaping, the results presented here seem to favor perception-centered theories of classical conditioning over response-centered theories — an outcome that is very compatible with Konorski's general approach. Further development of such theories will require particular attention to the formulation of response-selection rules, a problem that has consistently proved troublesome to S–S theorists of the past but that Konorski sought to handle in several ways, one of which is connected with his distinction between preparatory and consummatory reflexes.

4. Preparatory and Consummatory Reflexes

A previous section of this chapter reviewed one of the notable additions to Konorski's theory between 1948 and 1967: his distinction between consummatory and preparatory reflexes, which was based on the sequential occurrence of certain activities of organisms. Preparatory reflexes involve preliminary actions directed toward providing (preservative) or avoiding (protective) the appropriate stimuli for consummatory reflexes; preparatory reflexes involve relatively diffuse and complex movements, and would include approach–withdrawal and "searching" behavior. Consummatory reflexes, on the other hand, are typically precise and adaptive and are elicited by very specific stimuli. For example, food intake is preceded by (preparatory) responses involved in locating and seizing food, and the consummatory reflex does not start until (1967) "the moment when the edible substance makes contact with the mouth (p.11)". In considering the difference between consummatory and preparatory reflexes, one is reminded of Skinner's original distinction between unconditioned respondent and operant behaviors, based on whether or not the behavior is elicited by definite environmental stimuli (see Hearst, 1975a). Unlike Konorski, Skinner of course did not view such operants as reflecting the influence of some underlying drive.

As pointed out at several places in the preceding discussion, Konorski's suggestion that we should not limit classical conditioning to the study of consummatory CRs and URs seems convincing and heuristically valuable.

However, even if we grant the merit and advisability of examining a variety of responses in classical-conditioning arrangements, there remains the question of whether Konorski's distinction between preparatory and consummatory responses is sufficiently clearcut to be very useful in the theoretical analysis of classical conditioning. I am by no means sure of its value (cf. Mackintosh, 1974, pp. 21–23; Boakes, Chapter 9). It is certainly difficult to decide exactly when preparatory reflexes terminate and consummatory reflexes begin in a particular behavioral sequence; and Konorski (1967, p.401) seems to confuse the issue even further by the comment that for him the term "consummatory response" not only denotes the final biological component of a given unconditioned activity but also can apply, for example, to affectional behavior displayed when we meet a friend we have not seen for a long time. Furthermore, in laboratory research, measures of different preparatory CRs may often fail to covary consistently, when some independent variable is manipulated; and meaningful comparisons of preparatory and consummatory responses in terms of some common dimension or metric will probably prove very difficult to arrange—for example, in studying their relative rates of conditioning and extinction, a problem analogous to that encountered in so-called concurrent-measurement designs for analyzing interactions between Pavlovian and operant conditioning. (See Gormezano & Kehoe, 1975, pp. 160–161; Hearst, 1975a, p. 214.)

More specifically, is the distinction between preparatory and consummatory responses useful in the analysis of autoshaping? According to certain criteria offered by Konorski, the approach and key-pecking behaviors that arise as CRs during exposure to the standard autoshaping procedure would both be considered instances of preparatory responses, just as would the response of pecking at grain in the dispenser—these are motor activities that cause grain to reach the oral cavity and are presumably part of the last stage of preparatory feeding activity (1967, p.11). On the other hand, pecking is a precise and adaptive movement elicited by specific stimuli and could qualify in that respect as a consummatory response. My discussions with other contributors to this volume did not reveal any clear consensus as to how Konorski would classify the autoshaped key peck.

Perhaps the most interesting aspect of the phenomenon of autoshaping is the *directedness* of approach and contact behaviors toward the location of the CS rather than toward the location of the grain, an outcome that Konorski apparently would not have anticipated, despite his probable knowledge of Pavlov's findings in which dogs began to move toward and lick the lamp whose illumination signaled the delivery of food. As noted earlier, and unlike Pavlov's reports, Konorski's dogs regularly showed immobility and attention directed toward the food bowl during CS. Was this discrepancy between Pavlov's and Konorski's results due to differences in the degree of physical restraint under which dogs were placed in the two laboratories, to differences in the type and localizabiity of CS, its distance from the subject, or what? At any rate it seems unlikely that Konorski's distinction between consummatory and preparatory responses will help us to understand the most distinctive features of behavior in the autoshaping situation.

In connection with the analysis of variety of responses during the classical conditioning of restrained pigeons, the influential work of Staddon and Simmelhag (1971) deserves mention (see also Killeen, 1975). When grain was presented to hungry pigeons every 12 sec independently of their behavior and without any external signal preceding US delivery, certain responses such as jumping in the air, wing flapping, and head and limb movements were quite frequent throughout the early sessions of training. However, these behaviors were gradually replaced, particularly at times when US delivery was imminent, by such activities as pecking at the wall or at the continuously unilluminated key. Staddon and Simmelhag called the latter activities "terminal" and the other, more variable activities "interim". Because the more diffuse (interim) behaviors occurred early in training, then decreased, and eventually were unlikely to occur at times when the US was about to be delivered, their properties resembled those attributed by Konorski to preparatory activities. The temporal characteristics of pecking behavior, on the other hand, paralleled the properties of Konorski's consummatory activities, for they developed later and occurred in close conjunction with an expected US.

Research based on Staddon and Simmelhag's procedure of placing subjects in a situation where food is occasionally but predictably given, and then observing how various responses change, interact, and become ordered in a temporal sequence seems extremely valuable. Their approach should provide empirical data of the kind necessary to determine the existence of functionally similar and different classes of behavior. Whether categories determined in this fashion will correspond in any clear way to Konorski's distinction between consummatory and preparatory responses remains an open question. There is a definite need for more ethological and other behavioral work aimed at categorization of CRs appearing in unrestrained subjects exposed to various conditioning procedures (see Rescorla, 1978); but at the present time Konorski's dual classification does not seem particularly helpful in this respect.

5. The Classical-Instrumental Distinction

Recent work on the instrumental conditioning of autonomic responses (e.g., biofeedback experiments) and on the classical conditioning of directed movements of the whole organism (e.g., autoshaping experiments) has undermined one of the major criteria for justifying the existence of a fundamental distinction between classical and instrumental conditioning—that is, the autonomic- versus skeletal-response difference. Hearst (1975a; see also 1975b) has discussed and evaluated this point, as well as other empirical and theoretical issues relevant to the distinction. Because space is not available here to repeat or further develop those arguments, interested readers are referred to the earlier articles, which offered a generally negative view of the value of making a fundamental *theoretical* distinction between classical and instrumental conditioning. The two forms of conditioning do not seem to be clearly distinguishable on the basis of stimulus, response, and reinforcer characteristics; and the same basic phenomena and laws appear to hold for both types. Furthermore, with regard to Konorski's theory, how can one easily distinguish between directed movements involving *classical* preparatory "approach" or "searching" CRs and those involving "natural" or "parasitic" *instrumental* CRs?

In preparing the material for the present chapter, I was surprised to discover that Konorski's final views on the systematic utility of the Type I versus Type II distinction were apparently much less favorable than indicated in his writings over the 40 years since the appearance of his seminal paper on the topic with Miller. In 1969, he added a postscript to a translation (by B. F. Skinner) of the 1928 Miller-Konorski paper in which he commented (1969):" The sharp distinction between, not only the procedural side of Type I and Type II conditioned reflexes, but also between their physiological mechanisms, seems to me now largely exaggerated. In fact, further investigation shows with increasing clarity that both types can be explained on the basis of the same general principles of connectionistic processes (p. 189)."

Furthermore, in the last paper he published on the topic, Konorski (1973) stated that "both types of conditioning are based on the general laws of associations–connections between the centers involved. Whereas the experimental procedures of classical conditioning expose mainly the conditioned stimulus (CS)-unconditioned stimulus (US) connection, those of instrumental conditioning expose the conditioned stimulus (CS)-response (R) connection. Thus, the main differences between the two types of conditioning are those associated with the different centers involved in each, not the associative-connective laws themselves (p.2)."

However, certain inconsistencies with respect to the preceding statements appeared in Konorski's other remarks in the 1973 paper. He described (p. 5) the emergence of a classical CR as due to the familiar process of stimulus substitution, but (apparently combining views reminiscent of those proposed by Thorndike, Hull, Guthrie, and others) he analyzed the basis for instrumental movements in terms of drive reduction and retroactive inhibition. Doesn't this set of remarks imply the operation of different "laws of association" for the two types of conditioning? It is unfortunate that Konorski's final statements concerning the significance of the classical–instrumental distinction were not particularly well-developed or consistent.

V.   CONCLUDING COMMENTS

Konorski viewed classical conditioning as an especially useful arrangement for studying interperceptive (S–S) associations. According to his interpretation, the establishment and maintenance of a classical CR requires not only contiguity between CS and US but also the energizing effects of some background drive. Although his approach was perception-centered rather than response-centered, Konorski stressed that a variety of behaviors normally appear and interact during Type I conditioning; these reflexes may be classified as preparatory (drive) or consummatory, depending on their temporal location in basic sequences of organismic activity.

In this chapter Konorski's views concerning fundamental associative elements, contiguity, and reinforcement were compared with some other interpretations of classical conditioning, and then the applicability of his theory to recent research on the classical conditioning of directed movements (autoshaping) was explored. The principles of stimulus substitution and parasitic (superstitious) reinforcement, which Konorski frequently invoked in discussions of conditioning phenomena, cannot encompass some of the significant findings about autoshaping; but our view that CS–US associations or correlations play a more important role than response–US associations or correlations in the acquisition and maintenance of autoshaped behavior is generally compatible with Konorski's perception-centered interpretation of classical conditioning. Attempts to measure and categorize the various CRs and URs, autonomic and skeletal, that appear during classical conditioning seem needed in the future, although these responses will probably not fall neatly into the two categories, preparatory and consummatory, proposed by Konorski. The notion that classical conditioning and instrumental conditioning involve fundamentally different associative processes or laws— as Konorski maintained for 40 years but apparently abandoned in his final articles—seems especially hard to maintain in the face of data from autoshaping experiments that demonstrate the development of standard "operants" like key pecks or bar presses within classical-conditioning procedures.

Konorski anticipated many issues, methodological and theoretical, that are of considerable contemporary interest to students of animal and human learning: contextual conditioning, drive antagonisms, superstitious reinforcement of responses, expectancy disconfirmation, excitatory versus inhibitory control, and the possible role of imaginal processes—to name a few topics discussed in this volume. His willingness to speculate about the central mechanisms mediating overt behavior, and his attempts to relate psychological data to neurophysiological and clinical theories and findings, typified his general approach. Konorski was forthright in admitting his errors and he did not hesitate to radically modify his theoretical views when unfavorable data appeared. Other outstanding scientists may have had to admit error less often than he did, but few scientists can match his ingenuity, creativity, integrity, and enthusiasm in pursuit of the basic mechanisms of learning and motivation.

ACKNOWLEDGMENTS

The preparation of this article was supported by NIMH Research Grant No. MH-19300. I gratefully acknowledge the valuable suggestions and criticisms offered by participants in a graduate seminar on East–West Perspectives in Learning and Cognition, held at Indiana University in 1977. Edward Walker's comments on earlier drafts of this manuscript were very helpful.

REFERENCES

Bindra, D. A unified account of classical conditioning and operant training. In A. H. Black & W. F. Prokasy (Eds.), *Classical conditioning II: Current theory and research.* New York: Appleton-Century-Crofts, 1972.

Boakes, R. A. Performance on learning to associate a stimulus with positive reinforcement. In H. Davis & H. M. B. Hurwitz (Eds.), *Operant-Pavlovian interactions.* Hillsdale, N. J.: Lawrence Erlbaum Associates, 1977.

Boakes, R. A., & Halliday, M. S. *Inhibition and learning.* New York: Academic Press, 1972.

Boakes, R. A., & Ismail, R. B. *An effect similar to sensory preconditioning in autoshaping of key pecking in pigeons.* Unpublished manuscript, 1971. (Available from R. A. Boakes, Laboratory of Experimental Psychology, University of Sussex, Brighton, England).

Bolles, R. C. Reinforcement, expectancy, and learning. *Psychological Review,* 1972, *79,* 394–409.

Brown, P. L., & Jenkins, H. M. Auto-shaping of the pigeon's key-peck. *Journal of the Experimental Analysis of Behavior,* 1968, *11,* 1–8.

Browne, M. P. The role of primary reinforcement and overt movements in autoshaping in the pigeon. *Animal Learning and Behavior,* 1976, *4,* 287–292.

Champion, R. A. *Learning and activation.* Sydney, Australia: Wiley, 1969.

Davol, G. H., Steinhauer, G. D., & Lee, A. The role of preliminary magazine training in acquisition of the autoshaped key peck. *Journal of the Experimental Analysis of Behavior,* 1977, *28,* 99–106.

Deeds, W., & Frieman, J. *Latent learning of autoshaping: Stimulus-reinforcer learning in the nondeprived pigeon.* Paper presented at the meeting of the Midwestern Psychological Association, Chicago, May 1977.

Gormezano, I., & Kehoe, E. J. Classical conditioning: Some methodological-conceptual issues. In W. K. Estes (Ed.), *Handbook of learning and cognitive processes* (Vol. 2). Hillsdale, N. J.: Lawrence Erlbaum Associates, 1975.

Guthrie, E. R. Conditioning as a principle of learning. *Psychological Review,* 1930, *37,* 412–428.

Guthrie, E. R. Pavlov's theory of conditioning. *Psychological Review,* 1934, *41,* 199–206.

Guthrie, E. R. *The psychology of learning.* New York: Harper, 1935.

Hearst, E. The classical–instrumental distinction: Reflexes, voluntary behavior, and categories of associative learning. In W. K. Estes (Ed.), *Handbook of learning and cognitive processes* (Vol. 2). Hillsdale, N. J.: Lawrence Erlbaum Associates, 1975. (a)

Hearst, E. Pavlovian conditioning and directed movements. In G. Bower (Ed.), *The psychology of learning and motivation* (Vol. 9). New York: Academic Press, 1975. (b)

Hearst, E. Stimulus relationships and feature selection in learning and behavior. In S. Hulse, H. Fowler, & W. K. Honig (Eds.), *Cognitive processes in animal behavior.* Hillsdale, N. J.: Lawrence Erlbaum Associates, 1978.

Hearst, E., & Franklin, S. Positive and negative relations between a signal and food: Approach–withdrawal behavior to the signal. *Journal of Experimental Psychology: Animal Behavior Processes,* 1977, *3,* 37–52.

Hearst, E., & Jenkins, H. M. *Sign-tracking: The stimulus–reinforcer relation and directed action.* Austin, Texas: The Psychonomic Society, 1974.

Hebb, D. O. *The organization of behavior: A neuropsychological theory.* New York: Wiley, 1949.

Hogan, J. A. Responses in Pavlovian conditioning studies. *Science,* 1974, *186,* 156–157.

Hull, C. L. *Principles of behavior.* New York: D. Appleton-Century, 1943.

Hulse, S., Fowler, H., & Honig W. K. (Eds.). *Cognitive processes in animal behavior.* Hillsdale, N. J.: Lawrence Erlbaum Associates, 1978.

Jenkins, H. M. Effects of the stimulus–reinforcer relation on selected and unselected responses. In R. A. Hinde & J. S. Hinde (Eds.), *Constraints on learning.* New York: Academic Press, 1973.

Jenkins, H. M. Sensitivity of different response systems to stimulus–reinforcer and response–reinforcer relations. In H. Davis & H. M. B Hurwitz (Eds.), *Operant-Pavlovian interactions.* Hillsdale, N. J.: Lawrence Erlbaum Associates, 1977.

Jenkins, H. M., & Moore, B. R. The form of the auto-shaped response with food or water reinforcers. *Journal of the Experimental Analysis of Behavior,* 1973, *20,* 163–181.

Kamin, L. J. Predictability, surprise, attention, and conditioning. In B. A. Campbell & R. M. Church (Eds.), *Punishment and aversive behavior.* New York: Appleton-Century-Crofts, 1969.

Killeen, P. On the temporal control of behavior. *Psychological Review,* 1975, *82,* 89–115.

Konorski, J. *Conditioned reflexes and neuron organization.* Cambridge: Cambridge University Press, 1948.

Konorski, J. *Integrative activity of the brain.* Chicago: University of Chicago Press, 1967.

Konorski, J. Postscript. *Journal of the Experimental Analysis of Behavior,* 1969, *12,* 189.

Konorski, J. On two types of conditional reflex: General laws of association. *Conditional Reflex,* 1973, *8,* 2–9.

Konorski, J., & Miller, S. On two types of conditioned reflex. *Journal of General Psychology,* 1937, *16,* 264–272.(a)

Konorski, J., & Miller, S. Further remarks on two types of conditioned reflex. *Journal of General Psychology,* 1937, *17,* 405–407.(b)

Lorenz, K. Z. Innate bases of learning. In K. Pribram (Ed.), *On the biology of learning.* New York: Harcourt, Brace, & World, 1969.

Mackintosh, N. J. *The psychology of animal learning.* New York: Academic Press, 1974.

Miller, S., & Konorski, J. [On a particular form of conditioned reflex.] (B. F. Skinner, trans.). *Journal of the Experimental Analysis of Behavior,* 1969, *12,* 187–189. (Original publication: *Compte Rendu Hebdomadaire des Séances et Mémoires de la Societé de Biologie,* 1928, *99,* 1155–1157).

Mowrer, O. H. How does the mind work? Memorial address in honor of Jerzy Konorski. *American Psychologist,* 1976, *31,* 843–857.

Pavlov, I. P. *Conditioned reflexes.* London: Oxford University Press, 1927.

Pavlov, I. P. The reply of a physiologist to psychologists. *Psychological Review,* 1932, *39,* 91–127.

Pavlov, I. P. An attempt at a physiological interpretation of obsessional neurosis and paranoia. *Journal of Mental Science,* 1934, *80,* 187–197.

Peden, B. F., Browne, M. P., & Hearst, E. Persistent approaches to a signal for food despite food omission for approaching. *Journal of Experimental Psychology: Animal Behavior Processes*, 1977, *3*, 377-399.

Perkins, C. C. An analysis of the concept of reinforcement. *Psychological Review*, 1968, *75*, 155-172.

Peterson, G. B., Ackil, J., Frommer, G. P., & Hearst, E. Conditioned approach and contact behavior toward signals for food or brain-stimulation reinforcement. *Science*, 1972, *77*, 1009-1011.

Razran, G. *Mind in evolution*. Boston: Houghton Mifflin, 1971.

Rescorla, R. A. Pavlovian conditioning and its proper control procedures. *Psychological Review*, 1967, *74*, 71-80.

Rescorla, R. A. Informational variables in Pavlovian conditioning. In G. H. Bower (Ed.), *The psychology of learning and motivation* (Vol. 6). New York: Academic Press, 1972.

Rescorla, R. A. Some implications of a cognitive perspective on Pavlovian conditioning. In S. Hulse, H. Fowler, & W. K. Honig (Eds.), *Cognitive processes in animal behavior*. Hillsdale, N. J.: Lawrence Erlbaum Associates, 1978.

Rescorla, R. A., & Solomon, R. L. Two-process learning theory: Relationships between Pavlovian conditioning and instrumental learning. *Psychological Review*, 1967, *74*, 151-182.

Rescorla, R. A., & Wagner, A. R. A theory of Pavlovian conditioning: Variations in the effectiveness of reinforcement and nonreinforcement. In A. H. Black & W. F. Prokasy (Eds.), *Classical conditioning II: Current research and theory*. New York: Appleton-Century-Crofts, 1972.

Schwartz, B., & Gamzu, E. Pavlovian control of operant behavior. In W. K. Honig & J. E. R. Staddon (Eds.), *Handbook of operant behavior*. Englewood Cliffs, N. J.: Prentice-Hall, 1977.

Skinner, B. F. Two types of conditioned reflex: A reply to Konorski and Miller. *Journal of General Psychology*, 1937, *16*, 272-279.

Skinner, B. F. *The behavior of organisms*. New York: Appleton-Century-Crofts, 1938.

Skinner, B. F. "Superstition" in the pigeon. *Journal of Experimental Psychology*, 1948, *38*, 168-172.

Skinner, B. F. *Science and human behavior*. New York: Macmillan, 1953.

Staddon, J. E. R., & Simmelhag, V. L. The "superstition" experiment: A reexamination of its implications for the principles of adaptive behavior. *Psychological Review*, 1971, *78*, 3-43.

Stepien, I. The magnet reaction, a symptom of prefrontal ablation. *Acta Neurobiologiae Experimentalis*, 1974, *34*, 145-160.

Timberlake, W., & Grant, D. L. Auto-shaping in rats to the presentation of another rat predicting food. *Science*, 1975, *190*, 690-692.

Tolman, E. C. *Purposive behavior in animals and men*. New York: Century, 1932.

Tolman, E. C., & Brunswik, E. The organism and the causal texture of the environment. *Psychological Review*, 1935, *42*, 43-77.

Wagner, A. R. Stimulus validity and stimulus selection in associative learning. In N. J. Mackintosh & W. K. Honig (Eds.), *Fundamental issues in associative learning*. Halifax, Canada: Dalhousie University Press, 1969.

Wasserman, E. A. Pavlovian conditioning with heat reinforcement produces stimulus-directed pecking in chicks. *Science*, 1973, *181*, 875-877.

Wasserman, E. A., Franklin, S., & Hearst, E. Pavlovian appetitive contingencies and approach vs. withdrawal to conditioned stimuli in pigeons. *Journal of Comparative and Physiological Psychology*, 1974, *86*, 616-627.

Wasserman, E. A., Hunter, N. B., Gutowski, K. A., & Bader, S. A. Auto-shaping chicks with heat reinforcement: The role of stimulus–reinforcer and response–reinforcer relations. *Journal of Experimental Psychology: Animal Behavior Processes*, 1975, *104*, 158-169.

Williams, D. R., & Williams, H. Auto-maintenance in the pigeon: Sustained pecking despite contingent nonreinforcement. *Journal of the Experimental Analysis of Behavior,* 1969, *12,* 511–520.

Woodard, W. T., & Bitterman, M. E. Autoshaping in the goldfish. *Behavior Research Methods and Instrumentation,* 1974, *6,* 409–410.

Woodruff, G., & Williams, D. R. The associative relation underlying autoshaping in the pigeon. *Journal of the Experimental Analysis of Behavior,* 1976, *26,* 1–13.

Zener, K. The significance of behavior accompanying conditioned salivary secretion for theories of the conditioned response. *American Journal of Psychology,* 1937, *50,* 384–403.

3 Habituation and Memory

Allan R. Wagner
Yale University

I. INTRODUCTION

In his consummate work, *Integrative Activity of the Brain,* Konorski (1967), attempted to show how certain phenomena of "habituation" might be seen as relatively natural consequences of assumptions that he was otherwise inclined to make about the memory system. He presented a view of the long-term memory structure as a collection of representative elements (so-called gnostic units) interconnected via an associative network, and a view of short-term memory as that subset of the total collection of memory elements that is in a temporary state of "excitation." He then proposed that various indices of stimulus processing would be more or less pronounced, depending upon the likelihood that the stimulating event would "activate" an appropriate gnostic unit, which in turn would depend upon the latter's availability and state of excitation in the memory system.

Since that time, as a result of a variety of influences, much of this general point of view has become more or less common wisdom. As opposed to the attitudes of an earlier era, when, to use Estes' (1973) terminology, our thinking was more "response-oriented," we are inclined to follow Konorski in not treating a response decrement to iterated stimulation ("habituation")

53

as reflecting some isolated learning process, but in presuming that the evidence of memory that is involved should fit theoretically with other observable products of experience. There is also considerable acceptance of the kind of dual-state memory system that Konorski proposed, at least within so-called information-processing theory. The representative elements may be alternatively termed "ideas," "nodes," or "logogens," in different renditions; the elements may be assumed to be interrelated in ways that beg more precise characterization than "associative network;" and there may be interesting variation in what is intended by speaking of a state like "excitation." But, the core notions are very common. (For examples, see Anderson & Bower, 1973; Estes, 1975; Norman, 1968; Shiffrin & Schneider, 1977; Wagner, 1976, in press.) This being the case, we should not be surprised that Konorski's formulation bears a family resemblance to various proposals as to how variations in stimulus processing are keyed to the momentary state of a dual-memory system.

My assumption in this chapter, however, is that some of the important characteristics of Konorski's specific approach are less obvious than others, and that certain of these warrant current consideration in relationship to more recent data and theorizing. The brief text that Konorski explicitly devoted to "the problem of habituation" (1967, pp. 99–104) is in the context of a discussion of the *formation* of gnostic units. It is therefore possible to conclude that the relevant variation in stimulus processing in Konorski's formulation is essentially attributable to whether or not the subject has *available* an appropriate gnostic unit in long-term memory, and, because of other similarities as well, to conclude that Konorski's reasoning can be classified with that of Sokolov (e.g., 1969) as resting upon the presence or absence of a fitting "neuronal model." In fact, in Konorski's (1967) discussion of transient memory, of the interrelationship between perception and association, and of a variety of other issues, he painted a more complicated and interesting picture, which should be fully appreciated. I thus wish to try to bring some of the pieces of this picture together for inspection.

In regard to this constructive effort, I must apologize in advance for any errors that may result from biases in my vision. For what I shall go on to confess is that the expanded theory, as I see it, has a number of features that liken it to my own recent speculations on the consequences of "priming of short-term memory." And, I can then take the occasion to point to some of the kinds of research findings that might have been anticipated by Konorski, but that independently shaped our approach to some phenomena of "habituation."

Beyond this, I wish to emphasize how the formulation that I attribute to Konorski (1967) can deal reasonably with some phenomena not so explicitly treated in other theoretical schemes. Whether we find the overall formulation agreeable or not, there is incentive to present our theories at a commensurate level of detail.

II. KONORSKI'S THEORY OF HABITUATION

A. The Basic Formulation

The degree and kind of processing that a stimulus receives is most proximally determined in Konorski's theory by recurrent reflexes in the afferent system. A targeting reflex is said to occur to a stimulus in the form of adjustments of the peripheral and central components of the analyzers that lead to "better reception" of the event. An inhibitory perceptive recurrent reflex is said to occur to a stimulus in a form antagonistic to the targeting reflex that leads to a suppression of potential stimulus input (Konorski, 1967, pp. 66–67). The targeting reflex, according to Konorski, is furthermore generally accompanied by signs of somatic and autonomic arousal (the orienting response) and by EEG desynchronization, while the inhibitory perceptive recurrent reflex generally suppresses these signs, as well as other indices of stimulus processing.

To sketch a "hypothetical mechanism of habituation" that would indicate the occasions for the two kinds of reflexes and suggest the composite of behavioral effects involved, Konorski (1967) presented the diagram reproduced here as Fig. 3.1. It portrays the analyzer systems as being arranged with a sequence of "transit" levels between the receptor surface and the final representational (gnostic) level, and assumes that at each level the system is organized into "fields" that ultimately correspond at the gnostic level to the categories of perception. The portion labeled "a" is said to represent the sequence of events occasioned by a novel stimulus object (S-O) for which the subject has formed no gnostic unit in long-term memory. The stimulus pattern is shown to activate units of the receptor surface and consequent afferent, transit fields before being presented to a gnostic unit to which it has potential, but not actualized, connections. When the initial afferent field is activated, it is presumed to stimulate an "arousal system," the products of which are various autonomic indices of the orienting response (Sokolov, 1963), and the targeting reflex, the latter consisting of a facilitating bias to the relevant gnostic field, as indicated, as well as more peripheral receptor adjustments, not shown. Thus, the stimulus should produce notable signs of attention and arousal by the subject. In contrast, the portion of Fig. 3.1 labeled "b" is said to represent the sequence of events occasioned by a familiar stimulus object for which the subject has previously formed a gnostic unit in long-term memory. Basically, what is different in this case is that the stimulation is now assumed to *activate* an *actualized* gnostic unit (shown here as a filled, rather than open, circle in Fig. 3.1). This has two consequences. The pattern of stimulation is now "recognized," and the inhibitory perceptive recurrent reflex occurs. The latter is shown in Fig. 3.1 as the stimulation of a system that acts to inhibit the same arousal system that was provoked by the initial impact of the stimulus

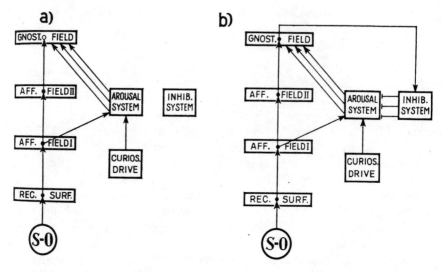

FIG. 3.1.   Konorski's depiction of a "hypothetical mechanism of habituation." The diagram labeled "a" represents the processing of a novel stimulus object, S-O, for which the subject has not actualized a gnostic unit. The diagam labeled "b" represents the processing of a familiar stimulus, for which the subject has actualized a gnostic unit. See text for further explication. Reprinted from Konorski (1967). ©1967, The University of Chicago Press.

pattern upon the afferent field. Thus, according to Konorski (1967), the stimulus should produce less persistent and less notable signs of attention and arousal by the subject, since the targeting reflex that is initiated will be "suppressed" and prevented from development, "cancelling it at its very outset (p. 101)."[1]

The variable link that determines the processing that a stimulus object receives is clearly the activation or not of a covering gnostic unit (or set of gnostic units) in memory. As previously noted, and as further shown in Fig. 3.1, Konorski discussed habituation primarily in reference to comparative instances in which the relevant gnostic units could be presumed to have been

---

[1]In this sketch and specific language one can draw parallels to Sokolov's (e.g., 1969) formulation. Sokolov assumes that any stimulus activating the receptive mechanism has consequent specific pathways to activate an "orienting reflex." And he assumes that the receptive mechanism also sends impulses to a "modeling structure." If the modeling structure has been previously excited by the same stimulus so as to contain an appropriate neuronal model, it will also have acquired a tendency to activate an inhibiting structure that in turn blocks the pathway between the receptive mechanism and the orienting reflex. Thus, Sokolov (1969) proposes the following: "Once this conditioned inhibitory link has formed and a stimulus begins to act upon the organism, it functions as a signal for the inhibitory process to develop. Consequently, the path which conducts excitation . . . (from receptive mechanism to the orienting reflex) . . . is already closed when the stimulus to which the organism has become accustomed begins to operate (p. 682)."

formed or not as a result of prior experience with the stimulus event. In his concern with a variety of problems of perception, recognition memory, and patterns of agnosia, he articulated a wealth of relevant assumptions about the formation of gnostic units. Thus, for example, it is important that gnostic units are assumed to be formed to represent particular *patterns* of stimulation, in a mapping that would generate a gnostic unit for each "unitary perception." The potentialities for actualization of gnostic units are presumed to vary with different analyzer systems and with the degree of elaboration and prior commitment of the gnostic fields involved. Furthermore, the likelihood of actualization is presumed to depend upon such training variables as the degree of arousal and the freedom from conflicting stimulation on the original occasions of the patterned stimulation. From the many assumptions made by Konorski (1967), which are only hinted at here, one can appreciate that an elaborate set of expectations concerning "habituation" is a ready consequence.

B. The Expanded Formulation

Konorski's formulation is complicated by his acknowledgement that whether or not a stimulus pattern will, in fact, activate a covering gnostic unit will depend upon memorial factors other than whether or not the nominal pattern of stimulation involved has been previously experienced so as to have permanently actualized its connections with a representative unit. There are two caveats. First, we should assume that the environment and/or the behavioral system is noisy so that one should anticipate occasions in which a familiar stimulus is not recognized. On such occasions, "when the convergence of afferent impulses from the proximate transit field upon the appropriate set of gnostic units is not complete because of the lack of some element normally participating in a given perception," the messages arriving at the gnostic units may "stimulate them subliminally" (Konorski, 1967, p. 185). Second, we should assume that there is some tendency for a relatively novel stimulus pattern to activate a gnostic unit in the absence of the prior experience that would permanently establish the representation. According to Konorski (1967): "potential synaptic contacts between the transmittent and recipient units are not completely impassable for incoming impulses, and in consequence, these impulses may produce the actual excitation of the recipient units if the latter are in a state of sensitization . . . In this way the sensitized gnostic units may be activated by the appropriate stimulus-pattern, although stable actual contacts linking them with the units of the lower level have not been established . . . [p. 493]."

This reasoning opens the way to two important influences upon stimulus processing not explicitly acknowledged in the abbreviated sketch of Fig. 3.1. The first is an associative influence. Since the critical gnostic units are the terms of the associative network of memory, they can presumably be provoked to activity, not only through the patterned external stimulation that they come to represent, but also through other gnostic units activated by previously associated stimulation. And, Konorski assumed that the input to a given gnostic unit from associative linkages can summate with the more or less appropriate patterned input from the periphery to produce gnostic activation that would be less reliably produced by the latter alone. Excitation from the associative network can, in effect, sensitize particular gnostic units. Konorski (1967) drew upon this assumption quite frequently in dicussions of perceptual and associative phenomena. It is the presumed basis for variability in the quality of perceptual reports depending upon the environmental context. It is what is presumed to allow for two of the four methods he discussed for studying associations: In the so-called expectation method, we may see that the recognition time for a stimulus object is less when the object is presented in its usual context than when in a novel context; in the so-called substitution method, we may see that signs of the targeting reflex and orienting response are less likely to occur to a stimulus when the latter is presented in a customary sequence of events than when in a changed sequence.

The second influence is a transient aftereffect of prior stimulation. Konorski added the discussion of this phenomenon in the last topic chapter of his 1967 book, and did not otherwise fully exploit it as he did the aforementioned associative influence. Indeed, the coverage is sometimes cryptic, requiring us to make some potentially important extrapolations on the basis of subtle choices in wording. However, Konorski was clear in asserting that whenever a given stimulus pattern is presented for which there is a potential gnostic unit (whether or not it is actualized), there is placed into operation a "corresponding" reverberating circuit "attached to" the gnostic unit. This circuit is presumed to bombard the gnostic unit with "circulating impulses." While the circuit is in operation, the gnostic unit is thus taken to be in a "state of sensitization," being more liable to activation from other influences.

Konorski (1967) assumed that the activity of the reverberating circuit is transient, that it "may die out spontaneously after a lapse of time or be suppressed by conflicting perceptions." He identified those gnostic units that are in a temporary state of "sensitization" by their respective reverberating circuits as being in short-term memory, or as he preferred to say, in "transient memory." It is worth underlining that the state of *excitation* that distinguishes the set of short-term memory elements in Konorski's system is not *activation*, but *stimulation toward activation*. Thus, for example, he allowed that although the "transient store of recent memory" may persist for extended periods of "hours or days" in the sense of persistent "sensitization," the representative gnostic unit will not be continually "activated" to influence such behaviors as perceptual reports. Such "activation" is only then more likely, as when the gnostic unit is the recipient of

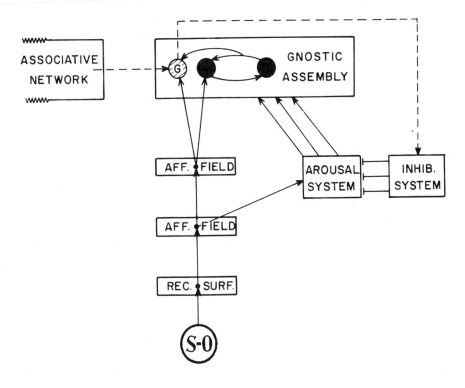

FIG. 3.2.   An expanded characterization of Konorski's hypothetical mechanism of habituation. It is identical with Fig. 3.1 except that it incorporates the additional influences on the gnostic unit of the associative network and a reverberating circuit of transient memory. See text.

input from its stimulus pattern (p. 493) from associated gnostic units (p. 497), or from an ill-specified attentional arousal of the gnostic field (p. 493).

Figure 3.2 shows how Konorski's sketch of the "hypothetical mechanism of habituation" reproduced in Fig. 3.1 might be elaborated so as to include the additional associative and reverberating influences, discussed earlier. It introduces no changes in the recurrent reflex components: Stimulation will consistently provoke the targeting reflex via the arousal system, unless the latter is suppressed by the inhibitory perceptive recurrent reflex, as previously described. The variable link that determines the processing that a stimulus object receives also remains the same as in Fig. 3.1. Does stimulation activate a covering gnostic unit, here indicated as a shaded circle labeled "G?" This illustration is meant to convey that (1) regardless of the previous history with respect to the pattern of stimulation, there will be *some* tendency for an appropriate gnostic unit to be activated, but that (2) whether or not it will be activated will depend not only on the degree of actualization of connections from the afferent field, but also on the concomitant excitation from other sources.

One of the latter sources is represented as originating in the remainder of the associative network. Presumably, input from this source would occur only in the case of gnostic units that have been actualized so as to have entered into associative relationship with other units. But then the stimuli that would excite the associated units may contribute to activation of the G unit, in summation with the pattern of stimuli particular to the latter unit.

The other source of input is from a closed reexciting circuit, here represented in kernel form by the connected, filled circles. To capture Konorski's proposal that each gnostic unit has such an "attached" circuitry for its own maintenance in transient memory, the collection of gnostic unit and closed loop is given an identity in Fig. 3.2 of "gnostic assembly." The wiring within the assembly, and between the components and the afferent field, is directed by Konorski's proposals that have been discussed earlier, which suggest that the gnostic unit is not integral to the reverberating circuit. Thus, Fig. 3.2 shows the gnostic unit as potentially receiving "sensitizing" input from the loop circuit, without activation of the gnostic unit being required for persisting excitement in the loop. Likewise, Fig. 3.2 shows the loop circuit as receiving excitation from the afferent field, independent of whether or not the gnostic unit is concomitantly activated.

Given this expanded formulation, it should be clear that Konorski's theory of habituation has richer predictive consequences than he explicitly anticipated in his concentration on the formation of gnostic units. Whether or not a stimulus will appear to be ignored or to be notably processed should vary with all of the regularities of transient memory and with all of the laws of associative learning and performance to which it is subject.

III. THE PRIMING OF STM

Since the writing of his 1967 book we have acquired a considerable amount of data that might have caused Konorski to shift his emphasis concerning the mechanism of habituation to be more in keeping with the sketch of Fig. 3.2 than the sketch of Fig. 3.1. The data have led me in a series of papers (e.g., Wagner, 1976, in press; Pfautz & Wagner, 1976) to suggest that variations in stimulus processing in a range of situations can be understood in terms of whether or not representation of the stimulus in question has been "primed" in short-term memory (STM). I have distinguished between an associatively generated priming[2] and a self-generated priming, to distinguish between

---

[2]Associatively generated priming has also been termed "retrieval-generated priming" (Wagner, 1976, in press) in the context of the common spatial analogy, in which an associated stimulus is said to retrieve representations from the long-term memory system into STM. The present terminology is employed for reasons of its better correspondence to Konorski's language, without other significance for the intended referent.

instances in which preexcitation of representation in the memory system can be attributed to the influence of previously associated stimuli versus instances in which the preexcitation can be attributed to recent exposure to the corresponding stimulus itself. But, I have been led to propose that priming by either route can have the same effect, of diminishing certain indices of stimulus processing. Now, one way to view Fig. 3.2 is as one particular mechanism that would accommodate these generalizations. The excitation from the associative network can be identified with associatively generated priming. The excitation from a previously initiated reverberating loop can be identified with self-generated priming. And it should be clear that either event, by favoring the prompt activation of the gnostic unit appropriate to a stimulus event, should lead to the prompt occurrence of the inhibitory perceptive recurrent reflex that is presumed to suppress various indices of stimulus processing.

In this section I summarize some of the major classes of observations that led to the aforementioned generalizations concerning the priming of STM, and which would appear to be agreeable to the characterization of Konorski's theory as presented in Fig. 3.2. In the process I draw some additional apparent parallels between our reasoning and that of Konorski (1967).

A. Associative Learning

Like Konorski, we can assume that the processing that a stimulus receives will be reflected in a variety of indices besides some species-typical unconditioned responses. One index should be the degree to which the stimulus can be shown to produce associative learning under conditions of pairing with some other stimulus event.

1. US Processing

a. Associatively generated priming.   Kamin (e.g., 1969) prompted our theorizing related to this category by his interpretation of the "blocking effect." It is now well documented that a CS–US pairing in Pavlovian conditioning will produce less evidence of an association, in conditioned responding to the CS, if the pairing is accompanied by another stimulus that has been previously trained to have association with the US. The accompanying stimulus may be another discrete CS (e.g., Kamin, 1969; Wagner, 1969), or it may involve more static contextual cues (e.g., Blanchard & Honig, 1976; Dweck & Wagner, 1970; Tomie, 1976). Kamin (1969) proposed that such a blocking effect might be attributed to a diminished processing of the US on occasions in which it was signaled by previously associated stimuli, as compared with occasions in which it was not so announced. Atkinson and Wickens (1971) and Wagner (1971) promulgated this view in the specific form of assuming that an associatively signaled stimulus is less likely to be "rehearsed in STM," and for this reason is less likely to become associated with other contiguous stimuli.

The general argument is that blocking can be attributed to the effects of associatively generated priming of the US representation in STM. It should be clear how this is allowed by the mechanism of Fig. 3.2. When a US occurs, it can more surely and/or more promptly activate its gnostic unit if the latter unit is concurrently excited via the associative network. In such cases the inhibitory perceptive recurrent reflex will more surely and/or more promptly be evoked to suppress the targeting reflex. The specific assumption that "rehearsal" is what is critically denied a signaled US would be tantamount to assuming that the essential ingredient is the suppression of the facilitating bias to the gnostic assembly that otherwise allows the reverberating circuit to continue after the US experience.

There are, of course, other possible interpretations of the blocking effect. It may be noted that there appears to be good reason (see, e.g., Mackintosh, 1975) to assume that more is involved in its production, in some situations, than a loss in effectiveness of a primed US, particularly if the critical CS–US training is carried out over a series of trials (e.g., Mackintosh & Turner, 1971) as opposed to a single pairing (e.g., Gillian & Domjan, 1977). Indeed, some extrapolations from the present notions (see Donegan, Whitlow, & Wagner, 1977) would lead us to expect that the target CS, as well as the US, might be differentially processed over a series of trials involving a primed as compared with a nonprimed US, and that this could importantly contribute to an overall blocking effect. However, rather than attempt here an exhaustive appraisal of this single phenomenon, I am inclined only to point to it as one instance of a more generally observable pattern.

b. Self-generated priming. What we would basically anticipate in Pavlovian associative learning is that whatever might be said about the potential loss of effectiveness of a US as a result of associatively generated priming (as in the case of blocking) could be said as well about the potential loss of effectiveness of the same US as a result of self-generated priming, and about the potential loss of effectiveness of a CS by either associatively or self-generated priming. These several anticipations have not been equally evaluated. But, the available data in each case have been encouraging.

William Terry, in a recent dissertation (1976) conducted in our laboratory, sought to evaluate whether or not a blockinglike effect would result in Pavlovian conditioning, if the CS–US episodes, rather than involving another CS that would signal the US, were shortly preceded by an instance of the US itself. The several experiments involved eyeblink conditioning in the rabbit with well-spaced CS–US pairings in which the CS was an 1100-msec visual stimulus and the US was a 100-msec paraorbital shock. In the first experiment all subjects were trained with two CSs. One of the stimuli, CS_N, was simply reinforced on each of its occasions—that is, was involved in the sequence CS_N–US. The other stimulus, CS_{US}, was similarly reinforced, but was also preceded 4 sec earlier by a US—that is, was involved in the sequence US–CS–US. What was anticipated from the priming notion was that in a within-subject comparison of the degree of conditioning to CS_N and CS_{US}, acquisition would be diminished to the latter cue. The pretrial US should have introduced US representation into STM, which could then be expected to persist over the time intervals involved (see Terry & Wagner, 1975; Wagner, Rudy, & Whitlow, 1973) to decrease the effectiveness of the reinforcing US when presented. In the language of Fig. 3.2, the pretrial US should have activated the reverberating loop that could be expected to sensitize the US gnostic unit and favor the recurrent inhibitory perceptive reflex on the subsequent occasion of the reinforcing US. The expected result was observed. There was considerably less conditioned responding to CS_{US} than to CS_N, both on the training episodes described and on special test trials with the CSs in isolation.

The within-subject comparisons of this study indicate that there is a decremental influence of a pretrial US on associative learning. Terry could not at this point, however, necessarily tie the source of decrement to the diminished processing of the subsequent US. Perhaps, for example, the priming US may have acted to mask or overshadow the following CS, or all of the subsequent stimulation without distinction. Terry commented upon the latter possibility in relationship to the priming notion by way of two further experiments. The studies were predicated on the general reasoning that a priming US should diminish the conditioning that results from a following CS–US pairing only to the degree that the pretrial and trial USs are the same, and only to the degree that the excitation of the US representation, presumably provoked by the priming stimulus, is allowed to persist until the trial occasion.

In the first of the two studies, Terry took advantage of the fact that in his experimental preparation he could arbitrarily select the area of the right or left eye to receive the conditioning US (and from which then to record conditioned responding), and he could equally choose to use as a priming US the exact same US or one that was as equivalent as possible in all respects except that it was delivered to the other location. If the priming US were effective only to the degree that it matched the subsequent conditioning US, it should have made a difference whether or not the two stimuli were delivered to the same location. Alternatively, if the priming US were effective by virtue of masking or overshadowing the subsequent CS, or by producing some more general processing disturbance, it should not have made any difference, everything else being equal, whether or not it was delivered to the same location as the subsequent conditioning US. The data from this study were clear. Using the same temporal US–CS–US sequence as in the preceding experiment, there was less evidence of conditioning, on training trials or on CS-alone test trials, for a group in which the priming US was the same as the terminal, reinforcing US as compared with a group in which the priming US was delivered to the alternate location.

Terry's (1976) second, follow-up study was somewhat more complex. Four groups of rabbits were conditioned. Group N had no pretrial stimuli scheduled, and simply received CS–US pairings on every trial. Group US received a priming US presentation 7 sec before the onset of each training trial—that is, the sequence US–CS–US. These two groups correspond to the CS_N and CS_{US} conditions, respectively, of Terry's first experiment, but in a between-groups design. Group US + D received a priming US 7 sec before each training trial as did Group US, but in addition received a "distractor" stimulus (D) beginning 2 sec after the priming US and 5 sec before the CS–US pairing — that is, the sequence US–D–CS–US. Finally, Group D received only the distractor stimulus in the same location as Group US + D — that is, the sequence D–CS–US.

The distractor stimulus was a 1-sec, irregular sequence of auditory clicks and vibrotactual stimulation designed to have a different pattern on each occasion. The intention was to make this stimulus sufficiently salient and novel so as to command the subject's attention in STM. To the degree that this was so, and to the degree that STM can be assumed to have a limited capacity, D could then be supposed to act to displace representation of the priming US in Group US + D. In this case a priming deficit should not be anticipated in Group US + D as it should be in Group US.

We have already hinted at the way in which Konorski's expanded formulation of Fig. 3.2 would make this same prediction. In general Konorski (1967) approached the notion of a limited capacity of STM as a "lateral inhibition" phenomenon, and specifically assumed that the reverberating circuit sensitizing a gnostic unit may be suppressed by "conflicting perceptions."

The experiment commented on alternatives to the priming notion in a manner similar to the preceding study. A deficit is expected not simply to the degree that *something* is represented in STM at the time of the CS–US trial, but to the degree that the *US* is represented. If US representation can be removed by a distractor interposed between the priming US and the CS–US trial, there should be an increase in the conditioning that results. The inclusion of Group D in the design was obviously to evaluate the unlikely possibility that the distractor itself would improve conditioning, in the absence of US priming.

Figure 3.3 depicts the percentage of conditioned responding observed in each of the 4 groups over the 8 training trials in each of 3 days (left panel), and over 2 test trials on the same days, when the CS was presented alone (right panel). The pattern of data is consistent on the two kinds of trials, and clear in its agreement with the priming expectations. Group US evidenced substantially less conditioning than Group N, with the magnitude of the difference shown being representative of other comparisons of these

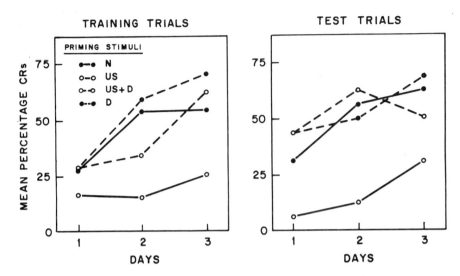

FIG. 3.3. Mean percentage eyelid conditioned responses in four groups of rabbits receiving different pre-CS–US priming events: no stimulation (N), a US alone (US), a US followed by a distractor (US + D), or the distractor alone (D). Plotted separately are the percentages on training trials as described and on test trials with no pre-CS–US stimulation in any group. Reprinted from Terry (1976).

treatments in Terry's (1976) experiments. In contrast, Group US + D was not suppressed in its level of conditioned responding. There were no reliable differences among Group N, Group D, and Group US + D, but all were reliably more responsive than Group US, on training and on test trials.

Terry's studies are the most analytical investigations available on the effects of US priming in Pavlovian conditioning. However, it should be added here that Konorski, Lubinska, and Miller (1936) conducted an investigation of salivary conditioning with dogs that was conceptually identical with Terry's (1976) first experiment, with similar outcome. More recently the phenomenon has come under investigation in the context of toxiphobia conditioning, with at least one additional notable finding: A priming US can diminish the conditioning that results from a subsequent CS–US pairing, even when the priming US occurs in temporal relationship to the following CS so that it would otherwise produce excitatory "backward" conditioning (Domjan & Best, 1977). In all, there are substantial data to indicate that a blockinglike effect is observable under conditions of self-generated US priming, as under conditions of associatively generated US priming.

2. CS Processing

When we turn to the question of potentially diminished *CS* processing under similar conditions of priming, the data are more fragmentary. There is one especially robust and reproducible effect that is congruent in the most general sense with Konorski's formulation concerning habituation—namely, the "latent inhibition" effect. Repeated preexposures to a CS alone retard the development of conditioned responding during subsequent CS-US pairings (see, e.g., Lubow, 1973). We assume that the CS comes to be less effectively processed for associative learning, rather than to have some specific behavioral tendency antagonistic to conditioned responding, because it is retarded in acquiring behavior indicative of "excitatory" or "inhibitory" associative relationships to either an aversive or an appetitive US (e.g., Best, 1975; Halgren, 1974; Reiss & Wagner, 1972; Rescorla, 1971). What is not clear is whether this latent inhibition effect should be attributed to some context-independent process, such as the *formation* of a covering gnostic unit in Konorski's schema, or to some context-dependent process, such as the priming notions captured in the expanded Fig. 3.2.

a. Associatively generated priming. The interesting possibility is that latent inhibition is like the blocking effect in being dependent, at least in part, upon an associatively generated priming of STM. When the CS is repeatedly presented in an experimental environment, it is possible that the environmental cues become associated with the CS so as to contribute to excitation of the latter's memorial representation, just as we have argued may be the case when a US is similarly preexposed in the environment of a blocking experiment (e.g., Blanchard & Honig, 1976; Dweck & Wagner, 1970; Tomie, 1976). There are, in fact, several studies of latent inhibition that indicate that the effect of CS preexposure is reduced if the environment is changed from the occasion of preexposure to the occasion of the CS-US pairings (e.g., Anderson, O'Farrell, Formica, & Caponegri, 1969; Lantz, 1973). Furthermore, the essential notion could be used to account for the findings of Mackintosh (1973) that a CS preexposed in sessions in which a US was also preexposed (in a random relationship) was later difficult to train as a signal for the training US, but not for another US. That is, we could assume that important contextual cues for the appearance of a latent inhibition effect were provided by the emotionally significant USs received in the same preexposure sessions. If the US were then changed, the diminished processing of the CS should have been less detectable.

The unique prediction from this point of view is that latent inhibition should be extinguishable. If the contextual stimuli have come to act as signals for the CS, associatively priming the CS representation in memory, we should be able to reduce this tendency by the conventional operation of presenting the signal without the customary consequence — that is, in this case by presenting the contextual stimuli in the absence of the CS.

Two relevant unpublished studies were conducted in our laboratory by Wagner, Pfautz, and Donegan (1977), one in eyelid conditioning with rabbits, the other in CER conditioning with rats. In the basic comparisons of interest, subjects were repeatedly exposed to a stimulus, later to be employed as a CS, in the experimental environment in which conditioning would occur, to an extent sufficient to produce a latent inhibition effect (100 stimuli per day for 1 or 2 days in the eyelid study; 4 stimuli per day for 6 days in the CER study). Half of the subjects thus pretreated were then returned for a like number of "extinction" sessions to the experimental apparatus, where they were confined without any stimulation. The remaining half of the subjects on these days were simply retained in their living cages. All subjects were then given their respective CS-US pairings. In both the eyelid-conditioning study and the CER study, the group that received the "extinction" sessions interposed between the CS exposures and the CS-US pairings showed substantially and reliably more conditioned responding than their nonextinguished counterparts.

These findings are clearly consistent with the notion that latent inhibition is, at least in part, a product of associatively generated priming. We must remain cautious, however, in that in the studies of latent inhibition that we have thus far conducted we have not evaluated the stimulus specificity of the extinction effect, as we have eventually done in the case of other related effects. This remains to be done, as it remains to be seen how generalizable is the extinction effect to the many situations in which latent inhibition can be demonstrated.

b. Self-generated priming. If some of the associative deficit accruing to a CS as a result of preconditioning exposure and grossly categorized as "latent inhibition" can be attributed to the priming of STM, we should see that not all preexposures have the same effect. In addition to a relatively permanent decrement in processing following a series of CS exposures, which might be attributed to associatively generated priming by contextual cues as discussed above, we could expect a relatively transient decrement in processing following a recent CS exposure, which might be attributed to self-generated priming.

In this instance we are especially encouraged by a study conducted by Best and Gemberling (1977), which is the more interesting because it comments upon a provocative observation originally reported by Kalat and Rozin (1973). The latter authors had compared the conditioned taste aversion produced by each of 3 treatments: One group (CS_{30}-US) received a taste CS 30 min before lithium chloride-induced toxicosis; a second group (CS_{240}-US) received the same taste CS 240 min before toxicosis; the third group (CS_{240}-CS_{30}-US) received 2 exposures to the taste CS, the first 240 min and the second 30 min before toxicosis. As would be expected from the differential temporal relationships, Group CS_{30}-US showed more of a subsequent taste aversion than did Group CS_{240}-US. The interesting observation, however, was that althouth Group CS_{240}-CS_{30}-US had the same final CS-US pairing as Group CS_{30}-US, it showed substantially less aversion to the CS, and indeed no more than was observed in Group CS_{240}–US. Kalat and Rozin (1973) interpreted these data as indicative of what they termed a "learned safety" effect: The longer the interval that the animal was granted following a novel taste CS in which a noxious event did *not* occur, the more it would presumably learn that the taste was "safe" and hence not to be avoided. A longer "safety" interval was provided in Group CS_{240}–CS_{30}–US (and in Group CS_{240}–US) than in Group CS_{30}–US.

Best and Gemberling (1977) argued instead that the data might reflect a CS-priming effect, with less conditioning produced in the double-CS case because of the persistence of the initial CS representation at the time of the terminal CS–US pairing. Among the data from several experiments reported by Best and Gemberling, the theoretically most decisive came from double-CS treatments in which the initial CS exposure was made increasingly more removed than 240 min from the final CS–US pairing (e.g., up to 450 min). When this was done there was not a systematic decrease in taste aversion, as would be expected from the learned-safety view, but rather a systematic *increase*, as would be expected from the priming notion. When it was made less likely that the initial CS would still be represented in STM at the time of the second CS, association of the latter CS and the US was apparently increased.

Other observations of Best and Gemberling (1977) are equally interesting. In the case of a CS–CS–US treatment, if the interval between the initial, priming CS and the terminal US is within the effective CS–US interval for conditioning, we would expect it to contribute directly to a CS–US association, as well as to reduce the associative effectiveness of the subsequent CS. In double-CS treatments in which the initial CS was placed at the same point (240 min) as in the Kalat and Rozin (1973) study or yet closer to the terminal US, it appeared that *all* of the conditioning that occurred was consistent with the interval between the initial CS and the US — that is, as though the more proximal CS were completely without effect. Only when the priming CS was quite removed from the terminal CS–US pairing did the latter relationship become apparent in conditioned aversion.

Such findings are consistent with the mechanism of Fig. 3.2. The assumption is that CS preexposure may transiently activate a reverberating loop that will persist to sensitize the subsequent activation of a covering gnostic unit, which will in turn favor the recurrent inhibitory perceptive reflex. We are cautious here again, only because we have not completed the kind of studies that made up the Terry (1976) investigation of US priming. Given a sequence of CS–CS–US, we would be more convinced by seeing that the depressed conditioning to the CS is greater when the priming CS is the same as the trial CS, as compared with when it is a different CS. And, we would wish to see that the effects of CS priming are reduced by placing a distractor between the priming and trial CSs.

B. Unconditioned Responding

Although we can interpret the studies of associative learning that have thus far been reviewed as consistent with the expansion upon Konorski's mechanism of habituation shown in Fig. 3.2, we would certainly wish to have supportive evidence from more typical measures of habituation—that is, in terms of some decrement in the usual responses to the stimuli in question. Our theoretical speculations have led us to ask whether or not self-generated and associatively generated priming of STM are evidenced in more typical measures of habituation, in specific relationship to the kinds of experimental subjects and stimuli employed in the preceding studies.

1. Self-generated Priming

There is ample evidence in studies of habituation to make one willing to entertain the notion of self-generated priming. At an empirical level it is relatively conventional (e.g., Castellucci & Kandel, 1976; Davis, 1970; Groves & Thompson, 1970; Horn, 1971) to distinguish between a transient habituation decrement that shows rapid "spontaneous recovery," versus a more persistent "long-term" habituation decrement that is detectable on tests remote from a habituation series. And, at a more analytical level, it is possible to show that, relatively independent of the level of "long-term habituation" that has occurred, one can see a systematic refractorylike effect, in which the response to the habituating stimulus is less likely the less time (within limits) since the immediately preceding stimulus (e.g., Davis, 1970; Leaton, 1976). Perhaps the major evidence to isolate the refractorylike effect as a separable phenomenon in habituation studies is that the massing of stimulus exposures leads to a more pronounced immediate response decrement but then to a less pronounced long-term decrement than does the spacing of stimulus exposures (e.g., Davis, 1970; Carew, Pinsker, & Kandel, 1972).

What must be particularly asked, however, about any transient response decrement to a stimulus is whether it might not be attributed to other than memorial processes — that is, to such factors as sensory adaptation or response system fatigue. Thus, our studies were calculated not only to look for evidence of a refractorylike habituation effect in the context of the kinds of experimental subjects and stimuli employed in our studies of association, but to do so in a fashion that would serve to identify any such effect as more or less consistent with presumptions already articulated about the priming of STM.

The two studies that best speak to this set of questions were part of a dissertation conducted by Jesse Whitlow (1975). The subjects were rabbits, under conditions of restraint identical with those in the preceding eyelid conditioning studies. The stimuli of interest were 1-sec tones comparable to those used as Pavlovian CSs. The behavior that was measured, however, was the peripheral vasconstriction response recorded from a photocell plethysmograph attached to the rabbit's ear. This response falls in the class of autonomic, "orienting responses" meant to be addressed by Konorski's formulation.

In his first study, Whitlow (1975) presented rabbits with a series of tones, equally often 530 Hz and 4000 Hz. The series was arranged with successive pairs of tones being separated by 150 sec, and the members within each pair (designated as S_1 and S_2) separated by 30, 60, or 150 sec in an irregular counterbalanced order. It was expected that a refractorylike response decrement would be observed in variation in responding to S_2 as a function of the S_1–S_2 interval. However, it was arranged that on half of the S_1–S_2 pairs, involving each interstimulus interval, the 2 stimuli were the same (equally often 530 Hz or 4000 Hz), while on the remaining pairs the 2 stimuli were different (equally 530 Hz followed by 4000 Hz, or 4000 Hz followed by 530 Hz). It was thus possible to evaluate the degree to which any decremental effect of S_1 upon the response to S_2 was specific to the occasions in which S_1 matched S_2 as suggested by the priming notion.

The study was run over 2 2-hr sessions and involved 48 S_1–S_2 pairs. There was an overall decrease in response to the respective stimuli in the pairs with repetitive stimulation (see Whitlow, 1975). But, the major data of interest can be seen in Fig. 3.4, which plots the averaged evoked vasoconstriction response at successive 5-sec intervals, following S_1 and then S_2 over all pairs in which the interstimulus interval was either 30, 60, or 150 sec, and separately for those pairs in which the stimuli were the same versus different tones. The principal finding is obvious and highly reliable. When S_2 followed S_1 by 30 sec, there was a substantial depression in the vasoconstriction evoked by the same as compared with the different tone. When S_2 followed S_1 by 60 sec, this difference was smaller, although still quite apparent. However, when S_2 followed S_1 by 150 sec, there was no longer a detectable depression in responding to S_2, as the subjects responded equivalently when S_2 was the same as or different from the preceding stimulus. There is, in this preparation, evidence of a stimulus-specific response decrement that dissipates over the tested intervals in a manner congruent with our priming assumptions.

Granted this finding, it was then possible for Whitlow to ask whether or not an extraneous stimulus that should compete for occupancy of STM would act to remove the stimulus-specific response decrement in a manner analogous to Terry's (1976) observation. The experiment was closely patterned after the previous study, with the exceptions that (1) the S_1–S_2 interval was consistently 60 sec, and (2) on half of each of the occasions in which S_2 was the same tone as S_1 or was different from S_1, the interval included a distractor stimulus 20 sec after S_1 and thus 40 sec before S_2. From the data

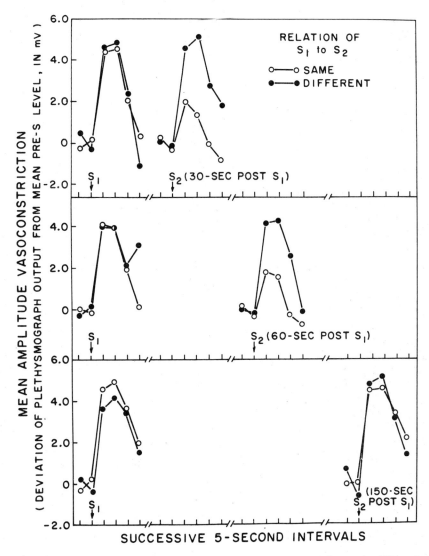

FIG. 3.4.   Mean average evoked vasomotor responses to the auditory stimuli S_1 and S_2 when the two were the same or different frequencies and separated by 30 sec (top panel), 60 sec (middle panel), or 150 sec (bottom panel). Reprinted from Whitlow (1975).

presented in Fig. 3.4, it could be expected that in the absence of the distractor stimulus there would be ample stimulus-specific response decrement to S_2 at this S_1–S_2 interval. The question was whether or not the distractor (which was a sequential compound of a 1-sec flashing light and a 1-sec electrotactile stimulation of the cheek) would remove the decrement, as extending the duration of the S_1–S_2 interval to 150 sec had been found to do.

Figure 3.5 summarizes the relevant data in the same manner as Fig. 3.4. When there was no intervening stimulation (see top panel), the response to S_2 was reduced on occasions in which S_1 was the same stimulus, as compared with occasions in which S_1 was a different stimulus. When the distractor was presented between S_1 and S_2 (see bottom panel), it produced an evoked response itself, but more importantly, it removed any differential effect upon S_2 responding of the same versus different S_1 events. These findings are what one would expect if the stimulus-specific response decrement were dependent upon a perseverating excitation of the test stimulus representation in STM, and the distractor acted to remove such excitation.

The outcomes of Whitlow's studies are particularly informative in not being liable to a sensory-adaptation or response-fatigue account. The stimulus specificity of the transient decrement argues against response fatigue, while the restorative effect of the distractor is outside of our notions of what might remove sensory adaptation. It is likewise important that the distractor did not appear to act as a *general sensitizer*. It is common to observe that some extraneous stimulus introduced in a habituation sequence will lead to an increase in responding (e.g., Humphrey, 1933; Thompson & Spencer, 1966). However, to the degree that the extraneous stimulation leads to an equivalent increase in the responding to nonhabituated stimuli as well as to habituated stimuli, there is no reason to assume that it has any action intrinsic to the mechanism of habituation (see Humphrey, 1933; Sharpless & Jasper, 1956; Thompson & Spencer, 1966; Castellucci & Kandel, 1976). The relatively innocuous but salient distractor employed by Whitlow (and presumably by Terry, 1976, as well) appears to have revealed a specific "dishabituation" effect that is important to our confidence in the phenomenon of self-generated priming as a memorial process in habituation.

2. Associatively Generated Priming

When we turn to the question of associatively generated priming in relationship to habituation, there is again some empirical encouragement. Ethological accounts often suggest that there is a context specificity to the waning of some species-typical behavior (see, e.g., Lorenz, 1965), and some scattered experiments make the same point. Peeke and Vino (1973) reported a relevant observation. They recorded the aggressive behavior of male

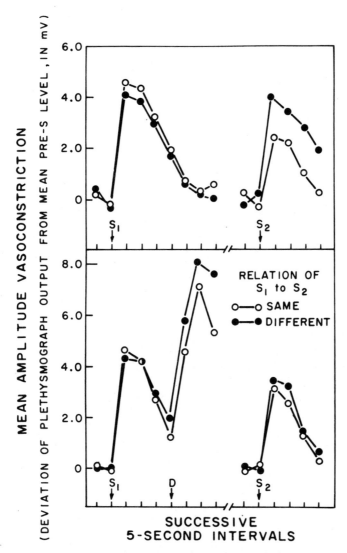

FIG. 3.5. Mean average evoked vasomotor responses to the auditory stimuli S_1 and S_2 when the two were the same or different frequencies and either involved no intervening stimulation (top panel) or were separated by a distractor, D (bottom panel). Reprinted from Whitlow (1975).

sticklebacks toward presentations of another male in two separate exposure sessions. They observed that there was more aggressive behavior (less evidence of habituation) in the second session when it was conducted in a different location in the aquarium from the first session, than when it was conducted in the same location as the first session.

One reason to be less than convinced that this kind of evidence of context specificity is due to associatively generated priming of STM is that it is possible to assume that the subject habituated to a total pattern, or Gestalt, which included not only the behavior-releasing stimulus, but also contextual cues. Changing the test environment would thus be equivalent to a stimulus-generalization test. Or, it is possible to assume that the subject habituated independently to both the releasing stimulus and the contextual cues. Changing the test environment might thus be viewed as a nonspecific sensitization manipulation.

In this light the most powerful test of the notion that the contextual cues may act via the associative network to excite stimulus representation in memory follows from the extinction manipulation employed in the aforementioned studies of latent inhibition by Wagner, Pfautz, and Donegan. A relevant habituation study was conducted in collaboration with Jesse W. Whitlow and Penn L. Pfautz (reported in Wagner, 1976), using the same vasomotor preparation employed by Whitlow (1975). In initial training, 2 groups of rabbits were subjected to a 2-hr habituation session during which a 1-sec tonal stimulus was presented on 32 occasions with a 150-sec interstimulus interval. The frequency of the habituation stimulus for half of each group was 530 Hz and for the remainder 4000 Hz. All subjects were returned for a retention test 2 days later. The only difference in treatment of the 2 groups was that on the intervening day one group (Control) was simply left in their home cages, while the other group (Extinguished) was placed in the experimental chamber, had the recording apparatus attached, and was treated exactly as during the previous habituation session except that no tones were presented. On the retention day all subjects received a series of 32 exposures to the previously habituated tone, as in session 1, and 2 test exposures to the alternate frequency tone.

Figure 3.6 summarizes the essential data from this experiment. In each panel is presented the averaged evoked response, at 5-sec intervals preceding and following the test stimulus, for each group. The left panel indicates the initial level of responding to the to-be-habituated stimulus (designated A) as it shows the averaged response to the first 2 tone presentations in the training session for both groups. The middle panel indicates the responding of the 2 groups to the comparable 2 tone presentations at the beginning of the retention test session. The major effect is obvious in these data. Both groups evidenced a habituation decrement in their responding on the retention day

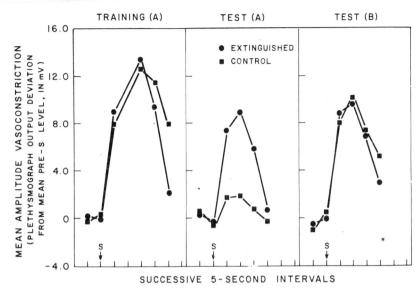

FIG. 3.6. Mean average evoked vasomotor response to auditory stimuli, S, in 2 groups of subjects. The successive panels present the response to the first 2 of 32 presentations of tone A during the habituation training session (left panel), the response to the first 2 exposures to the same tone in a retention test session 2 days later (middle panel), and finally the response to 2 exposures to a novel tone, B, in the same retention session. Between training and testing, 1 group (Extinguished) was exposed to the experimental environment in the absence of stimulation, while the other group (Control) was not. Reprinted from Wagner (1976).

as compared with the initial stimulation. However, whereas there was a very sizable habituation decrement (an 85% reduction in response magnitude) seen in the case of the Control group, there was a much smaller habituation decrement (a 34% reduction in response magnitude) seen in the Extinguished group.

The test trials to the nonhabituated tonal stimulus (designated B) were included in the retention session as a check that the extinction manipulation had not been effective by producing some general change in vasomotor responsivity. The right panel of Fig. 3.6 presents the evoked response to the novel tone. It is apparent that the two groups did not differ in their response to this stimulus, whereas they did differ in their response to the previously habituated tone.

C. Perceptual Judgments

From the studies that I have reviewed in the previous sections, and from the major emphasis of this chapter on relating the phenomena involved to

Konorski's (1967) treatment of habituation, one might be left with the impression that the priming of STM consistently leads to a decrement in measures of stimulus processing. This is clearly not the case. In some tasks the operation of preceding a stimulus pattern with a previously associated stimulus or with an instance of the same pattern leads to an enhancement of discriminative performance to the target event.

Consider the studies of Meyer, Schvaneveldt, and Ruddy (1975). They required human subjects to judge whether a string of letters was a word or a nonword in a speeded reaction-time task. Judgmental reaction time was faster to a word (e.g., Butter) when it was immediately preceded in the stimulus list by another word from the same semantic category (e.g., Bread) than when similarly preceded by a word from another category (e.g., Nurse). Similar observations have been made by Neely (1977) when the initial stimulus was a category name and the target stimulus was either a member of that category or not and by Posner and Snyder (1975) when the initial stimulus was either a component of the second stimulus or not. The common interpretation that has been given to these findings is that the initial stimulus (it is called a "priming stimulus" in this literature also) excites a memorial representation of the target stimulus prior to its presentation either as a result of the physical identity or of the associative linkages between the two. With this prior excitation, it is assumed that the judgmental behavior required to the target, when it occurs, can proceed faster. The experimental technique is presently of theoretical interest as it may comment on such issues as the organization of memory and on a distinction between an "automatic spread of activation" versus a "controlled processing." (See, e.g., Neely, 1977; Posner & Snyder, 1975; Shiffrin & Schneider, 1977.) But, in the present context, the critical fact is simply that self-generated or associatively generated priming of STM can facilitate performance in such instances.

The obvious theoretical challenge is to be able to specify what class of behavioral measures will be decremented by operations of priming and what class will be facilitated by the same operations. I have indicated earlier in this chapter that many of the decremental effects of priming that have been observed in our laboratory could be attributed to a decreased likelihood or duration of poststimulation *rehearsal* of the target stimulus. One possibility is that the decremental effects of priming are in fact restricted to those behavioral measures that reflect differences in rehearsal. Thus, I have suggested (Wagner, in press) that "refreshing" a stimulus representation previously primed in STM might cause the representation more surely or more promptly to reach some "threshold" necessary to generate some behavior than would de novo excitation, while the same "refreshment" would be less likely to occasion the sustained activity in STM (i.e., rehearsal) that is critical to other indices of stimulus processing, such as associative learning or a protracted orienting response. Another possibility is that the representational processing of a primed stimulus is more generally decremented, but that behavior sequences initiated by the priming stimulus may persist to summate with behavior actually occasioned by the target stimulus, making the processing of the latter stimulus per se only appear to be facilitated in some instances. Consider, for example, the likely results of CS-US pairings when the CR that develops to the CS mimics the UR to the US. Even though US processing (and thus UR magnitude) might be generally diminished when the stimulus is associatively primed in STM by the CS, just the opposite might appear to be the case when the CS provokes a CR that mimics the UR and what is measured at the time of the US is a combination of CR and UR.

At the moment, one might best conclude that additional data are necessary to guide our assumptions. We do not, for example, know what range of responses will evidence the kind of contextual control of the habituation decrement seen in our aforementioned extinction study (Wagner, 1976) of vasomotor responding. Nor do we know in relationship to this range what specific conditions of training and testing with explicit signals for associatively generated priming will produce a "conditioned diminution of the UR" (e.g., Kimble & Ost, 1961; Kimmel, 1966), versus an apparent facilitation of the UR (e.g., Hupka, Kwaterski, & Moore, 1970), versus a more complex mixture of facilitation and diminution (e.g., Wagner, Thomas, & Norton, 1967).

Any provisional account, however, can be measured in usefulness against that of Konorski (1967). The operations of priming are presumably effective, as we have noted, because they increase the likelihood that the target stimulus will activate its corresponding gnostic unit. But, Konorski explicitly assumes that activation of a gnostic unit from peripheral stimulation has two separable effects. First, it leads to the recognizable perception of the stimulus. Second, it leads to the recurrent inhibitory perceptive reflex that suppresses the several components of the targeting reflex, including the facilitating bias to the gnostic assembly. It is, I think, within the obvious sense of these distinctions to anticipate that priming operations would thus facilitate speeded perceptual judgments, as observed by Meyer et al. (1975), while at the same time depressing associative learning and vasomotor responsiveness, as illustrated in the studies I have reviewed.

IV. CONCLUDING COMMENTS

It is with great respect for the scope and subtlety of Konorski's theorizing that I have attempted to show how his views on habituation and memory put forth in 1967 could accommodate more recent data and speculations concerning the priming of STM. Having apologized in advance for whatever unwarranted remarks I might make about Konorski's approach to habituation, I need now only to try to excuse what I have left out.

In emphasizing associative and transient influences on habituation phenomena, I have said little of substance about that process that Konorski (1967) emphasized. That is, I have ignored the role of the *formation* of gnostic units. In this regard I would mainly note that Konorski's (1967) explicit account of this process hardly requires similar interpretation. However, it is also true that to the degree that we must acknowledge the influence of self-generated and associatively generated priming of STM on habituation, we must depreciate the fact of habituation as being presumptive evidence of the formation of gnostic units, as Konorski (e.g., 1967, p. 100) sometimes implied. It remains to be seen what instances of habituation, if any, will truly require appeal to the formation of new representative elements in long-term memory, as opposed to the formation of new associative relationships among the given elements of the memory system.

In pursuing the particular subject of this paper I have also ignored other recent advances in our understanding of habituation that are equally deserving of being related to Konorski's formulation. I have, of course, emphasized those phenomena that have occupied my own interests. But important findings have resulted from the investigations of habituation in "simple systems," such as the gill-withdrawal reflex of the *Aplysia* (see, e.g., Castellucci & Kandel, 1976), the tail-flip response of the crayfish (e.g., Zucker, 1972), and the neuronal activity of the isolated spinal cord of the frog (see, e.g., Thompson & Glanzman, 1976). A major result of this work is the appreciation that certain instances of response decrement to iterated stimulation that would be classified as habituation (e.g., Thompson & Spencer, 1966) can be attributed to what Horn (1967) has generically termed "self-generated depression," — that is, a passive decrease in sensitivity intrinsic to the fact of prior activity. In some cases (e.g., Castellucci & Kandel, 1976) it is clear that the decrement can be attributed to presynaptic depression with neuronal activity decreasing the subsequent availability and release of transmitter substance.

Konorski's formulation concerning habituation does not involve the notion of self-generated depression. Habituation is, instead, consistently attributed to the occurrence of active excitatory and inhibitory processes. This omission is presently a detraction. The formulation must, of course, be evaluated primarily in terms of its own aims. It is a theory about stimulus processing in a complex system that is otherwise assumed to enjoy a variety of mechanisms, including an elaborate dual-memory system. And, it is intended to account for an intricate pattern of variations in behavior, including receptor adjustments, autonomic indices, associative learning, and perceptual judgments under a complex set of historical variables, including those I have emphasized as involving the priming of STM. It meets these aims admirably. However, the absence of appreciation of self-generated depression limits its generality, or, to say this more correctly, limits the ease with which it appears possible to form integrative links with simpler formulations that now appear sufficient and appropriate for other habituating systems.

My assumption is that Konorski's formulation could be recast to advantage by exploiting the reality of self-generated depression in neural networks, as well as its present notions of excitatory and inhibitory influences. Indeed, my original views about the priming of STM (Wagner, 1976), although inspired by conceptions of information-processing theory (e.g., Atkinson & Shiffrin, 1968), were voiced so as to be generally compatible with the notion of self-generated depression (e.g., Horn, 1967). That is, it seemed reasonable to suppose that when some stimulus representation had recently been activated in STM, whether by occurrence of its corresponding stimulus or by an associated retrieval cue, then critical units in a rehearsal circuit would be in a temporary state of depression and thus be less susceptible to activation than they otherwise would be. Unfortunately, there is a substantial gap between such generalities and a concrete formulation, equivalent to Konorski's schema, as presented in Fig. 3.2.

ACKNOWLEDGMENTS

Preparation of this chapter was supported in part by National Science Foundation Grant BMS 74-20521. It benefited from the knowledgeable criticism of Nelson H. Donegan, Penn L. Pfautz, and Judith Primavera.

REFERENCES

Anderson, D. C., O'Farrell, T., Formica, R., & Caponegri, V. Preconditioning CS exposure: Variation in the place of conditioning and presentation. *Psychonomic Science*, 1969, *15*, 54-55.

Anderson, J., & Bower, G. *Human associative memory.* New York: Winston, 1973.

Atkinson, R. C., & Shiffrin, R. M. Human memory: A proposed system and its control processes. In K. W. Spence (Ed.), *The psychology of learning and motivation* (Vol. 2). New York: Academic Press, 1968.

Atkinson, R. C., & Wickens, T. D. Human memory and the concept of reinforcement. In R. Glaser (Ed.), *The nature of reinforcement.* New York: Academic Press, 1971.

Best, M. R. Conditioned and latent inhibition in taste-aversion learning: Clarifying the role of learned safety. *Journal of Experimental Psychology: Animal Behavior Processes*, 1975, *2*, 97-113.

Best, M. R., & Gemberling, G. A. The role of short-term processes in the CS preexposure effect and the delay of reinforcement gradient in long-delay taste-aversion learning. *Journal of Experimental Psychology: Animal Behavior Processes*, 1977, *3*, 253-263.

Blanchard, R., & Honig, W. K. Surprise value of food determines its effectiveness as a reinforcer. *Journal of Experimental Psychology: Animal Behavior Processes,* 1976, *2,* 67–74.

Carew, T. J., Pinsker, H. M., & Kandel, E. R. Long-term habituation of a defensive withdrawal reflex in *Aplysia. Science,* 1972, *175,* 451–454.

Castellucci, V., & Kandel, E. R. An invertebrate system for the cellular study of habituation and sensitization. In T. J. Tighe & R. N. Leaton (Eds.), *Habituation: Perspectives from child development, animal behavior, and neurophysiology.* Hillsdale, N. J.: Lawrence Erlbaum Associates, 1976.

Davis, M. Effects of interstimulus interval length and variability on startle-response habituation in the rat. *Journal of Comparative and Physiological Psychology,* 1970, *72,* 177–192.

Domjan, M., & Best, M. R. Paradoxical effects of proximal unconditioned stimulus preexposure: Interference with and conditioning of a taste aversion. *Journal of Experimental Psychology: Animal Behavior Processes,* 1977, *3,* 310–321.

Donegan, N. H., Whitlow, J. W., & Wagner, A. R. Posttrial reinstatement of the CS in Pavlovian conditioning: Facilitation or impairment of acquisition as a function of individual differences in responsiveness to the CS. *Journal of Experimental Psychology: Animal Behavior Processes,* 1977, *3,* 357–376.

Dweck, C. S., & Wagner, A. R. Situational cues and correlation between CS and US as determiners of the conditioned emotional response. *Psychonomic Science,* 1970, *18,* 145–147.

Estes, W. K. Memory and conditioning. In F. J. McGuigan & D. B. Lumsden (Eds.), *Contemporary approaches to conditioning and learning.* Washington, D.C.: Winston, 1973.

Estes, W. K. Structural aspects of associative models for memory. In C. N. Cofer (Ed.), *The structure of human memory.* San Francisco: Freeman, 1975.

Gillian, D. J., & Domjan, M. Taste-aversion conditioning with expected versus unexpected drug treatment. *Journal of Experimental Psychology: Animal Behavior Processes,* 1977, *3,* 297–309.

Groves, P. M., & Thompson, R. F. Habituation: A dual process theory. *Psychological Review,* 1970, *77,* 419–450.

Halgren, C. R. Latent inhibition in rats: Associative or nonassociative? *Journal of Comparative and Physiological Psychology,* 1974, *86,* 74–78.

Horn, G. Habituation and memory. In G. Adam (Ed.), *Biology of Memory.* Budapest: Publishing House of the Hungarian Academy of Sciences, 1971.

Humphrey, G. *The nature of learning.* New York: Harcourt, Brace, 1933.

Hupka, R. B., Kwaterski, S. E., & Moore, J. W. Conditioned diminution of the UCR: Differences between the human eyeblink and the rabbit nictitating membrane response. *Journal of Experimental Psychology,* 1970, *83,* 45–51.

Kalat, J. W., & Rozin, P. "Learned safety" as a mechanism in long delay taste-aversion learning in rats. *Journal of Comparative and Physiological Psychology,* 1973, *83,* 198–207.

Kamin, L. J. Predictability, surprise, attention and conditioning. In B. Campbell & R. Church (Eds.), *Punishment and aversive behavior.* New York: Appleton-Century-Crofts, 1969.

Kimble, G. A., & Ost, J. W. P. A conditioned inhibitory process in eyelid conditioning. *Journal of Experimental Psychology,* 1961, *61,* 150–156.

Kimmel, H. D. Inhibition of the unconditioned response in classical conditioning. *Psychological Review,* 1966, *73,* 232–240.

Konorski, J. *Integrative activity of the brain.* Chicago: University of Chicago Press, 1967.

Konorski, J., Lubinska, L., & Miller, S. Elaboration des reflexes conditionnels dans l'ecorce cerebral a l'etat d'inhibition inducte. *Acta Biologiae Experimentalis,* 1936, *10,* 297–330.

Lantz, A. Effect of number of trials, interstimulus interval, and dishabituation during CS habituation on subsequent conditioning in a CER paradigm. *Animal Learning and Behavior,* 1973, *4,* 273–278.

Leaton, R. N. Long-term retention of the habituation of lick suppression and startle response produced by a single auditory stimulus. *Journal of Experimental Psychology: Animal Behavior Processes,* 1976, *2,* 248–260.

Lorenz, K. *Evolution and modification of behavior.* Chicago: University of Chicago Press, 1965.

Lubow, R. E. Latent inhibition. *Psychological Bulletin,* 1973, *79,* 398–407.

Mackintosh, N. J. Stimulus selection: Learning to ignore stimuli that predict no change in reinforcement. In R. A. Hinde & J. Stevenson-Hinde (Eds.), *Constraints on learning.* London: Academic Press, 1973.

Mackintosh, N. J. A theory of attention: Variations in the associability of stimuli with reinforcement. *Psychological Review,* 1975, *82,* 276–298.

Mackintosh, N. J., & Turner, C. Blocking as a function of novelty of CS and predictability of UCS. *Quarterly Journal of Experimental Psychology,* 1971, *23,* 359–366.

Meyer, D. E., Schvaneveldt, R. W., & Ruddy, M. G. Loci of contextual effects in visual word recognition. In P. M. A. Rabbitt & S. Dornic (Eds.), *Attention and performance V.* London: Academic Press, 1975.

Neely, J. H. Semantic priming and retrieval from lexical memory: The roles of inhibition-less spreading activation and limited-capacity attention. *Journal of Experimental Psychology: General,* 1977, *3,* 226–254.

Norman, D. A. Toward a theory of memory and attention. *Psychological Review,* 1968, *75,* 522–536.

Peeke, H. V. S., & Vino, G. Stimulus specificity of habituated aggression in three-spined sticklebacks *(Gasterosteus Aculeatus). Behavioral Biology,* 1973, *8,* 427–432.

Pfautz, P. L., & Wagner, A. R. Transient variations in responding to Pavlovian conditioned stimuli have implications for the mechanisms of "priming." *Animal Learning and Behavior,* 1976, *4,* 107–112.

Posner, M. I., & Snyder, C. R. R. Facilitation and inhibition in the processing of signals. In P. M. A. Rabbitt & S. Dornic (Eds.), *Attention and performance V.* New York: Academic Press, 1975.

Reiss, S., & Wagner, A. R. CS habituation produces a "latent inhibition effect" but no "conditioned inhibition." *Learning and Motivation,* 1972, *3,* 237–245.

Rescorla, R. A. Summation and retardation tests of latent inhibition. *Journal of Comparative and Physiological Psychology,* 1971, *75,* 77–81.

Sharpless, S., & Jasper, H. Habituation of the arousal reaction. *Brain,* 1956, *79,* 655–680.

Shiffrin, R. M., & Schneider, W. Controlled and automatic information processing: II. Perceptual learning, automatic attending, and a general theory. *Psychological Review,* 1977, *84,* 127–190.

Sokolov, E. N. *Perception and the conditioned reflex.* New York: Pergamon, 1963.

Sokolov, E. N. Modeling properties of the nervous system. In I. Maltzman & M. Cole (Eds.), *Handbook of contemporary Soviet psychology.* New York: Basic Books, 1969.

Terry, W. S. The effects of priming US representation in short-term memory on Pavlovian conditioning. *Journal of Experimental Psychology: Animal Behavior Processes,* 1976, *2,* 354–370.

Terry, W. S., & Wagner, A. R. Short-term memory for "surprising" versus "expected" unconditioned stimuli in Pavlovian conditioning. *Journal of Experimental Psychology: Animal Behavior Processes,* 1975, *1,* 122–133.

Thompson, R. F., & Glanzman, D. L. Neural and behavioral mechanisms of habituation and sensitization. In T. J. Tighe & R. N. Leaton (Eds.), *Habituation: Perspectives from child development, animal behavior, and neurophysiology.* Hillsdale, N. J.: Lawrence Erlbaum Associates, 1976.

Thompson, R. F., & Spencer, W. A. Habituation: A model phenomenon for the study of neuronal substrates of behavior. *Psychological Review,* 1966, *197,* 16–43.

Tomie, A. Interference with autoshaping by prior context conditioning. *Journal of Experimental Psychology: Animal Behavior Processes,* 1976, *4,* 323–334.

Wagner, A. R. Stimulus selection and a "modified continuity theory." In G. H. Bower & J. T. Spence (Eds.), *The psychology of learning and motivation.* Vol. 3. New York: Academic Press, 1969.

Wagner, A. R, Elementary associations. In H. H. Kendler & J. T. Spence (Eds.), *Essays in neobehaviorism: A memorial volume to Kenneth W. Spence.* New York: Appleton-Century-Crofts, 1971.

Wagner, A. R. Priming in STM: An information processing mechanism for self-generated or retrieval-generated depression in performance. In T. J. Tighe & R. N. Leaton (Eds.), *Habituation: Perspectives from child development, animal behavior, and neurophysiology.* Hillsdale, N. J.: Lawrence Erlbaum Associates, 1976.

Wagner, A. R. Expectancies and the priming of STM. In S. H. Hulse, H. Fowler, & W. K. Honig (Eds.), *Cognitive processes in animal behavior.* Hillsdale, N. J.: Lawrence Erlbaum Associates, in press.

Wagner, A. R., Rudy, J. W., & Whitlow, J. W. Rehearsal in animal conditioning. *Journal of Experimental Psychology,* 1973, *97,* 407–426. (Monograph)

Wagner, A. R., Thomas, E., & Norton, T. Conditioning with electrical stimulation of motor cortex: Evidence of a possible source of motivation. *Journal of Comparative and Physiological Psychology,* 1967, *64,* 191–199.

Whitlow, J. W. Short-term memory in habituation and dishabituation. *Journal of Experimental Psychology: Animal Behavior Processes,* 1975, *1,* 189–206.

Zucker, R. S. Crayfish escape behavior and central synapses: II. Physiological mechanisms underlying behavioral habituation. *Journal of Neurophysiology,* 1972, *35,* 621–637.

4 Conditioned Inhibition and Extinction

Robert A. Rescorla
Yale University

From the beginning of the study of Pavlovian conditioning, the notion of inhibition has played a major role. In both his theorizing and his experimental work, Pavlov (1927) devoted his principal efforts to its exploration. Indeed, it is fair to say that the study of inhibition, rather than of excitation, was responsible for the bulk of Pavlov's ideas. Consequently, despite the historical skepticism of American psychology, there has never been any serious question elsewhere that inhibition, particularly learned inhibition, deserves a primary role in our thinking about conditioning.

Nevertheless, it is really with Konorski (1948) that most modern thinking about learned inhibition originates. One can identify three important ways in which Konorski influenced our thinking about inhibition. First, he clearly articulated measurement procedures by which one can reliably identify the presence of inhibition. Although Pavlov had used a variety of identification procedures, Konorski's writings seem much clearer on the fundamental opposition of excitation and inhibition, an opposition that has produced the currently popular assessment techniques of summation and retarded excitatory conditioning. Second, Konorski developed several theoretical accounts of conditioned inhibition. In constructing those theories, he took the important step of viewing inhibition as an associative phenomenon, coordinate with an associative excitatory process; moreover he specified alternative forms that those associations might take. Finally, Konorski and his coworkers at the Nencki Institute gave us what was, until the 1960s, the only really detailed data available on Pavlovian learned inhibition. It would not be an exaggeration to say that in most important respects, Konorski provided the basis upon which modern thinking about inhibition is simply an elaboration.

The present chapter has several objectives. First, it reviews Konorski's thinking on inhibition. Particular attention is paid to his views as represented in the 1948 book, because those views seem less widely appreciated than his later ideas. Second, it discusses several recurring issues in the study of inhibition, the role of inhibition in extinction, and the degree to which learned inhibition and excitation are parallel processes. It is argued that Konorski's views provide a profitable framework within which to consider these issues and to guide their future exploration.

I.   KONORSKI'S VIEWS

In his 1948 book, Konorski put forth his first general theory of conditioned inhibition. That theory, like all that have followed it, grappled with three separate issues: What is learned when a stimulus becomes a conditioned inhibitor? What conditions produce that learning? How does that learning evidence itself in the organism's behavior?

Konorski's answers to these questions came from his view of Pavlovian excitatory conditioning, in each case deriving from the answer to a parallel question about conditioned excitation. Konorski made explicit a framework for excitatory conditioning that many others have subsequently adopted: that excitatory conditioning involves the formation of an association between an internal representation of the signaling CS and of the signaled US. In the language of his 1948 book, the CS center developed excitatory connections with the US center. The formation of those excitatory connections was described as depending on the activation of the CS center at a time when the US center undergoes an increase in its own activity, an increase normally produced by the occurrence of the US itself. As those excitatory connections accumulate, the presentation of the CS, which activates the CS center, also becomes capable of activating the US center, thus producing a conditioned response. Konorski proposed a particular function relating the level of activation so induced in the US center to the number of CS–US connections. That performance function was given a negatively accelerated shape, a decision based primarily on physiological considerations.

Konorski's treatment of conditioned inhibition was largely analogous. It, too, was viewed as the learning of connections between the CS and US centers, but in this case the nature of the connections is inhibitory rather than excitatory. Likewise, the establishment of these inhibitory connections depends on the CS being present when the activation of the US center undergoes a change; in this case, the fall in the activation of that center is responsible for establishing the connection. Finally, conditioned inhibition was viewed as influencing performance by modifying the relationship between the number of excitatory associations and the activation of the US center. In this view a conditioned inhibitor acts to shift the performance function so as to permit a given level of excitatory connections less impact on the US center and thus on performance.

The relationships between inhibition and excitation in generating performance are shown in Fig. 4.1, which redraws Konorski's curves. The negatively accelerated performance function describes the characteristic of a particular US center at a particular point in time. It begins at a point greater than zero CS–US connections to suggest the importance of a threshold level of excitation in generating performance; it grows in a negatively accelerated fashion to suggest that although performance is monotonically related to strength of association, the relationship is one of diminishing returns. Moreover, because the lateral position of this function changes with various treatments, the same level of excitatory association will not always generate the same activation of the US center. In particular, the role of a conditioned inhibitor is temporarily to shift the function to the right, thus raising the threshold and reducing the behavioral output of a given excitatory stimulus, such as A.

This performance scheme makes it clear why summation and the retarded development of performance under excitatory acquisition should be good indices of a conditioned inhibitor. In the former case, an inhibitory stimulus, X, reduces the level of activation that an excitatory A would otherwise produce by shifting the function on AX trials compared with that on A alone trials (as in curves 1 and 2 in Fig. 4.1). In the latter case, an in-

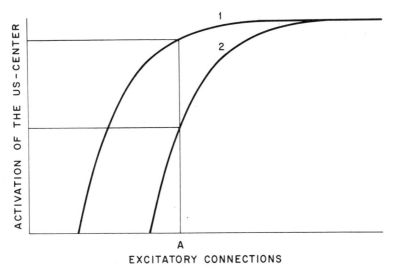

FIG. 4.1. Konorski's representation of how excitation and inhibition interact to activate the US center. The two functions show different levels of inhibition. A given level of excitation (A) would be mapped, via these different functions, into different activations of the US center.

hibitory X produces a shifted function that would then require more excitatory conditioning of X to generate a given activation of the US center. Similarly, it clarifies the failure to observe any particular behavior when an inhibitor is presented separately, an observation that has historically plagued the concept of inhibition. Within this view, an inhibitor has no behavioral output of its own, but instead serves to modulate the success of excitatory stimuli in producing their output.

According to this theory, conditioned inhibition is formed when a CS accompanies a fall in excitation in the US center. Konorski argued that therefore, the optimal procedure should be to pair a CS with the termination of a US. When he wrote there was only sketchy data with this procedure, but it was supportive. To this day there are few relevant data and the available results remain ambiguous (cf. Rescorla, 1969). But Konorski noted that other, more frequently employed, procedures should also produce conditioned inhibition according to this theory. For instance, the standard conditioned inhibition technique, in which A is reinforced except when accompanied by X (A+, AX-) ensures the presence of X at a time when the removal of A allows the US center to decrease in activity. In this way, the inhibitor can be associated with a particular US center, forming inhibitory connections with it. Moreover, the magnitude of the fall in US center activity, and thus the magnitude of inhibition conditioned, should depend on A's ability to activate the US center. Consequently, the acquisition of inhibition should follow a negatively accelerated course. This fits with the observations normally made in the special case of no X at all—namely, extinction of the previously excitatory A. Related applications of the notion of a fall in the US center may be made to other known procedures for the generation of inhibition.

In summary, according to Konorski's view in 1948, a conditioned inhibitor had the power to raise the performance threshold because it had inhibitory associations with the US center due to its past presence when that center had undergone a fall in its level of activation.

By 1967, however, Konorski had largely abandoned this view in favor of an account of inhibition with a quite different flavor. Two changes in Konorski's general thinking were central to this modification of his views on inhibition. First, he developed the notion of "gnostic units" to replace that of CS and US centers. Although he continued his commitment to a physiological basis for behavior, these gnostic units are more psychological in nature. They correspond to what might fairly be called psychological representations of individual events, such as the CS and the US. Although wishing to leave the detailed mechanisms vague, Konorski clearly intended that these gnostic units be partly memories built up from experience with such events. Moreover, he distinguished between two types of gnostic units that were constructed from the presentations of USs. He noted that such events have both consummatory and drive aspects; so he suggested that separate gnostic units might represent those different aspects. For instance, the presentation of food was viewed as activating the "taste" gnostic unit and the "hunger-drive" gnostic unit.

The second change that importantly influenced Konorski's discussion of inhibition was his placing these gnostic units within a theoretical scheme that envisioned a network of facilitatory and inhibitory connections among the units themselves (see Dickinson & Dearing, Chapter 8). Thus the activation of one gnostic unit might have either excitatory or inhibitory effects on the units identified with other events. Of particular relevance was the assumption that the failure of food to occur could under some circumstances activate "no-taste" and "antidrive" gnostic units antagonistic to those activated by food presentation (see Dabrowska, Chapter 12).

Within this new framework, the notion of an inhibitory association was abandoned. Instead, conditioned inhibition was viewed in the broader context of antagonistic relationships among various gnostic units. A conditioned inhibitor was seen simply as a stimulus that had an excitatory association with a gnostic unit antagonistic to the unit of the US in question. For the case of shock, conditioned inhibitors were described as stimuli with excitatory associations to a "no-shock" gnostic unit, which in turn can inhibit the "shock" gnostic unit and thus reduce the CR one would otherwise see when the latter unit was activated. "Inhibitory" associations of this sort were constructed exactly like any others: by the repeated co-occurrence of the stimulus with the "no-shock" gnostic unit. In the case of food USs, for which Konorski more clearly separated the consummatory and drive aspects of the US, this gnostic unit structure was replicated for each aspect. Thus, conditioned exciters and conditioned inhibitors were viewed as entirely comparable in their conditions of formation and function; they differ only in which gnostic units are involved.

In describing this new theory of inhibition, Konorski argued that it better accords with the experimental data accumulated in his laboratory since the 1948 book. However, that argument does not seem very convincing. Rather, it seems likely that this shift in views of inhibition was a consequence of Konorski's desire to integrate conditioned inhibition into his new framework (see Dickinson & Dearing, Chapter 8).

In a general way, one can see that this view of inhibition does agree with much of the available empirical data. For instance, inhibition should be measurable by the techniques of summation and retardation because the inhibitor activates a gnostic unit antagonistic to that of the "excitatory" US. That activation should interfere with the activation of that US unit by other stimuli and thus reduce the response they produce in a summation test. Moreover, the previous association with an antagonistic unit should retard the development of excitation for the same reason that transfer of excitatory conditioning across other antagonistic USs is slow.

Most of the recognized procedures for establishing conditioned inhibition are claimed by Konorski to fit this new description of pairing with the "no-US" gnostic unit. Although his views could have used some clarification on this point, he argued that this unit is activated by the nonoccurrence of the US only in a context that otherwise contains the US. Thus conditioned inhibition should not result from simple repeated presentation of an event. Rather inhibitors should be US-specific precisely because they demand the activation of a specific "no-US" unit for their formation.

Perhaps the strongest claim that Konorski made for the superiority of this view to his older one concerns its ability to deal with variability. He argued that the opposing-unit view of inhibition demands an instability when two units are simultaneously provoked either by different stimuli or, in the case of extinction, by the same CS. But one may wonder why that instability is any more inherent in two excitatory associations activating opposing centers than it is in simultaneous inhibitory and excitatory associations acting on the same center. Indeed, one may question whether the simple presence of any sort of opposition should particularly demand high variability.

Overall, the 1967 theory, although using quite a different vocabulary from the 1948 theory, really turns out to be quite similar to it in general content. Both theories assume conditioned inhibition to be associative; both attribute its occurrence to conditions parallel to those generating excitation; and both assume that antagonism between excitation and inhibition is central to their effects on performance. There are some differences in detail of prediction, but those are relatively minor. The major difference is one of flavor and perhaps heuristic value.

Just how influential Konorski's work on inhibition has been can be appreciated by noting that most modern views of conditioned inhibition accept his general thinking. There really are no monolithic general theories of inhibition, but most workers in the field seem to share certain views with each other and with Konorski.

Although differing in detail, most discussions describe a conditioned inhibitor as a stimulus that provides a sort of negative information about the US. Conditioned inhibitors are commonly thought to result when a stimulus predicts decreases in the probability, intensity, or duration of the US. Such emphasis on prediction of change has clear antecedents in Konorski's 1948 view that only those CSs that accompany increases or reductions in the activation of the US center become conditioned (as exciters and inhibitors, respectively). It also agrees with Konorski's desire to permit contextual determination of the activation of the "no-US" gnostic unit in the 1967 theory. This mode of thinking is captured in perhaps the best-specified modern theory of conditioned inhibition, that of Rescorla and Wagner (1972). That theory describes the conditions generating inhibition in more formal language, as the occurrence of a negative discrepancy between the current associative strength of a stimulus compound and the strength that the ensuing reinforcer will support. But negative discrepancies can readily be identified as circumstances that result in a fall in the activation of the US center. Indeed, although he stopped short of predicting phenomena like blocking, for which the Rescorla–Wagner theory was designed, Konorski suggested the essentials of that theory in 1948. Thus modern descriptions of the circumstances generating inhibition have strayed very little from those specified by Konorski.

There has been substantially less modern concern over describing the nature of the learning produced by the application of these procedures. Most theorists have apparently been willing to adopt Konorski's views that inhibition involves some sort of association with a US representation. We examine that view in more detail later.

Finally, there was historically considerable concern over the way in which inhibition interacts with excitation to produce behavior. But many theorists have apparently acceded to the view that inhibition does more than "reset" or "erase" excitation (although theories like the Rescorla–Wagner model seem to disagree). The two primary alternatives to resetting are both represented in Konorski's own theories: that an inhibitor elicits a competing behavior (or activates a competing center) and that an inhibitor only functions so as to mask excitation. However, there has been only scattered empirical and theoretical work on the question of how inhibition acts on performance. We return to this issue later.

II.   TWO RECURRING ISSUES

Konorski not only suggested particular theories, but also was influential in phrasing most of the important conceptual issues that have occupied the attention of subsequent workers. We here deal with two of those issues in somewhat more detail. The first is the degree to which inhibition and excitation should be viewed as parallel processes. The second concerns the role of inhibition in the extinction of an excitatory CS. We illustrate certain features of these issues by reference to some recent experiments conducted in our laboratory.

A. Parallels to Excitation

Accounts preceding Konorski (1948) often described inhibition in terms quite different from those applied to conditioned excitation. Thus Pavlov (1927) viewed the development of inhibition as an inevitable consequence of the repeated presentation of stimuli, regardless of the reinforcement conditions. That view might fairly be called a nonassociative account in which inhibition is a property of the CS. Moreover, the common notion that inhibition is a broadly depressing state of the organism, not necessarily tied to antecedent stimulus events, was partly responsible for its lack of acceptance in Western psychology.

However, at least since Konorski (1948) and Asratyan (1961), discussions of inhibition have increasingly emphasized its parallels to conditioned excitation. Both of Konorski's own theories repeatedly appeal to such parallels. Similarly, one attractive feacture of both the contingency description of conditioning and the molecular Rescorla–Wagner model is their derivation of both excitatory and inhibitory conditioning from a common set of principles.

We may consider the parallels between inhibition and excitation in terms of the three issues identified earlier: the description of what is acquired, the rules of its acquisition, and the mapping of that learning into performance.

Discussions of the nature of what is learned in either conditioned excitation or conditioned inhibition are not as popular as they once were. As noted earlier, most modern theorists appear to adhere to some form of an S–S theory of excitatory conditioning. Or more properly stated, they appear unwilling to enter into what they consider a meaningless debate in the historical S–S versus S–R format. Consequently, theorists will speak of excitatory associations with the "reinforcer," ".US," "US representation," "US center," or "gnostic unit" — without apparent commitment to which aspects of those events (e.g., stimulus or response) are involved in the learning. Similarly, most of us have come to view conditioned inhibitors as bearing associative relationships to the same ill-defined entity. At this level of discussion, Konorski's distinction between inhibitory associations with the same " gnostic unit" and excitatory associations with a built-in antagonistic "gnostic unit" seems unimportant. Both views assume a basic parallel in what is learned in excitatory and inhibitory conditioning.

This view that conditioned excitation and inhibition produce parallel associative learnings is encouraged by many of the preparations designed to establish inhibition. Thus descriptions of such preparations as negative correlation between the CS and US, pairing a CS with the termination of the US, discriminative conditioning, and even long-delay conditioning all make reference to the experimenter-arranged relationship between the inhibitor and the reinforcer. Under those circumstances, it is natural to think that this relationship generates the learning—that is, that conditioned inhibition involves an association between those events.

However, the frequently employed "conditioned inhibition" procedure, which some authors have described as the paradigm case of inhibition, actually suggests a number of other possibilities for conditioned inhibition. Expanding on Konorski's comments, Rescorla (1975) noted that there are three obvious (but surely not exhaustive) alternatives to an A–US association that also might account for the inhibition that X gains from an A+, AX−paradigm. Thus X might develop an inhibitory association with the exciter, A, becoming capable of reducing A's effectiveness altogether; it might develop an association with the conditioned response that A produces, becoming capable of suppressing a particular behavior; or it might develop an inhibitory connection to the A–US association, instead of a connection to any particular event representation. A view of excitation as some form of A–US association would find none of these alternatives a parallel process.

Just as Konorski identified most of these alternative possibilities, he also proposed the kinds of experimental tests that would distinguish among them. He noted that if one could successfully transfer an inhibitory X to a new exciter based on the same US but not to one based on a different US, then any inhibitory associations would appear to be US-specific but not CS-specific. Such an outcome would suggest that the inhibitor is associated with either the original reinforcer or with the response elicited by exciters based on that reinforcer. Moreover, Konorski reported some salivary conditioning data in support of these transfer predictions (Konorski & Szwejkowska, 1956).

Since Konorski's report, there have been several confirmatory studies for conditioned inhibitors generated within an instrumental-learning paradigm (Brown & Jenkins, 1967; Hearst & Peterson, 1973) and some additional work with Pavlovian preparations (e.g. Marchant, Mis, & Moore, 1972; Wagner, 1971).

I describe here one study (Rescorla & Holland, 1977) from our own lab because it illustrates the logic while simultaneously providing additional evidence on several alternatives. This study used an observational technique recently developed by Holland (1977). In this procedure, rat subjects receive auditory and visual stimuli paired with food in a Pavlovian fashion. The unique feature of the situation is that CSs in different modalities tend to produce different overt behaviors as a result of those pairings. A light paired with food comes to produce primarily a rearing response followed by passive sitting in front of the food magazine. A clicker so paired, however, elicits a sharp startle reaction followed by rapid lateral head movements (head jerking). These behaviors can be readily observed and reliably scored at a rate of 2 times in each 10-sec CS.

The idea of Holland's study was to establish both the light and clicker as excitatory stimuli, evoking these different responses, and to simultaneously endow a tone with inhibitory properties by nonreinforcing it in conjunction with one of these CSs. The question of interest is whether an inhibitor established in conjunction with one exciter will transfer its inhibitory effect to another exciter based on the same US but eliciting a different response pattern. Successful transfer would not only argue against X's having inhibitory action on A or on the A–US association, but would also provide evidence separating the usually confounded other two possibilities, inhibitory action on the US or on the response indexing excitation. Because most Pavlovian conditioning preparations observe similar responses whenever the same US is used with different CSs, that separation has not previously been possible. To evaluate the degree of transfer of inhibition, a control group received excitatory conditioning with both the light and clicker, but the tone was presented separately rather than in a conditioned-inhibition paradigm.

Figure 4.2 shows the data from a test session in which all animals received the excitatory clicker and light, as well as each compounded with the inhibitory tone. The dependent variable is the percentage of all observations falling into a particular behavior category. The top panels show the reaction to the clicker in the three groups. When presented alone, the clicker provoked a startle reaction and head jerking in each group. However, the presence of the tone reduced those behaviors both in the group with prior

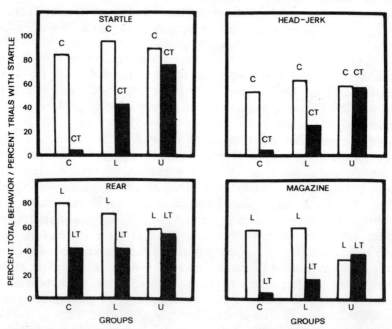

FIG. 4.2.   Results from a test session in which clicker (C) and light (L) are exciters evoking different behaviors. Tone (T) is either an inhibitor established in conjunction with the light (Group L), or the clicker (Group C), or is a control stimulus (Group U). The top two panels display the responses produced by the clicker and the clicker-tone compound in the various groups; the bottom two panels show the responses produced by the light and the light-tone compound in those same groups. From Rescorla & Holland (1977).

CT- trials (Group C) and in that with prior LT- trials (Group L). The ability of the tone to inhibit clicker-induced behaviors in Group CT- simply represents the success of the conditioning treatments. However, its ability to inhibit clicker-induced behaviors in Group LT- represents transfer to the clicker of inhibition trained in conjunction with the excitatory light. The control group, which had received separate T- trials (Group U, for unpaired), showed little reduction in the clicker-generated behaviors as a result of the tone's presence. The bottom panels show the light-produced rearing and magazine behaviors in the three groups. Those behaviors, too, were reduced by the presence of the tone in both Groups C and L, but not in Group U. Here Group C shows evidence of transfer across exciters of the tone's inhibition. Thus, whichever exciter had been used to establish the tone as an inhibitor, the tone acted to suppress the response to both excitatory stimuli. In a subsequent phase of the experiment, Holland paired those exciters with a shock US, which enabled them to provoke freezing. The inhibitory tone, however, had no ability to decrease the likelihood of that new response.

This pattern of data is in good agreement with that reported by Konorski. Inhibition transferred from one exciter to another within US modality, but not between US modalities. Moreover, that transfer was not dependent on the exciters' evoking a common response. This suggests that the inhibitory association is not with a particular response, a finding that seems to weaken a competing response theory while supporting the view that inhibitors are associated with the US representation or gnostic unit.

It may be noted that this view is also supported by the normal outcome of the other major detection procedure for inhibition — the retardation of excitatory acquisition test. In that test, inhibition is evidenced by especially slow development of an excitatory CR when the stimulus is paired with the US. The absence of any other explicit exciter during that test makes less plausible any appeal to associations involving A for describing X's inhibitory associations. Moreover, the available data suggest at least the kind of gross US specificity mentioned earlier; inhibitors based on food do not show retardation in pairings with shock, and vice-versa (Konorski, 1967). One may note, however, that such retardation assessments are actually quite rare for inhibitors generated by an A+, AX− procedure.

Despite these generally confirmatory pieces of evidence, it should be cautioned that we still have data on these issues for relatively few conditioning preparations. Because of the importance of inhibitory learning in the life of the organism, it would not be too surprising if multiple kinds of inhibitory connections were formed with different preparations and paradigms. But at the present time there is little reason to question Konorski's suggestion of a basic parallel between conditioned excitation and conditioned inhibition in terms of the content of the learning.

A second way in which conditioned inhibition and excitation seem parallel is in the rules that describe their acquisition and removal. We have already noted that most theories have sought such parallel descriptions, and that such a search has been reasonably successful. For instance the Rescorla–Wagner version of Konorski's change in the US center description seems to do a good job of detailing the rules for the building up of inhibition in terms that parallel those for the accumulation of excitation. In both cases the discrepancy between the current associative strength and that supportable by the consequent reinforcer is argued to govern the building of associative strength, whether inhibitory or excitatory (see Wagner & Rescorla, 1972).

The rules for extinction of excitation and inhibition are more problematical, however. On the one hand, current models seem agreed that the activation of excitatory associative strength greater than that supportable by nonreinforcement should lead to the extinction of excitatory CSs. On the other hand, rules for the extinction of inhibitory CSs have been less frequently suggested. Historically, only the use of an excitatory conditioning procedure seems to have been considered for extinguishing inhibition. But at least one modern theory has suggested that inhibition should be extinguishable by a procedure parallel to that successful with excitation—simple nonreinforced exposure. According to the Rescorla–Wagner model, a separately presented inhibitor should evoke a negative value discrepant from the zero strength supportable by nonreinforcement. Somewhat more generally, because nonreinforcement establishes neither the excitatory nor inhibitory association, one might anticipate that its application should also lead to a reduction in each of those associations. That is, separate presentation should extinguish both excitation and inhibition.

There are presently very few data available on this degree of parallel in conditions producing extinction of excitatory and inhibitory CSs. But one series of experiments from our laboratory repeatedly failed to find any loss of inhibition as a result of such an extinction procedure, despite substantial extinction of excitation under comparable conditions (Zimmer-Hart & Rescorla, 1974). In those studies the repeated separate presentation of an inhibitory X, established in an A+, AX− paradigm, did not reduce its power to inhibit the responses either to A or to another exciter. Once established by the arrangement of a negative discrepancy, an inhibitor may no longer need that discrepancy for its maintenance. Clearly we need more data, but that preliminary outcome implies a lack of parallel in the description of what eliminates inhibition and excitation.

However, it is in the mapping of learning into performance that most theorists have seen the principal lack of parallel between inhibition and excitation. Historically, the most glaring empirical discrepancy has also been the observation that originally made many skeptical of the existence of conditioned inhibition: The presentation of exciters leads to the production of a response, but the presentation of inhibitors seldom provokes a noticeable behavior change. Although Konorski has reported some cases in which inhibitors generated responses antagonistic to those produced by exciters, most investigators have not been able to identify any such regular responses. For instance, the observational study described earlier included separate presentations of the inhibitor; but even that systematic, but relatively openended inspection of an inhibitor could identify no regularly produced behavior. Such observations are, of course, most difficult for a simple competing-response notion of conditioned inhibition. They may be less embarrassing for a more sophisticated "no-US" center notion for which one has difficulty knowing whether or not to anticipate a response.

The rather widespread use of the summation and retardation assessment techniques for detecting the presence of inhibition has served to alleviate temporarily the practical difficulty raised by this apparent lack of parallel. By using techniques that ensure the presence of excitation at a time when an inhibitor is presented, we have been able to reliably assess the presence of inhibition and consequently study its properties. And the success of those techniques has encouraged the view that the failure to observe a regular response to an inhibitor is a technical one. It has permitted us to assume that inhibitors always affect the organism but that the effect goes unnoticed unless we arrange for concurrently present excitation to provoke a response that can be measurably attenuated.

However, the use of those techniques has also led us to ignore the important theoretical issue of why inhibition must be so detected, whereas the supposedly parallel excitation needs no such techniques. One extreme possibility, in particular, must be considered: that inhibition does not always affect the organism but instead requires the simultaneous presence of excitation to be activated. Because our detection procedures all arrange for excitation to be present, we cannot readily distinguish this from the more commonly accepted alternative. However, the possibility that exciters activate inhibitors would imply a substantial difference between the ways inhibition and excitation affect the organism.

Some information on this extreme possibility can be gathered if we use a detection procedure that does not require the presence of excitation. One may interpret several experiments on second-order conditioning as doing just that. Experiments by Volbroth (see Pavlov, 1927) and Lindberg (1933) and more recently from our own laboratory (Rescorla, 1976) have paired a neutral CS with a previously established inhibitor; these pairings were carried out in the absence of any excitation, so that the inhibitor provoked no noticeable response. However, the subsequent testing of that second-order CS in a retardation and summation procedure revealed it to have acquired inhibitory properties as a result of those pairings. Those findings suggest that even in the absence of excitation, the inhibitor had sufficient impact on the organism to endow an antecedent stimulus with some inhibitory power. Whether or not the inhibitor was fully active cannot readily be determined; but excitation was not necessary for it to have some effect.

Nonetheless, the elimination of this extreme form of asymmetry still leaves the problem of accounting for the lack of parallel action on performance generally. It remains true that separately presented exciters produce performance, whereas separately presented inhibitors do not.

However, the model of conditioning suggested by Konorski in 1948, may provide a useful framework within which to think about this asymmetry. The mode of interaction between excitation and inhibition shown in Fig. 4.1 suggests that they play separate roles in performance. That figure describes a function, presumably unique to each US center, that governs the mapping of excitation into performance. In Konorski's view, that function demanded a minimum level of excitation and then described a negatively accelerated relationship, such that changes in the level of excitation are differentially effective at different levels. The role of inhibition is to shift that function, so as to "add" to or "subtract" from excitation.

The differential effects of the separate presentation of inhibitors and exciters are quite natural in such a framework. Excitation activates the US center, resulting in the changes in level of activation of that center that are presumed responsible for both performance and learning. However, separately presented inhibitors do not activate the US center; instead they shift the function mapping excitatory associations into activation of the US center. In the absence of exciters, that shift should not change the level of activation in the center and consequently should neither generate performance nor support the associative changes necessary to extinguish inhibition.

As Zimmer-Hart and Rescorla (1974) noted, one may incorporate such a threshold view of the action of inhibition into formal models of conditioning like that suggested by Rescorla and Wagner (1972). That is, one may use a scheme like that of Fig. 4.1 to determine the net associative output of a stimulus compound. In that case, discrepancies between such an output and the output that the ensuing US will support would determine the changes in excitation and inhibition that occur on a trial. In fact if the function shown in Fig. 4.1 were linear, then such a view would differ from the original Rescorla–Wagner model only in those cases for which stimulus compounds are net inhibitors. For all those cases, the output of the function would be zero, the same as that of a neutral CS. Because zero is presumably the output supportable by a nonreinforcement, separate presentation of any inhibitor (or neutral CS) would not be discrepant from its consequences and there would be no associative change. That is, we would not anticipate the extinction of inhibition. One may more fully incorporate the details of Konorski's particular function if one is willing to assume that associative strengths summate exponentially rather than linearly.

This kind of description may also provide a useful way to think about the effects of variables other than inhibition on learning and performance. For instance, Konorski argued that the effects of motivational changes may be viewed as shifting this function in much the same way as do inhibitors. Grice (1971) has made a similar suggestion in a different context. The next section presents other performance phenomena that may be thought of in the same way.

In sum, one may argue that Konorski provided us with just the proper mixture of parallel and lack of parallel between inhibition and excitation. On the one hand, the 1948 model provides for the processes to be opposing in nature as a result of their associations with the same reinforcing event. Furthermore, it provides a common framework within which to describe the conditions for their acquisition and extinction. On the other hand, it gives them sufficiently different roles in the determination of performance and subsequent learning to make understandable some presently available asymmetries.

B. Inhibition in Extinction

For both Pavlov and Konorski the phenomenon of extinction was the starting point for discussion of inhibition. Both were impressed by the loss of performance to a previously trained stimulus when it was repeatedly nonreinforced. Both referred to such a stimulus as "inhibitory," apparently equating inhibition with response decrement in their initial discussions.

But the evidence for the participation of learned inhibition in extinction has always been different from that for its presence in other contexts. Although Pavlov reported some evidence that extinguished stimuli function as inhibitors in summation and retardation test procedures, few others have made similar observations. Indeed, both those procedures have been described instead as techniques for revealing the continued excitatory power of an extinguished CS (e.g., Reberg, 1972; Rescorla, 1969). Consequently, there remains very little evidence that the nonreinforcement carried out during extinction converts exciters into predominately inhibitory stimuli. The issue of inhibition's involvement in extinction has thus not been whether extinction produces net inhibitors but whether it generates sufficient inhibition to compensate for the previously trained excitation. Is it the presence of inhibition that is primarily responsible for the decrement involved in extinction?

In most of his discussions Konorski was acutely aware of this distinction, often separating primary inhibitory stimuli (which were net inhibitors) from secondary inhibitory stimuli (which were initially excitatory and then extinguished). Most of his experimental generalizations about inhibition were derived from the former. Unfortunately, in some places Konorski continued to refer to both extinguished stimuli and net inhibitors as "inhibitory stimuli," without differentiation. On occasion, this usage led to misleading conclusions. For instance Konorski's (1948, p. 168) account of positive induction and excitatory responding resulting from the summation of two "inhibitory" stimuli is really applicable only to extinguished CSs. The analysis depends on the "inhibitory" stimulus' having concurrent excitation and would not apply to pure inhibitors.

Techniques other than summation and retardation have been developed, of course, for the detection of inhibition after extinction. These techniques have depended on the assumption that after extinction the CS evokes both excitatory and inhibitory tendencies. They then intend to unveil the remaining excitation by temporary or permanent removal of the presumed inhibition. The logic is that if such unveiling is successful, then excitation must have survived extinction: but if excitation survived, something must have been preventing its appearance in the absence of that unveiling. Conditioned inhibition has been the obvious candidate for that role. The array of techniques used in the service of this argument is largely a familiar one: spontaneous recovery, disinhibition, rapid reconditioning, and reinstatement by US presentation.

However, one may question this logic on several grounds (cf. Mackintosh, 1974). First, the use of some of these techniques appears to attribute to conditioned inhibition special properties that are not comparable to those of conditioned excitation, and for which there is little independent evidence. Many accounts of spontaneous recovery and disinhibition, for instance, depend on inhibition's being especially susceptible to the disruptive effects of the passage of time or to the presentation of other neutral stimuli. Of course, one may agree with Konorski that the susceptibility may not be a special property of inhibition but instead related to the "depth"or the "age" of the association. The fact that the inhibitory connection is normally younger might make it more vulnerable. Nevertheless, it is not at all clear that any "inhibition" identified by such procedures is the same as the "inhibition" identified by procedures employed in the detection of net inhibition.

More importantly, these postextinction techniques are really directed more at the question of whether excitation survives extinction than at the question of whether inhibition is present. What they apparently demonstrate is that extinction does not wipe out all of the trained excitation; what they fail to demonstrate is that it is conditioned inhibition that prevents the exhibition of that excitation after extinction.

1. A sample of other masks of excitation

This logical point becomes especially clear when one considers one of the less frequently used unveiling techniques: reinstatement. Both Pavlov (1927, p. 59) and Konorski (1948, p.185) report that the response to an extinguished CS can be substantially restored by the simple unpaired presentation of the original US. Recent evidence from our laboratory both substantiates those observations and suggests that various uninteresting interpretations are inapplicable. Using a conditioned suppression procedure with rat subjects, Rescorla and Heth (1975) found that US presentation produced substantial reinstatement whether or not the US was signaled by another CS and even if 24 hours intervened betweeen US presentation and testing of the previously conditioned and extinguished CSs. Rescorla and Heth argued that these reinstatement results could not easily be interpreted in terms of the US aftereffects, reinstituting conditions of training, the excitatory conditioning of situational cues, or stimulus generalization. They suggested that US-induced reinstatement is not specific to the removal of associative inhibition between any particular CS and the US, but rather involves a modification of the US representation itself. In the language of Konorski, it changes the state of the US center, or of the gnostic unit, which in turn modifies performance. During extinction, deterioration of the US representation may occur, reducing performance even when the excitatory association remains partially intact. The reoccurrence of the US may restore the representation and thus unveil the excitatory association. However, in this instance it does not appear to have been the presence of conditioned inhibition that was masking the excitation.

Further evidence for this view of reinstatement comes from some recent experiments, which identify a procedure capable of attenuating reinstatement (Rescorla & Cunningham, 1977). If extinction attenuates a US representation associated with a CS, and if the reoccurrence of the US restores the representation, then perhaps further extinction of another CS would be capable of "erasing" that restoration. That is, the reinstating effect of a US upon one extinguished CS should be attenuated if another excitatory CS is nonreinforced between that US presentation and the testing of the target CS.

Figure 4.3 shows the results of one conditioned suppression experiment, which illustrates the phenomenon of reinstatement and tests that proposition. This figure displays performance during terminal extinction and postreinstatement testing of a tone that had previously been paired with shock. The results are expressed in terms of a suppression ratio of the form $A/(A+B)$, in which A and B are respectively the response rates during the CS and pre-CS periods. The rats in all 3 groups had initially received training to bar-press for food, followed by pairings of both a 2-min flashing

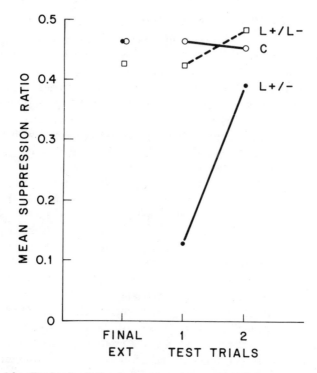

FIG. 4.3.   Final extinction and reinstatement test performance to a tone CS. Groups
L+/− and L+/L− received reinstating shocks between final extinction and testing
of the tone. Group C received no such shocks. Group L+/L− additionally received
nonreinforced light presentations during the test session, before tone presentation.
Adapted from Rescorla & Cunninhgam (1977).

houselight and a 2-min 1800 Hz tone with a 1/2-mA, 1/2-sec shock. Each
animal then received nonreinforced presentations of the tone until its fear
reached the level shown in the left-hand side of Fig. 4:3. Animals in Groups
L+/− and L+/L− then received a reinstatement treatment consisting of 4
presentations of the shock, each preceded by the light; animals in Group C
received no reinstatement. On the following day, all animals received 2 test
trials with the tone. Before those test trials, the animals in Group L+/L−
received 4 nonreinforced trials with the light; the other groups did not.
    It is clear that intervening shocks produced substantial reinstatement
(Group L+/− is more suppressed than is Group C), but it is also clear that
intervening nonreinforcement of the light removed that reinstatement
(Group L+/− is also more suppressed than Group L+/L−). Other experiments have indicated that this erasure effect occurs only when the light
itself is fear-eliciting, but does not depend on the light's having been present
during the reinstating shocks. Moreover, this particular light and tone do
not show any evidence of associative stimulus generalization. Consequently,

one interpretation of these findings is that nonreinforced light presentations reversed the effects of US presentation, effectively erasing the boost that had been given to the US representation. That is, this particular unveiling technique may act on the event representation rather than on associative inhibition. Learned processes other than conditioned inhibition may mask any excitation that remains after extinction.

This conclusion suggests that some of the more commonly employed unveiling techniques may also act partly on a depressed US representation rather than on associative inhibition. For instance, spontaneous recovery might reflect changes in the nonassociative US representation with time, in addition to (or instead of) changes in inhibitory–associative connections. That alternative is encouraged by the results of several instrumental learning experiments (e.g., Boakes & Halliday, 1975; Liberman, 1944), which suggest that when the US continues to be presented in a noncontingent fashion during extinction then spontaneous recovery is reduced. That outcome would be anticipated if a portion of the spontaneous recovery observed in standard extinction procedures resulted from improvement in a depressed US representation; the continued occurrence of the US would act to prevent that depression and so attenuate its contribution to the decrement observed during extinction.

The idea that changes in the US representation might contribute to spontaneous recovery has also received support in a recent experiment from our laboratory. In that experiment we attempted to erase spontaneous recovery in the way we had erased reinstatement; successful erasure would point to a role of the US representation. For this purpose, 2 groups of rats received acquisition and massed extinction with a 2-min tone CS in a conditioned suppression procedure using a 1/2-mA, 1/2-sec shock US. One group had previously also been given fear conditioning with a light, the other had not. Following tone extinction, both groups were given a 7-day rest period during which they remained on deprivation in their home cages. Finally, they were tested for spontaneous recovery of the fear to the tone. All animals received a 2-hr test, during the second half of which 2 nonreinforced tones were given; but all animals also received 2 nonreinforced lights in the first half of the test session. For Group E (Erased), those lights were fear-eliciting and so would be expected, on the basis of reinstatement experiments, to erase. For Group C, the lights were not fear-eliciting, permitting these animals to serve as a control in which spontaneous recovery should be exhibited.

Figure 4.4 shows the course of tone extinction and test. Extinction to the tone was similar in the two groups. Moreover, Group C showed substantial but short-lived spontaneous recovery. More interestingly, the prior nonreinforced presentation of the excitatory light effectively attenuated that recovery in Group E. Those results suggest that spontaneous recovery is sensitive to the same erasure procedures as is reinstatement. They are thus con-

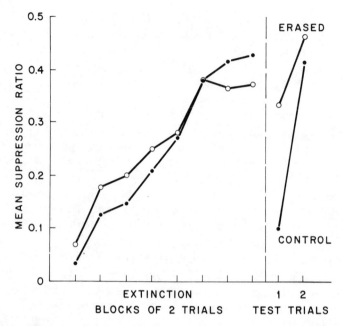

FIG. 4.4.   Extinction and spontaneous recovery of the fear response to a tone CS.
Both groups received two nonreinforced light presentations before testing of the tone.
For the Erased Group, the light was fear-eliciting; for the Control Group, it was not.

sistent with the view that spontaneous recovery does not solely involve the
loss of a CS-specific conditioned inhibition. Its occurrence and removal are
importantly affected by the treatment of other CSs in a manner consistent
with the conclusion that it partly involves changes in the US representation.

These observations emphasize the flaws in a logic that attempts to infer
the presence of conditioned inhibition to an extinguished CS by manipulations revealing that underlying excitation remains. They suggest that there
are also important learned nonassociative changes taking place during extinction that are viable alternatives for the role of hiding remaining excitation.

One can find in Konorski's writings three ways of thinking about results
such as these. At one level, Konorski (1967) emphasized the learned formation of gnostic units representing individual events. Although Konorski did
not detail the mechanism, one might easily imagine that the organism functions to bring his gnostic unit into line with the appropriate event in the
manner of Sokolov's (1963) neuronal model. In that case, the US presentation would strengthen the gnostic unit, whereas the activation of that unit in
the absence of the US (as in the nonreinforced excitatory CS presentation)
would depress the unit. If the integrity of the gnostic unit is important for
the production of a response, then its manipulation would yield response

changes, despite unchanged excitatory associations.

At a different level of discussion, one might describe effects such as these in terms of the conditioning of drive states to situational cues. Konorski's 1967 theory emphasized the multiple conditioning effects of a US presentation. In addition to conditioning discrete CRs to explicitly presented CSs, the occurrence of USs might condition drives to situational stimuli. Konorski envisioned those drives as facilitating the learning and performance of more discrete responses, although their sole presence was not sufficient for performance. Moreover, although again lacking in precision, his discussion indicates that the rules for the establishment and elimination of those connections between situational stimuli and drives might differ from those describing associations between sporadic CSs and responses. For that reason, this account might survive arguments against reinstatement's being the simple conditioning of excitation to situational stimuli (see Rescorla & Heth, 1975).

Finally, one may profitably discuss these reinstatement effects in terms of Konorski's 1948 performance function. In that context, Konorski emphasized the ability of various treatments to shift temporarily the function mapping excitation into performance. In addition to describing conditioned inhibition in this way, he noted that motivational variables might shift the function. To describe these later effects, he spoke of the excitability of the US center as modulating the performance impact of a given level of conditioned excitation. This language may also be used to describe reinstatement and erasure: The US occurrence shifts the function leftward, lowering the threshold for performance, whereas nonreinforced activation of the US center shifts it to the right, raising the threshold. Moreover, one might view stimuli present during the latter, erasing operation as acquiring the conditioned ability to raise the function—that is, to be conditioned inhibitors. In that case the present manipulations partly produce shifts in the US center excitability, which themselves admit of conditioning.

This account of reinstatement suggests that separate US presentation might affect not only performance but also the course of further associative learning and extinction. Earlier it was suggested that the output of the performance depicted in Fig. 4.1 might enter into equations like those of Rescorla and Wagner for the determination of changes in associative strength. Because that output might be affected by US presentation following extinction, as well as by conditioned excitation and inhibition, we might expect it to modify subsequent associative changes. Some recent experiments in our laboratory (Rescorla & Cunningham, 1977) suggest that this may be true, at least for the course of further extinction. Extinction carried out after reinstatement makes especially difficult the retraining of a CS, suggesting that extinction with a strong US representation produces an especially large associative change. Moreover, the interposition of erasure after that reinstatement, but before the additional extinction, attenuates that effect. Although one may entertain a variety of accounts for such findings, including attributing them to the importance of drive conditioning of situational cues, they are also consonant with the threshold notion Konorski offered in 1948. The modulating effects of different performance levels on subsequent associative learning may be influenced by not only the associative but also the nonassociative factors affecting that performance.

2. An alternative to unveiling for detecting inhibition

In the present context an important consequence of the preceding discussion is that one must acknowledge that learned decremental processes other than CS-controlled associative inhibition might conceal . the continued presence of excitation during extinction. That observation greatly reduces the value of most of the unveiling procedures for providing evidence of the presence of associative inhibition during extinction. The issue is then whether we can find alternative evidence of its role in extinction.

We have elsewhere suggested one possible alternative for detecting the presence of inhibition in extinction (Rescorla, 1975). The difficulty in the straightforward use of summation and retardation techniques stems from the apparent failure of simple extinction to produce sufficient inhibition to make the CS a net inhibitor. The decrement normally attributed to inhibition is usually claimed to just balance the excitation that remains. However, one might try to separate out some of that inhibition by arranging for a neutral stimulus to be concurrently present when an established exciter undergoes extinction. Such a stimulus would have no initial excitatory value but, at least according to some theories of conditioning, it might share in the inhibition developed during extinction. If so, one might separately test *that* stimulus in summation and retardation procedures, to reveal any inhibition it gained. Identification of that stimulus as a net inhibitor would at least imply that inhibition developed during extinction and would be consonant with the claim that the extinguished CS also gains inhibition.

There are several reasons for suspecting that such a detection procedure would be successful. First, notice that this technique is an $A+,AX-$ procedure, in which the two kinds of trials are applied sequentially. Appplication of those same treatments in an intermixed fashion constitutes the standard conditioned inhibition procedure, known to endow X with inhibitory power. Of course, the failure to continue the reinforcer in extinction, as distinct from the conditioned inhibition procedure, might well make the kinds of nonassociative processes described earlier so powerful as to prevent the development of inhibition to X in the case of extinction. Second, Kamin (1969) has reported some evidence that an X added during extinction does become inhibitory. After $A+$ conditioning and then $AX-$ extinction in a conditioned-suppression procedure, he retested A alone. He found that A still evoked a substantial response despite the absence of a response to AX.

This comparison between A and AX constitutes a summation test that implicates X as inhibitory. However, the summation test of X against the extinguishing CS itself may not be entirely convincing to those who worry about generalization decrement or configural learning.

A somewhat more powerful use of this paradigm was recently made in our laboratory, in conjunction with David Furrow. Using a conditioned suppression procedure, we initially paired a 30-sec low (250 Hz) tone with a 1/2-mA, 1/2-sec shock for 3 groups of rats; this stimulus was later used as a standard exciter in a summation test procedure. Then a 30-sec high (1800 Hz) tone was paired with shock in 2 of those groups but not in a third. Subsequently, 1 of the fear-conditioned groups (Group A−) received simple extinction of the high tone (A) for 14 trials. The other 2 groups received nonreinforced tone presentations accompanied by a flashing light (X). For 1 group (A$^+$X−), A had previously been conditioned, and so X accompanied an exciter during its extinction. For the other group (A^{0}X−), A had not been fear-conditioned, and so X was presented in the context of a neutral, rather than an extinguishing, stimulus. After extinction, all groups received 2 assessments of X to determine whether it had developed conditioned inhibition in Group A$^+$X−. The first test involved summation, in which the standard low tone first received additional conditioning and was then presented both alone and in conjunction with X. The second test was a retardation procedure in which the X was presented alone and followed on 50% of its 12 trials with the foot shock. The anticipated results were that X would prove more inhibitory in both tests in Group A$^+$X−.

Figure 4.5 suggests that those anticipations were borne out. During extinction, Groups A− and A$^+$X− both showed initial suppression which extinguished rapidly over trials. As would be expected from most theories, the latter group lost its fear-eliciting power earlier. Although the animals in Group A^{0}X− showed some moderate level of suppression throughout, they did not change substantially in extinction. During the summation test, all three groups showed substantial fear of the low tone alone. Of more interest, they showed differential attenuation of that suppression by the concurrent presentation of the light. Group A$^+$X− showed evidence of greater inhibition controlled by X. A similar pattern emerged during the retardation test, in which the light conditioned more slowly in Group A$^+$X− than in the other two groups.

These results provide some support for the proposition that a neutral stimulus present during the extinction of an exciter develops conditioned inhibition. Such evidence appeared in two assessment techniques, but its magnitude was not especially large. However, that is perhaps to be expected because X must presumably share with A the total inhibition gained in extinction. That necessity for sharing differs from most conditioned inhibition procedures, in which continued intermixed A+ trials guarantee that A will

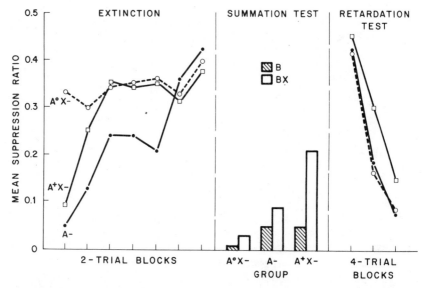

FIG. 4.5.   Extinction, summation, and retardation testing of a light (X). Group A+X− received nonreinforcement of a previously fear-conditioned high tone (A) with X present; Group A°X was treated identically, except that A had never been conditioned; Group A− received prior fear conditioning of A, but its nonreinforcement occurred in the absence of X. In each group, X was tested in summation by comparing the response to another low-tone fear elicitor (B) with the response to BX, and in a retardation procedure.

remain fully excitatory, thus forcing X to control all the inhibition. Thus these data do provide relatively direct support for the idea that in extinction some inhibition develops.

It is worth mentioning one additional implication of these results. In a standard extinction procedure, situational stimuli are always present during the nonreinforced CS presentations. One might consequently anticipate that those situational stimuli would acquire conditioned inhibition during the course of extinction. Historically, that possibility has seemed implausible because those situational stimuli are also present during acquisition and during the intertrial interval in extinction. But some of the evidence noted earlier suggests that neither of those circumstances may prevent the situational stimuli from developing inhibition during extinction. During initial acquistion, situational stimuli may indeed become excitatory, but most modern evidence suggests that they lose strength by the later stages of acquisition because they are repeatedly nonreinforced during the intertrial interval. Those stimuli may then be expected to be relatively neutral at the start of extinction of the discrete CS. But if the aforementioned results apply to situational stimuli, they may gain inhibition because of their presence during the nonreinforcement of an explicitly presented excitatory stimulus. Moreover, at least some results suggest that the presentation of inhibitory stimuli alone (i.e. the occurrence of the intertrial interval) may not lead to the extinction of their inhibition (Zimmer-Hart & Rescorla, 1974). Consequently, at the end of every straightforward extinction procedure, the situational stimuli may be net inhibitors. (See Wagner, Chapter 3, for a relevant account of US representational changes in acquisition.)

This possibility that situational stimuli become inhibitory may also provide an alternative account of the US representation changes previously described as "nonassociative." When interpreting those effects as involving shifts in the performance function shown in Fig. 4.1, we earlier noted that conditioned inhibitors may be described as producing a similar shift. Thus, for example, the phenomenon of reinstatement might be thought of as the removal of situational stimulus inhibition through reconditioning. That is, this particular unveiling technique might in the end reveal an excitation hidden by conditioned inhibition. But the stimulus controlling that inhibition may be situational rather than the originally trained CS.

The viability of this interpretation cannot be fully discussed here, but it is worth noting that it is not simply verbally different from viewing reinstatement as the result of establishing excitation to situational stimuli. One of the interesting features of reinstatement has been its failure to produce changes in the general level of fear in the absence of CSs, despite its ability to reinduce substantial fear in explicitly presented CSs. That feature would be understandable if reinstating USs acted simply to remove inhibition rather than to establish excitation.

One final aspect of the potential involvement of inhibition in extinction must be discussed. Even if we accept these results as suggesting that inhibition occurs during extinction, there remains the question of whether excitation escapes nonreinforcement unscathed. That some excitation remains seems demanded by the various unveiling procedures; whether it all remains is difficult to evaluate from the kinds of data presently available. Historically, discussions of extinction have emphasized two extreme possibilities—that excitation is entirely lost or that it is entirely preserved but masked by inhibition. Consequently, the apparent identification of the presence of some inhibition has sometimes led to the faulty conclusion that the excitation has been entirely preserved. In the laboratories of Pavlov and Konorski this conclusion was bolstered by the observation of what they describe as complete spontaneous recovery from extinction. However, modern studies of that phenomenon have been substantially less successful. Even when spontaneous recovery is large, as in the control group in Fig. 4.4, the occurrence of rapid reextinction undermines the conclusion that excitation was fully preserved. It is difficult to know whether the classical observations were really different or whether the criteria for completeness are not the same. Certainly had we a procedure for fully recovering a stimulus to its preextinguished level of performance (perhaps such as the patience to wait long enough for full spontaneous recovery), it would provide powerful evidence for the full preservation of excitation. Unfortunately, there is presently no evidence of that sort available.

Finally, it is of interest to note that Konorski's use of the threshold notion of inhibition anticipates not only that all of the originally trained excitation is preserved but also that an extinguished CS remains somewhat more excitatory than it was before training and extinction. If one views excitatory and inhibitory associative changes as governed by the performance output of a function like that of Fig. 4.1, then the growth of inhibition will stop when the function shifts just enough to drop the output to zero. That is, extinction will become stable when the threshold just equals the level of excitation previously conditioned. However, if one also agrees with Konorski's reasonable assumption that before any conditioning, the threshold is above zero and the CS has no excitation, then the levels of excitation and inhibition are closer after extinction than they were before training. That is, a novel CS has a threshold with no excitatory strength, whereas an extinguished CS has a level of excitation just at threshold level. Consequently, an extinguished CS in addition to having both excitatory and inhibitory tendencies, would also have a net excitatory value when compared with untrained CSs. One would then expect retraining to proceed more rapidly than initial conditioning, even if that retraining did not act to especially remove inhibition.

In summary, the involvement of inhibition in extinction remains uncertain. Traditional demonstrations seem flawed, at least as identifiers of associative inhibition controlled by the extinguished CS. Other procedures, more derivative from general conditioned inhibitory procedures, may provide a more promising approach to identifying the presence of inhibition in extinction. But we are still uncertain about the degree to which excitation survives extinction and the relative contributions of associative inhibition and other decremental processes in masking any surviving excitation. However, it may be that the view of inhibition suggested by Konorski in 1948 will again provide a useful framework. At least it gives a language in which to discuss currently available data.

III. CONCLUSION

We have attempted an exposition of Konorksi's views of conditioned inhibition and have given some detailed discussion of two particular problems on which those views bear. Over the past 10 years we have seen a striking reawakening of interest in the problem of conditioned inhibition. But the remarkable thing is the degree to which both our theoretical questions and our data collection have been influenced by Konorski's own views. There is very little in our current thinking about inhibition that was not anticipated by Konorski years earlier.

Some have claimed that the highest tribute a science can pay to one of its students is to name a phenomenon in his honor. And indeed it would not seem inappropriate to me to speak of Konorskian conditioned inhibition. But in some ways we have already paid Konorski an even higher tribute. So widely accepted have been Konorski's ideas in our thinking about inhibition that many of them have attained the status of essentially universally accepted assumptions. They are so ingrained in our conceptions that we no longer feel it necessary even to cite Konorski as their original author. That it seems to me, is the highest compliment a science can give, and one that Konorski richly deserves.

ACKNOWLEDGMENTS

The research reported in this chapter was supported by grants from the National Science Foundation. Peter C. Holland and Christopher L. Cunningham contributed importantly to its content.

REFERENCES

Asratyan, E. A. The initiation and localization of cortical inhibition in the conditioned reflex arc. *Annals of the New York Academy of Sciences,* 1961, *92,* 1141–1159.

Boakes, R. A., & Halliday, M. S. Disinhibition and spontaneous recovery of response decrements produced by free reinforcements in rats. *Journal of Comparative and Physiological Psychology,* 1975, *88,* 436–446.

Brown, P. L., & Jenkins, H. M. Conditioned inhibition and excitation in operant discrimination learning. *Journal of Experimental Psychology,* 1967, *75,* 255–266.

Grice, G. R. A threshold model for drive. In H. H. Kendler and J. T. Spence (Eds.), *Essays in neobehaviorism: A memorial volume to Kenneth W. Spence.* New York: Appleton-Century-Crofts, 1971.

Hearst, E., & Peterson, G. Transfer of conditioned excitation and inhibition from one operant response to another. *Journal of Experimental Psychology,* 1973, *99,* 360–368.

Holland, P. C. Conditioned stimulus as a determinant of the form of the Pavlovian conditioned response. *Journal of Experimental Psychology: Animal Behavior Processes* 1977, *3,* 77–104.

Kamin, L. J. Selective association and conditioning. In N. J. Mackintosh & W. K. Honig (Eds.), *Fundamental issues in associative learning.* Halifax: Dalhousie University Press, 1969.

Konorski, J. *Conditioned reflexes and neuron organization.* Cambridge: Cambridge University Press, 1948.

Konorski, J. *Integrative activity of the brain.* Chicago: University of Chicago Press, 1967.

Konorski, J., & Szwejkowska, G. Reciprocal transformations of heterogeneous conditioned reflexes. *Acta Biologicaeexperimentalis,* 1956, *17,* 141-165.

Liberman, A. M. The effect of interpolated activity on spontaneous recovery from experimental extinction. *Journal of Experimental Psychology,* 1944, *34,* 282-301.

Lindberg, A. A. The formation of negative conditioned reflexes by coincidence in time with the process of differential inhibition. *Journal of General Psychology,* 1933, *8,* 392-419.

Mackintosh, N. J. *The psychology of animal learning.* New York: Academic Press, 1974.

Marchant, H. G., Mis, S. W. & Moore, J. W. Conditioned inhibition of the rabbit's nictitating membrane response. *Journal of Experimental Psychology,* 1972, *95,* 408-411.

Pavlov, I. P. *Conditioned reflexes.* Oxford: Oxford University Press, 1927.

Reberg, D. Compound tests for excitation in early acquisition and after prolonged extinction of conditioned suppression. *Learning and Motivation,* 1972, *3,* 246-258.

Rescorla, R. A. Pavlovian conditioned inhibition. *Psychological Bulletin,* 1969, *72,* 77-94.

Rescorla, R. A. Pavlovian excitatory and inhibitory conditioning. In W. K. Estes (Ed.), *Handbook of learning and cognitive processes (Vol. 2).* Hillsdale, N. J.: Lawrence Erlbaum Associates, 1975.

Rescorla, R. A. Second-order conditioning of Pavlovian conditioned inhibition. *Learning and Motivation,* 1976, *7,* 161-172.

Rescorla, R. A., & Cunningham, C. L. The erasure of reinstated fear. *Animal Learning and Behavior,* 1977, *5,* 386-394.

Rescorla, R. A., & Heth, C. D. Reinstatement of fear to an extinguished conditioned stimulus. *Journal of Experimental Psychology: Animal Behavior Processes,* 1975, *1,* 88-96.

Rescorla, R. A., & Holland, P. C. Associations in Pavlovian conditioned inhibition. *Learning and Motivation,* 1977, 8, 429-447.

Rescorla, R. A., & Wagner, A. R., A theory of Pavlovian conditioning: Variations in the effectiveness of reinforcement and nonreinforcement. In A. H. Black & W. F. Prokasy (Eds.) *Classical conditioning II: Current research and theory.* New York: Appleton-Century-Crofts, 1972

Sokolov, Y. N. *Perception and the conditioned reflex.* Oxford: Pergamon press, 1963.

Wagner, A. R. Elementary associations. In. H. H. Kendler & J. T. Spence (Eds.), *Essays in neobehaviorism: A memorial volume to Kenneth W. Spence.* New York: Appleton-Century-Crofts, 1971.

Wagner, A. R., & Rescorla, R. A. Inhibition in Pavlovian conditioning: Application of a theory. In R. A. Boakes & M. S. Halliday (Eds.), *Inhibition and learning.* New York: Academic Press, 1972.

Zimmer-Hart, C. L., & Rescorla, R. A. Extinction of Pavlovian conditioned inhibition. *Journal of Comparative and Physiological Psychology,* 1974, *86,* 837-845.

5 Brain Processes and Conditioning

John W. Moore
University of Massachusetts—Amherst

I. INTRODUCTION

The name Jerzy Konorski stands not only for a set of theoretical insights, hypotheses, and techniques, but also for an idea—namely, that an enduring understanding of psychological processes, such as those underlying conditioning, requires a multitiered approach combining sophisticated psychological theory and technique with equally rigorous and sophisticated concepts and criteria from neurobiology. Such an approach has been successfully applied to many areas of psychology — sensory, motivation, personality, even clinical—but it has met with only indifferent or dubious success when applied to learning and conditioning. Here the pattern has been for two groups of scientists to proceed independently with little regard of each other's efforts. The reasons for the failure of physiology to gain more than token acceptance within the mainstream of psychological thought concerning learning are many. As Halliday points out (Chapter 1), learning theory became the province of behaviorism, which in some camps expressed open hostility toward physiology. Physiology was seen as a threat to behaviorism's independence as a science. For the past few decades, students of learning theory, the author included, were trained in such a way that the potential relevance of physiology was obscure to say the least. The physiologists were little help, as their approach to learning seemed too simple and theoretically barren.

Jerzy Konorski saw earlier than most that the way to bring about a rapprochement between sophisticated behavioral theorizing and up-to-date neurophysiology was to get the two disciplines "under the same hat" — to get behaviorists and physiologists to speak the same language. Whether by design or by accident, Konorski's influence has been a bit one-sided. His books and articles have been more successful in getting Western behaviorists to think in physiological terms than in getting physiologists outside the Nencki Institute directed toward fertile areas of research that make contact with learning theory. This chapter is written primarily for physiological psychologists. It addresses questions such as the whereabouts and nature of the engram and the role of various brain structures and systems in the acquisition and performance of a CR with little regard for readers whose interests lie elsewhere. Its inclusion in this volume, along with the other physiological contributions, may serve as an inducement for physiologists to take note of how physiological techniques can be brought to bear on questions of interest to learning theorists and, more importantly, how hard-won behavioral principles can be used in connection with physiological experiments to clarify questions of brain function. With these motives, this chapter is dedicated to Jerzy Konorski the neuroscientist.

As a preview, the reader is led to conceive of conditioning as occurring initially and most critically within the deepest portion of the brain, anatomically very near to the final efferent pathway of the CR. Conditioned inhibition emerges as having properties, portrayed by Konorski (1948; see also Rescorla, Chapter 4), of actively suppressing the CR via an anatomically distinct circuit that interrupts the flow of information from the CS to the response by convergent interactions within the narrow confines of the brain stem — at the point where conditioning occurs. The flow of information from the CS to the response is also subject to "higher" influences of an "attentional" nature under the control of the hippocampus and perhaps neocortex. The neocortex is pictured as helping to channel the CS to the critical zone of CR initiation, but the evidence for this is weak and merely suggestive. Far more convincing is the unmistakeable contribution of the hippocampus in "tuning out" irrelevant or redundant stimuli. Just exactly how this may happen is spelled out in a theoretical model that attempts to reconcile all the available evidence.

II. BACKGROUND

Nemesius, a Byzantine bishop of Emesa (Syria), first proposed in the fourth century A.D. that psychological processes are differentially distributed within the brain (Rosner 1974). Since then, the question of the physical basis of learning and memory has come to be appreciated as one of incredible complexity. Satisfactory answers regarding the engram, perhaps an unfortunate term suggesting that learning can be understood as a unitary process, have been slow in coming despite the explosive growth of the neurosciences. The interdisciplinary approach may be the only fruitful course, for it has become increasingly obvious that pinning down an engram requires converging threads of evidence from many traditional areas of knowledge. From psychology, the search for engrams waits on the resolution of tough questions as to precisely what is meant by learning and memory. It has been an article of faith that characterizing the engrams of animal conditioning will rate high marks in both scientific precision and in relevance to processes fundamental to human learning and memory. Jerzy Konorski appreciated these problems, and his culminating work, *Integrative Activity of the Brain* (1967), represents a hallmark toward their solution.

Traditionally, the search for the neural substrates of learning has spawned a host of strategies and techniques that were often founded on antithetical assumptions. For example, the approach embodied by the lesion technique assumes that functional localization can be inferred from observations following destruction of various brain structures or systems. Any resulting loss of function may be transient or it may be linked to such things as reduced sensorimotor capabilities, altered states of arousal, altered vegetative functions, hormone imbalances, and so forth. If these possibilities can be eliminated, one can then argue that these structures or systems are implicated in the lost or reduced function.

The problem of diaschisis usually prevents the stronger conclusion that a given structure or system is the critical locus of integrative transactions. In the case of functional loss following brain damage, the destroyed tissue might include nerve fibers that convey essential information to or from regions far from the lesioned area. Alternatively, the lesion might interrupt normal neural traffic in such a way as to alter the physiological properties of neurons in the critical zone. An example would be the phenomenon of denervation supersensitivity, in which the physiological response of neurons to stimulation of one afferent pathway is enhanced as a result of interrupted tonic inflow along a second pathway.

Chemical diaschisis can be a problem in psychopharmacological studies because the principal neurotransmitter systems of the brain tend to act reciprocally in such a way that a drug or lesion that affects one system— norepinephrine, for example — can produce compensatory adjustments in some other system — 5-HT or serotonin, for example. Without additional evidence, it is often difficult to tell which neurotransmitter system is the more intimately involved in a particular function.

Electrophysiological techniques, though insufficient in themselves for establishing functional localization in the brain (see Gabriel, 1976), are nevertheless invaluable in tracking the flow of neural information through the nervous system. When combined with brain stimulation, electrophysiology can reveal the synaptic organization within particular structures and systems through the combination of orthodromic and antidromic stimulation of nerve fibers. Together with autoradiographic and older fiber-tracing techniques, it becomes possible to establish the afferent–efferent relationships among neurons.

Electrophysiological investigations of learning often avoid the question of localization altogether, by emphasizing the necessity of understanding the brain in terms of the component processes that go into the formation of an engram and into the retrieval or performance of acquired behavior. This approach is familiar and requires no elaboration except to note that its logical underpinnings are close to those of contemporary psychology, which also strives to understand learning and memory in terms of information processing.

Evidence regarding the physical basis of learning also comes from investigations of what is usually referred to as brain function. Until recently, the emphasis here has not been on how the various parts of the brain contribute to, say, classical conditioning. In seeking the function of brain systems through experimentation and clinical observation, the stress has been on sensorimotor capabilities or vegetative functions, on the one hand, or on complicated and subtle perceptual and intellectual capabilities, on the other. Nevertheless, a few parts of the telencephalon have been judged to be especially relevant for learning — the frontal and prefrontal areas of neocortex, basal ganglia, and the septohippocampal complex of the limbic system, to mention the more obvious examples.

These observations are meant only to emphasize the number of options available for pursuing questions of brain processes and conditioning. The approach described in this chapter is one in which the contributions of various levels, structures, and systems of the brain are assessed in terms of mechanisms of interest for understanding not only conditioning but also brain function.

III. MODEL SYSTEMS OF LEARNING

With the discovery of similar mechanisms of reflex habituation in mollusks and mammals (see Thompson, 1976), the trend toward ever-simpler preparations in which to investigate behavioral plasticity has been replaced by renewed interest in associative learning in mammals. But even here the engram question must be hedged in light of evidence and intuition that argues that answers depend on the particular species, cue, reinforcement, response, setting, and so forth that one has in mind. What has been needed is a fresh consensus as to which preparations are the most promising, from the viewpoint not only of their amenability to neurobiological research, but also of their suitability in terms of producing phenomena and revealing processes deemed to be of fundamental importance by behavior theorists. In principle, the model-systems approach simplifies the task of coordinating research findings among laboratories and across a variety of techniques.

My aim is to review evidence concerning the location and nature of engram systems in a few of the mammalian preparations currently in vogue, principally the rabbit nictitating membrane response (NMR) developed by I. Gormezano. Thompson (1976) has summarized the considerations that have combined to push the rabbit NMR preparation to the forefront of contemporary research. Summarizing his arguments, the rabbit NMR preparation possesses a number of features that make it ideal for investigation of the neuronal basis of learning in animals: (1) The CR can be acquired within a single training session, but requires a substantial number of trials to become fully developed; (2) the CS does not produce a CR ("alpha" response) before training; (3) presentation of the CS and US does not produce sensitization or pseudoconditioning; (4) the parametric features of the CR are well characterized, and the role of such variables as CS intensity, US intensity, and intertrial and interstimulus intervals is known; (5) the behavioral CR is robust and discrete, with easily measured features such as latency and amplitude; (6) the efferent system controlling the behavioral response is known; (7) the preparation permits precise control and specification of stimuli; (8) the preparation is physiologically normal and remains essentially motionless during training, and intertrial responses are infrequent. To Thompson's list we might add the fact that the rabbit NMR preparation has substantiated relevance for research of interest to contemporary theorists (e.g., Dickinson & Dearing, Chapter 8; Wagner, Chapter 3).

Eye-blink conditioning in cats may hold similar promise, and Patterson, Olah, and Clement (1977) have recently demonstrated NMR conditioning in cats. The rabbit and cat share similar neural mechanisms for NM and blink reflexes. Neocortical development is more advanced in the cat than in the rabbit, but because the two preparations can be anchored to a common point on the motor or output side, meaningful comparative investigations become a real possibility. The rabbit NMR is the more extensively investigated of the two, and most of the ensuing discussion of experimental findings concerns this preparation.

For both species, the somas of motoneurons controlling extension of the NM are located, along with those innervating the lateral rectus muscles of the eyes, in the abducens nucleus (N. VI.). These motoneurons are often tightly coupled so as to permit electronic synaptic interactions. Discharge of one can result in the discharge of its neighbor without the need of additional synaptic drive. Axons project to the retractor bulbi muscles, which form bands of "white" or "twitch" muscle fibers encapsulating the optic nerve at the base of the orbit.

Lorente de No (1933) thought that extension of the rabbit's NM is a passive consequence of eyeball retraction involving the retractor bulbi and the six other extraocular muscles as well. More recent evidence indicates that the retractor bulbi muscles, under sole innervation of the abducens nerve, are responsible for most, if not all, of the NMR in rabbits (Cegavske, Thompson, Patterson, & Gormezano, 1976). Mis (1977) reports normal-appearing conditioned and unconditioned NMRs following transection of all extraocular muscles save for retractor bulbi.

Motor control of the extraocular muscles in the rabbit and cat, including retractor bulbi, is a bit simpler than is the case for spinally innervated skeletal muscles, in that they are not provided with muscle spindle proprioceptors or a gamma-efferent system. Nor do the axons of motoneurons projecting to extraocular muscles give off recurrent collaterals. The NMR, then, lacks many of the feedback systems we typically associate with voluntary or instrumental behavior.

Defensive eye reflexes in rabbits and cats are usually accompanied by extension of the obicularis oculi muscles, which are innervated by a branch of the facial nerve (n. VII). These, too, are devoid of proprioceptors and gamma-efferents. Feedback of eye-ball retraction and blinking must arise from other routes, such as from changes in visual input.

IV.   FOREBRAIN STRUCTURES AND CLASSICAL CONDITIONING

Perhaps too much discussion in this chapter is devoted to the question of where the initial and most crucial events in conditioning occur. Nevertheless, with perhaps a few exceptions, engrams for conditioning remain as elusive as ever. One approach has been to determine the minimal or most caudal portions of the vertebrate nervous system capable of supporting conditioned responding. A recent review by Patterson (1976) can leave little doubt that instances of classical conditioning that satisfy stringent behavioral criteria have been demonstrated in spinal preparations. Impressive and informative as such demonstrations are, they may have little bearing on where conditioning occurs in an animal whose nervous system is essentially intact. Buchwald and Brown (1973) have recently surveyed the literature on the effects of forebrain destruction on classical conditioning. From the period covered in their review, there have been some significant developments concerning this question in studies using defensive eye reflexes in rabbits and cats.

Oakley and Russell (1972, 1974, 1975, 1976, 1977) investigated rabbit NMR conditioning following extensive, and in some cases virtually complete, neodecortication. The surgery involved pial stripping and aspiration of gray matter performed in two stages, one hemisphere at a time with several weeks allowed for recovery after each stage. Survivors appeared healthy and were able to maintain their weight, but at a below-normal level. Neodecorticate rabbits conditioned to a tone or light with an eye-shock US conditioned normally, except that initiation of the process required a greater number of trials than was the case for unoperated control rabbits. Russell (personal communication) believes this retardation of the onset of conditioning may involve an attentional rather than associative deficit because, once the CR starts to appear, its rate of growth and terminal level of performance matches that of normal animals.

The form or topography of the CR was essentially the same for both normal and neodecorticate rabbits, except that the latency of the CR following CS onset tended to be longer by 50 to 75 msec in neodecorticates and showed little inclination to shorten as conditioning progressed. A progressive decrease of CR latency is usually seen in rabbit NMR conditioning, and this was observed in the intact animals.

Neodecorticate rabbits showed better differential conditioning between tone and light CSs than normals. Most of this difference was due to extremely low response rates to CS− by neodecorticates. These animals may have had an exaggerated ability to ignore or "tune out" stimuli, or, equivalently, an exaggerated difficulty in tuning in. This would be consistent wth the idea that they suffered an attentional deficit that somehow postponed the onset of conditioning. This same deficit would work to the animal's advantage as far as differential conditioning is concerned by better enabling it to ignore the nonreinforced CS and in this way achieve a low rate of responding.

An attentional deficit could also account for the longer CR latencies observed in neodecorticate rabbits, because a bias to ignore the CS would have to be overcome before the CR could occur. Overcoming a bias to ignore would introduce a lag in the system that would preclude CRs of short latency, even in well-conditioned animals.

Alternative interpretations of these studies come to mind. In the case of differential conditioning (Oakley & Russell, 1975, 1976, 1977), reduced responding to CS− by neodecorticates, whether initially or in later reversal training, could have been due to an increase in "active" or conditioned inhibition. That is, the animals might have had an exaggerated ability to suppress CRs when they were not appropriate. A number of considerations argue against this interpretation — frontal neocortex is closely linked to inhibitory function, for example. In the absence of appropriate behavioral tests for conditioned inhibition, there is no way of knowing whether CR-suppressing mechanisms contributed to these observations.

In a similar vein, the postponement of the onset of conditioning and increased CR latency in neodecorticates, rather than reflecting an attentional deficit, could instead indicate a tonic bias favoring CR-suppressing mechanisms. Finally, elevated sensory thresholds to CSs could play a part in extending the number of trials before observing CRs, in longer CR latency, and in reduced responding to a nonreinforced CS. Resolution of these questions awaits further research.

The significance of Oakley and Russell's findings, nevertheless, is in the support they provide for the idea that engrams for classical conditioning can be formed and stored subcortically. In this regard their research is consistent with other literature reviewed by Konorski (1967, p. 296), Buchwald and Brown (1973), Woody, Yarowsky, Owens, Black-Cleworth, and Crow (1974), and, most thoroughly, by Enser (1976). Moreover, their evidence suggests that, apart from an attentional deficit, the formation and storage of the engram for rabbit NMR conditioning proceeds normally in the absence of neocortex. This point is made most forcefully in their most recent report (Oakley & Russell, 1977), which describes tests of CR retention following neodecortication. A group of 10 intact normal rabbits received differential conditioning to tone and light CSs. Half of these animals were then designated for neodecortication, and the 3 that survived were retrained in the differential-conditioning situation as before and compared with the 5 remaining normal rabbits, which were retrained at the same time. Retention by the neodecorticates was almost as good as in normals, a result that clearly implies subcortical storage of the engram in the normal intact brain. If storage of the engram were limited solely to neocortex, a profound retention loss would have been evident following recovery from surgery.

This result is particularly significant because it argues against the possibility that animals conditioned de novo following surgery make use of subcortical circuits not normally involved in conditioning. Alternatively, it could be argued that subcortical structures spared following neodecortication undergo functional (perhaps structural) reorganization and therefore have capabilities for processing and storing information that bear little relationship to the normal brain. By demonstrating retention of the CR following neodecortication, Oakley and Russell (1977) have reduced the force of these arguments.

Oakley and Russell's (1977) retention study showed that conditioned responding to CS−, reasonably high before neodecortication, dropped to zero during retraining, whereas normal rabbits maintained the same modest rate observed in the initial training phase. Also of interest is that CR latency of neodecorticated rabbits shifted immediately into the same range, 50 to 75 msec greater than normals, which is characteristic of this preparation. Accepting the attentional interpretation of neocortical involvement in conditioning, these observations reinforce the idea that neodecortication adversely affects one of the processes involved in conditioning—namely, the ability of the animal to tune in a reinforced CS—while at the same time providing an exaggerated ability to tune out nonreinforced stimuli. The associative component of conditioning is formed and survives intact within a subcortical store.

Another experimental assessment of the role of forebrain in rabbit NMR conditioning was carried out by Enser (1976) in I. Gormezano's laboratory at the University of Iowa. Enser compared neodecorticate and sham-operated rabbits in acquisition, differentiation, and reversal, using tone and light CSs. He found, as did Oakley and Russell, retarded conditioning in neodecorticate rabbits in comparison with controls, but a high level of performance was eventually achieved by both groups. Unlike Oakley and Russell's experiments, however, differential conditioning, both initially and during reversal, was somewhat poorer for neodecorticates than for controls. The difference in this case was due to a combination of slightly lower conditioned responding to CS+ and a slightly higher level of responding to CS−. Enser did, however, bear out the earlier findings regarding CR latency, which was 50 to 75 msec longer on the average among neodecorticates than among controls.

Whatever discrepancies exist between the two sets of experiments can probably be laid to differences in the pattern of forebrain destruction inflicted by surgery. In the Oakley and Russell experiments, particularly those described in the more recent reports, neocortical destruction was virtually complete while subcortical stuctures remained intact with little indication of tissue loss. In Enser's study, as much as 25% of neocortex was spared in some rabbits, while subcortical tissue loss was described as substantial, involving compression and degeneration of basal ganglia, limbic structures, and thalamus.

In his research, Enser also produced hemidecorticate and thalamic preparations. The thalamic animals had basal ganglia and hippocampus aspirated away in addition to neocortex. These animals showed acquisition performance described as being 30 to 40% below that of neodecorticates. Thalamic rabbits were also capable of differential conditioning and reversal.

The fact that thalamic animals were significantly impaired in comparison with neodecorticates — which were, in turn, impaired in comparison with hemidecorticate and sham-operated animals—led Enser to interpret his data in terms of a mass-action principle. Here, this hypothesis states that acquisition and performance of the CR is inversely related to the total volume of brain tissue destroyed; precisely which structures are lost or damaged presumably makes little difference so long as sensorimotor and vegetative functions remain intact. Enser's results support this hypothesis in that his surgical groups were neatly ordered on all of the usual measures of CR strength (frequency, amplitude, latency) in the following way: sham-operates, hemidecorticates, neodecorticates, thalamics. Correlational analyses, relating the weight of spared brain tissue to individual performance measures, also supported a mass-action interpretation of the data.

Oakley and Russell (e.g., 1972) compared hemidecorticate rabbits with normal and bilateral neodecorticates, but found no differences between hemidecorticates and normals in terms of speed of conditioning. They considered, but rejected, a mass-action interpretation of conditioning—at least as far as the neocortex is concerned.

The two sets of studies can be reconciled, I think, by recognizing that Enser's bilateral neodecorticates and hemidecorticates sustained varying amounts of subcortical damage. It therefore seems likely that a mass-action principle applies, not to neocortex as Lashley originally proposed, but rather to subcortical sparing and loss. But even on this basis it is difficult to accept mass action in the sense Lashley used the term. "Surgically simplified" preparations like Enser's thalamic rabbits tend to be bizarre in so many ways that the fact that any conditioning can be demonstrated has greater significance than the fact that it may be slow to occur and labile once established. Conditioning entails a number of processes and mechanisms that in all likelihood have different neural substrates. The finding that conditioning seems to conform to a mass-action principle does not necessarily mean that the engram is diffusely represented as a sort of hologram in neural tissue. What it probably means instead is that the formation of the engram must overcome successively more rigorous obstacles as more and more of the forebrain is removed.

Recent experiments with drastically "simplified" preparations by Norman, Villablanca, Brown, Schwafel, and Buchwald (1974) and by Norman, Buchwald, and Villablanca (1977) have demonstrated eye-blink conditioning in cats, based on an eye-shock US, following decerebration as deep as the boundary between midbrain and pons. Lower brain-stem transection even left intact one animal's ability to differentiate between tones of 1 and 2 kHz. Given the degree of stimulus generalization one might expect between tones in this frequency range, this feat probably required active suppression of the CR to the nonreinforced CS, rather than mere tuning out.

These reports suggest that the conditioning obtained in decerebrate cats, though fragile and slow to develop, entails formation and storage of the engram below the level of the colliculi — probably within the reticular core of the brain stem. The possibility that discriminative functions involving CR suppression can also exist at this level is one of the more exciting developments to come out of the decerebration approach.

V. LIMBIC STRUCTURES AND RABBIT NMR CONDITIONING

Limbic-system structures, particularly the septohippocampal complex, have been implicated in a general way in processes presumed to be important for conditioning — attention, inhibition, and the formation of the associative engram itself. (See Grossberg, 1975; Isaacson & Pribram, 1976, Iverson, 1976, for a sampling of recent developments.) Evidence from a number of laboratories is beginning to bring into focus the precise roles of the septal nuclei and hippocampus in rabbit NMR conditioning.

A. Acquisition Processes

Bilateral lesions of the septal area or dorsal hippocampus do not impair simple acquisition or performance of the conditioned NMR (Lockhart & Moore, 1975; Maser, Dienst, & O'Neal, 1974; Schmaltz & Theios, 1972; Solomon, 1977; Solomon & Moore, 1975). These observations at first glance seem inconsistent with two other lines of research — one involving electrophysiological recording and the other involving posttrial brain stimulation.

Berger, Alger, and Thompson (1976; see also Thompson, 1976) have identified a "neural substrate" of conditioning while recording unit activity from dorsal hippocampus during paired presentation of a tone CS and a corneal air-puff US. Significant activity correlated with behavioral CRs appeared in pyramidal cell layers and from the granule cell layer of the dentate. The firing pattern of these units appeared to match the topography of the response, except in that the neuronal response consistently anticipated the behavioral response by 25 to 35 msec.

Signs of conditioning in hippocampal units appeared very early in training, and from the published figures conditioning could be detected after only a few trials in some cases. At this point the neuronal response did not anticipate the US, as would a criterion CR, but instead appeared as a miniature or shadow of the UR. This early replica of the UR, the first sign of conditioning, also anticipated the behavioral response by 25 to 35 msec. From here both the neuronal and behavioral response grew in strength — decreasing in latency and increasing in amplitude — until by the end of training, unit responses were easily extracted from background activity.

Hippocampal units in animals that received only unpaired presentations of the tone and air puff failed to show increased activity either to the tone, the US, or the UR. These observations rule out the possibility that correlated activity observed when the two stimuli were paired was the result of pseudoconditioning, sensory artifacts, or motor potentials. The observed changes were clearly the result of conditioning.

Another line of evidence implicating the hippocampus in simple acquisition has been reported by Salafia, Romano, Tynan, and Host (1977), who found that bilateral stimulation of the dorsal hippocampus immediately after conditioning trials drastically impaired the rate of CR acquisition. The level of electrical stimulation in these experiments was below that which would induce convulsions but high enough to evoke EEG afterdischarges, or "seizure activity." Rabbits given posttrial stimulation of the hippocampus required over 160 trials to attain a CR rate greater than 10%. Normal rabbits and neocortically stimulated controls required 50 trials, on the average, to achieve this rate. Once CRs began to appear, hippocampal stimulation produced no further disruptions and conditioning proceeded normally. Salafia et al (1977) suggested that posttrial hippocampal stimulation interfered retroactively with engram consolidation. Subsequent research (Salafia, personal communication) indicates that the onset of conditioning can be delayed just as severely if brain stimulation is below a level that triggers EEG seizure activity or if it is delivered to the amygdala instead of the dorsal hippocampus.

B. Differential Conditioning

Lockhart and Moore (1975) compared normal rabbits and rabbits with septal lesions in 2-tone differential conditioning (700 versus 1600 Hz, counterbalanced; 76 db SPL). Conditioning of the NMR progressed normally in septal animals, but their rate of responding to the nonreinforced tone over the course of extended training was consistently elevated by about 15% compared with controls. Tests of stimulus generalization in extinction failed to uncover any greater tendency for conditioning to generalize to other tones in septal animals compared with normals. (Septal rabbits were, overall, more resistant to extinction.) Nor did the 2 groups differ in their rates of spontaneous responding. Threshold tests eliminated the possibilities that the differences between the 2 groups could be laid to greater sensitivity of the septal animals to the nonreinforced CS or to the eye-shock US.

What is not so easily resolved is the question of precisely which mechanism of differential conditioning was affected by the lesion. Septal animals might have been less able than normals to suppress CRs to CS – because of deficits in active or conditioned inhibition. Alternatively, they might have been less able than normals to tune out or ignore the nonreinforced CS. Because this experiment employed two stimuli from the unidimensional scale of audiofrequency, it seems likely that active suppression of the CR would be required to counter generalization from CS+ to CS–.

On the other hand, deficits in an attentional process — tuning out a nonreinforced stimulus — could also contribute to both the higher rate of responding to CS – and the greater resistance to extinction reported by Lockhart and Moore (1975). This interpretation is supported by Powell, Milligan, and Buchanan (1976), who compared normal rabbits and rabbits with septal lesions in 2-tone differential conditioning (304 versus 1216 Hz; 80 db SPL). Responding to CS – was in excess of 15% in septals compared with controls, even after extensive training. Powell et al, (1976) monitored cardiac-decelerative reactions concurrently with eye-blink conditioning and found septal rabbits to be more reactive than normals to both CSs. Because cardiac deceleration is typically taken as a sign of "orienting," these authors suggested that their septal rabbits were, in fact, impaired in their ability to tune out the CS–. Habituation of the orienting response may be involved in the tuning-out process. In this case septally lesioned animals would be at a disadvantage in establishing inhibitory stimulus control.

C. Latent and Conditioned Inhibition

Because simple differential conditioning and extinction can be accomplished either by tuning out the nonreinforced signal or by active suppression of the CR — or by a combination of the two — we have abandoned these tests in favor of others that permit independent assessment of these two inhibitory mechanisms.

Latent inhibition — that is, the observation of retarded conditioning because of preexposure to the CS (see Lubow, 1973) — is generally conceded to be an attentional phenomenon (e.g., Reiss & Wagner, 1972). Briefly, repeated preexposure to the CS results in a loss of "salience," and this attentional bias must be overcome in order for conditioning to occur. The need to recover lost salience once conditioning begins implies slower acquisition than would occur had the CS not been preexposed (Mackintosh, 1975).

Conditioned-inhibition training — that is, differential conditioning in which one stimulus is reinforced and the same stimulus in the presence of an added cue is not — requires active suppression of the CR (e.g., Marchant, Mis, & Moore, 1972). For the purpose of the present discussion, one need only note that in order for the animal to successfully discriminate the reinforced CS from the nonreinforced compound stimulus, it must pay attention to the added cue.

Solomon and Moore (1975) investigated the effect of bilateral destruction of dorsal hippocampus on latent inhibition of the rabbit NMR. The experimental design involved 3 surgical treatments: dorsal hippocampectomy, bilateral removal of overlying neocortex, and nonoperated controls. Each group was then divided in 2, with half the animals given 450 preexposures to the to-be-conditioned stimulus—a tone—and half which merely remained in the conditioning chambers for a comparable period of time.

The effect of hippocampectomy on latent inhibition was dramatic: Normal and neocortical controls required an average of 65 trials to achieve a criterion of 5 consecutive CRs, but after 450 preexposures to the tone, this figure increased to 107 trials. For hippocampectomized rabbits, controls required 86 trials to criterion, which was not significantly different from controls in the other surgical treatments, while those preexposed to the CS required only 51 trials. Hippocampectomy not only eliminated latent inhibition, but actually reversed it.

Threshold tests were carried out to ensure that hippocampectomized rabbits were not more sensitive to the CS than controls. Other tests ruled out the possibility that these animals were hyperreactive to the eye-shock US.

Our interpretation of this experiment was much along the lines of

Douglas's (1972) view that the hippocampus is somehow responsible for tuning out nonreinforced stimuli. To pursue this further, Solomon (1977) tested whether nonreinforcement was indeed the key to understanding the role of the hippocampus in conditioning. In one experiment, naive hippocampectomized rabbits were compared with neocortical and nonoperated control groups for conditioned inhibition between a light (CS+) and the light compounded with a tone (CS−). Following training, the tone was paired with the US to further assess its acquired inhibitory potential. The point of this experiment is that conditioned inhibitors may have adaptive significance even though they are correlated with nonreinforcement. Would the hippocampus conspire to tune out such a signal?

Evidently not, for Solomon (1977) found no difference among the three groups either in differential conditioning or in the subsequent resistance-to-reinforcement test. The results of this experiment are summarized in Fig. 5.1.

The hippocampus evidently plays no part in conditioned inhibition of the rabbit NMR; and Douglas's (1972) idea, although sound as far as latent inhibition is concerned (and this could also apply to simple differential conditioning and extinction), requires amending if it is to incorporate Solomon's (1977) conditioned-inhibition results. As Solomon (1977) puts it, the hippocampus seems to be involved in tuning out irrelevant stimuli. Conditioned inhibitors, although nonreinforced, are decidedly relevant.

D.   Blocking of the Rabbit NMR

This distinction between nonreinforcement and adaptive significance is made with greater force in a second experiment reported by Solomon (1977). Previous research in our laboratory had established procedures for demonstrating blocking in Kamin's (1969) two-stage design (Marchant & Moore, 1973). Kamin's (1969) design involves establishing a CR to one CS and then continuing the conditioning procedure to this same CS, plus a new stimulus compounded with the original CS. In comparison with controls, the acquisition of conditioning to the added stimulus is blocked by earlier conditioning to the other element of the compound, presumably because of its redundancy as a predictor of the US.

Solomon adopted these procedures and compared naive hippocampectomized rabbits with the usual neocortical and normal controls. Briefly, animals in the blocking groups received 300 to 400 pairings of a tone CS and eye-shock US—enough to ensure a CR rate greater than 90% for 200 consecutive trials. For each surgical treatment, this stage was matched by control animals, which were restrained in the conditioning chambers for a comparable period of time, but which experienced neither the tone nor the eye shock.

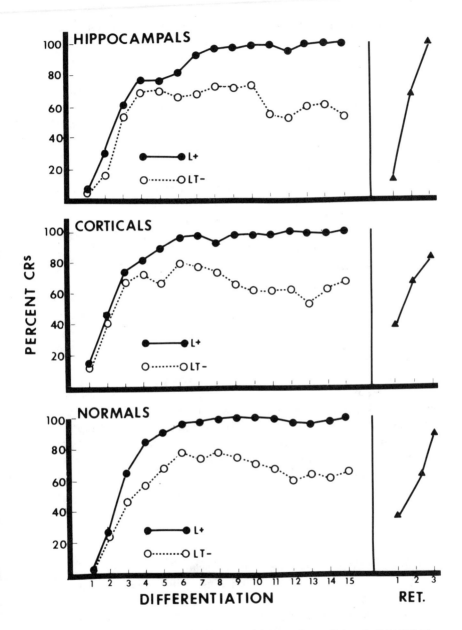

FIG. 5.1. Left panels: Mean percentage of conditioned responding to the light (CS+) and the light plus tone (CS−) for hippocampals, corticals, and normals over 15 days of conditioned inhibition training. Right panels: Mean percentage of conditioned responding to the tone over the 3 days of retardation testing. From Solomon (1977).

All animals were treated alike in the second stage of training, receiving 500 reinforced trials to a compound CS consisting of a light and the same tone employed in the first stage. The test for blocking consisted of 100 randomly ordered presentations of the light (blocked CS) and the tone (blocking CS) administered over 2 daily extinction sessions. The 2 stimuli were never compounded at this stage.

The results can be expressed most conveniently by comparing the percentage of CRs to the light in the various treatments. Combining the normal and neocortical groups, those animals conditioned to the tone in the first stage gave an average of 19 CRs to the light out of 100 opportunities. Their controls, which had no conditioning before the second stage, gave an average of 46 CRs to the light. This difference was significant and a clear indication that blocking occurred.

Hippocampectomized rabbits conditioned to the tone before compound CS training gave an average of 60 CRs to the light. Their controls actually gave fewer, 45, although the difference was not significant. In short, dorsal hippocampectomy completely eliminated the blocking effect that was so evident in the other surgical groups.

Solomon's (1977) interpretation of these results was along lines that by now should be familiar. Normal blocking, according to some views (e.g., Kamin, 1969; Mackintosh, 1975), is basically an attentional phenomenon. Presumably the redundant cue, although reinforced, tends to be ignored, and conditioning is therefore considerably weaker than it would be if the stimulus were part of a novel stimulus compound. Destruction of the dorsal hippocampus evidently takes with it the animal's ability to tune out or ignore the added cue. Solomon's (1977) final word is that the hippocampus appears to be involved in tuning out redundant or irrelevant stimuli — namely, stimuli with no adaptive significance, regardless of whether or not they are reinforced.

VI.   NEUROPHARMACOLOGY OF ATTENTIONAL PROCESSES

Solomon's research, as we have seen, implicates the hippocampus in certain aspects of attention related to conditioning. Still to be determined are the physiological transactions involving the hippocampus that lie behind these processes. One approach that seems to hold promise involves the administration of drugs that selectively affect the various neurotransmitter systems of the brain. All the brain amines (norepinephrine, dopamine, serotonin) and acetylcholine have been mentioned in one context or another as substrates of attention, inhibition, learning, and so forth.

Carlton (e.g., 1969) may have been the first to suggest that central cholinergic systems are essential for inhibition of nonreinforced behaviors.

Cholinergic blockade often mimics the effect of hippocampal lesions, and anatomical evidence indicates a major cholinergic pathway originating in the midbrain and projecting to the septohippocampal complex (Kuhar, 1976; Lewis & Shute, 1967).

These considerations led us to investigate the role of central cholinergic systems in latent inhibition of the rabbit NMR (Moore, Goodell, & Solomon, 1976). Rabbits were injected with 1.5-mg/kg scopolamine hydrobromide, scopolamine methylbromide, or isotonic saline. Scopolamine hydrobromide competes with acetylcholine at muscarinic receptor sites and therefore blocks the normal synaptic transmission in both the central and peripheral nervous systems. Scopolamine methylbromide does not pass the blood–brain barrier and therefore serves as a control by which the peripheral action of the drug can be separated from its central effects.

In their experiment, Moore et al. (1976) tested the reactivity of the unconditioned NMR over a series of habituation–dishabituation trials. Neither form of scopolamine affected UR reactivity or sensitivity to the US. The next experiment consisted of assessing latent inhibition under each drug treatment by comparing subgroups preexposed to the to-be-conditioned stimulus—a tone—with those that were merely confined in the conditioning chambers but not preexposed to the tone. Subsequent conditioning over a series of tone–eye-shock pairings indicated that central cholinergic blockade by scopolamine profoundly retarded acquisition. As is so often the case, however, all animals ultimately acquired the CR, and their terminal performance matched that of controls. A distinct but statistically unreliable latent inhibition effect was seen in these animals, but subsequent tests indicated that the drug significantly raised the threshold of the tone for eliciting the CR by an average of over 10 db in comparison with controls.

The question of whether cholinergic systems are involved in tuning out a latent inhibitor was resolved only when a light was substituted for the tone as the CS. Scopolamine did not affect the CR threshold to the light, but once again the drug produced a marked retardation of the onset of conditioning. Nevertheless, latent inhibition was obtained. Drugged rabbits preexposed 450 times to the light before conditioning required an average of 693 trials to achieve a criterion of 5 consecutive CRs, whereas nonpreexposed rabbits required only 328 trials to attain this criterion. Scopolamine may adversely affect the process of tuning in a CS (as would appear to be the case with neodecortication; e.g., Oakley & Russell, 1972), but unlike hippocampectomy, the drug had no effect on tuning out a latent inhibitor.

Because latent inhibition could be demonstrated in the rabbit NMR preparation despite central cholinergic blockade, we have turned our investigations to the other mesolimbic chemical pathways. The serotonergic pathway originating in the median raphe nuclei of the brain stems and projecting to the septohippocampal complex, among other forebrain structures, is presently under investigation in our laboratory. Para-chlorophenylalanine (pCPA), which depletes brain serotonin by inactivating an enzyme, trypto-phan hydroxylase, essential for its synthesis, is being injected into rabbits for tests of latent inhibition. Although preliminary results suggest that pCPA may eliminate latent inhibition where scopolamine did not, data are too fragmentary at this point to support a firm conclusion.

More convincing support for the hypothesis that serotonergic systems mediate latent inhibition has been obtained by Solomon, Kiney, and Scott (1977) in a 2-way active-avoidance task with rats. Animals injected with pCPA were given 30 or 0 preexposures to the warning signal—a tone. Injec-tion control groups showed significant latent inhibition of avoidance condi-tioning, but pCPA-injected rats preexposed to the tone were actually slightly faster to acquire the avoidance CR than their 0-preexposed controls. Obser-vations on spontaneous shuttling and escape-avoidance latencies argue against the possibility that the elimination of latent inhibition by pCPA was due to spurious factors.

Serotonin may act in combination with other neurochemical systems, and the crucial sites of action remain to be specified. Nevertheless, it would appear that serotonergic interactions in the septohippocampal complex may be an essential component in the tuning-out function of this part of the brain.

VII.   A MIDBRAIN–BRAIN STEM CIRCUIT FOR CONDITIONED INHIBITION OF THE RABBIT NMR

Recalling Solomon's (1977) failure to implicate the hippocampus in condi-tioned inhibition, the question arises as to the locus of the inhibitory engram. The question is posed in a simpleminded way in order to dramatize the fact that a clearly defined substrate for inhibitory learning has been just as elusive as that for excitatory learning. Mis (1977), working in our laboratory, has recently identified a critical link in the system responsible for active suppression of the conditioned NMR by a conditioned inhibitor.

Inhibitory postsynaptic potentials (IPSPs) have been recorded from ab-ducens motoneurons in response to stimulation of various regions of the brain stem (e.g., vestibular nuclei). Although physiological investigations have not specified whether abducens motoneurons showing IPSPs are those that innervate the retractor bulbi muscles or those that innervate the lateral rectus muscles, this and other evidence pointed toward the existence of brain-stem systems capable of inhibiting the NMR.

In his first experiment, Mis (1977) applied electrical stimulation (ESB) to various midbrain and brain-stem loci through chronically implanted elec-trodes. The basic procedure was simply to condition the implanted rabbits to a tone CS and eye-shock US and then observe the effect on CR topography of ESB applied either before or after CS onset. Each of the five electrodes in an array was tested in turn. Mis was most interested in identifying those loci that when stimulated produced a diminution or inhibition of the CR.

Some striking instances are illustrated in Fig. 5.2. Panel A shows a normal CR topography with no ESB. Panel B shows what happened when ESB preceded the CS. In this case the CR was completely inhibited. Panel C shows that this inhibition occurred even when the ESB was applied after initiation of the CR. These results illustrate that ESB acted on the output stage of the system rather than masking or overshadowing the CS.

Panels D, E, and F of Fig. 5.2 show the same pattern as Panels A, B, and C, the difference being that these tracings were taken following transection of all the extraocular muscles (four recti, two obliques) except the retractor bulbi. Because the same pattern of ESB-induced inhibition was observed in all such tests, it seems safe to conclude that the inhibition was central and not due to stimulation of muscles that could oppose the retractor bulbi. Videotape records failed to reveal any spurious ocular or facial reactions that could be construed in terms of peripheral antagonism of the NMR.

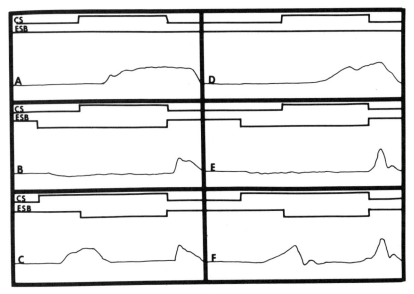

FIG. 5.2.   Tracings of polygraph records depicting the substantial reduction in the CR amplitude produced by stimulation of Electrode 1 in rabbit FM88 following transection of the extraocular muscles. Panels A and D depict the normal CR topography, while Panel B depicts the change in CR topography observed when the onset of the brain stimulation preceded the onset of the CS. Panel C depicts the change in CR topography observed when the onset of the CS preceded the onset of the ESB. The parameters of ESB employed during these tasks were 1000 pps, .25 msec pulse duration, and $300\,\mu$ A pulse amplitude. From Mis (1977).

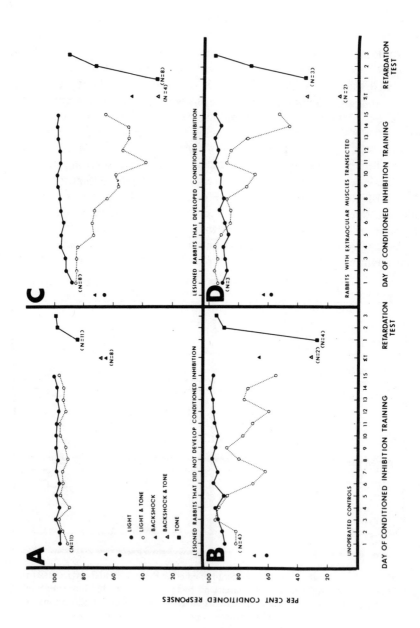

The effect of ESB on the unconditioned NMR was tested in 6 rabbits. Fourteen of the 30 electrodes represented by these animals produced a significant diminution of the CR, and 9 of these also produced a consistent reduction of UR amplitude amounting to 10 to 20% in most cases.

Histology indicated that the greatest and most consistent inhibitory effects were obtained from electrodes with tips located in ipsilateral regions immediately rostral to the oculomotor complex (N. III). The implicated structures are the accessory oculomotor nuclei, the nucleus of Darkschewitsch and the interstitial nucleus of Cajal, and adjacent parvocellular portions of the red nucleus on the same side—that is, the side of the conditioned NMR.

Using naive rabbits, Mis (1977) then produced small ipsilateral lesions in and around the suspected regions of the midbrain by electrocoagulation. After recovery, these animals, plus some normal rabbits and other non-lesioned rabbits with severed extraocular muscles, were conditioned to a light CS and then given 15 100-trial sessions of conditioned-inhibition training. The light was reinforced, and the same light compounded with a tone was the conditioned inhibitor. Some rabbits were also conditioned to a tactile CS (mild back shock) in the first stage for the purpose of carrying out independent summation tests for conditioned inhibitory properties of the tone. All animals received retardation testing — that is, tone acquisition following training or summation testing.

Figure 5.3 summarizes the results of this experiment. Panel B shows the development of conditioned inhibition in normal rabbits, and Panel D indicates that transection of extraocular muscles did not prevent this development. The results of summation testing, carried out in extinction, are depicted at the points labelled ΣT on the abscissa. Inhibitory summation is indicated by the lower level of extinction responses to the back shock-tone compound compared with back shock alone.

Panels A and C show the results for lesioned animals, and two populations are clearly defined: those that developed conditioned inhibition (Panel C) and those that did not (Panel A). Histological comparisons of the location of the lesions in these two clusters of animals confirmed the expectations from Mis's (1977) first experiment: Lesions that invaded the ipsilateral

FIG. 5.3. Mean percentage of CRs during the initial-acquisition, conditioned-inhibition training, summation-testing, and retardation-testing phases of this experiment. Panel A depicts the mean percentage responses for the lesioned rabbits that did not develop conditioned inhibition, while Panel C depicts the mean percentage responses for the lesioned rabbits that did develop conditioned inhibition. Panel B depicts the mean percentage of CRs for the unoperated controls, while Panel D depicts the mean percentage of CRs for the rabbits with their extraocular muscles transected. From Mis (1977).

nucleus of Darkschewitsch, the interstitial nucleus of Cajal, or the parvocellular red nucleus effectively destroyed the animal's ability to process
the tone into an inhibitory CS. As the retardation test results suggest, these
animals had no difficulty acquiring a CR to this stimulus, an observation
that eliminates the possibility that the lesion affected their ability to tune in
the tone.

A follow-up experiment extended these results by showing that a lesion in
che critical zone disrupted previously established conditioned inhibition. Differential conditioning could not be reestablished over a training period comparable to that administered before lesioning. The disruption did not
depend on the conditioned inhibitor's being a tone, because other conditioned inhibitors—light and back shock—could also be rendered ineffective
by the lesion.

Mis's (1977) research suggests that a number of midbrain and brain-stem
nuclei and fiber tracks play a role in modulating the output of abducens
motoneurons innervating the retractor bulbi muscles. However, the accessory oculomotor nuclei would appear to be crucial integrating sites for
inhibition of the conditioned NMR. Precisely how this region of the brain
influences the output of abducens motoneurons remains to be established. It
is possible that information originating in more rostral parts of the brain—
in the frontal cortex, for example—flows into this region and that the lesion
does nothing more than block the transfer of inhibiting neural traffic to
lower centers. All of these possibilities are being investigated. However,
anatomical and physiological evidence suggests that neurons in the critical
zone project to other cell groupings, principally the red nucleus, and do not
synapse directly into abducens motoneurons. Although the red nuclei send
numerous projections into the brain stem of the rabbit, their nearest terminations are among reticular formation neurons located lateral and ventral
to the abducens nucleus.

VIII.   THE ENGRAM FOR RABBIT NMR CONDITIONING

One implication of a midbrain–brain-stem circuit for inhibition of the conditioned NMR is that the most likely point of contact between the opposing
processes of conditioned excitation and inhibition is the reticular core of the
brain stem—specifically, among neurons ventral to abducens motoneurons.
This propinquity, if proven, would support the hypothesis that the crucial
events in conditioning occur in this part of the brain (see Gastaut, 1958b).

The idea that classical conditioning consists of informational transactions
within the reticular formation is by no means new (Eccles, 1953; Gastaut,
1958a). That the initial and crucial events in conditioning normally occur in
the more caudal portions of this system — in the brain stem or bulbar
reticular formation — has gained support only recently (e.g., Halas,
Beardsley, & Sandlie, 1970; Oleson, Ashe, & Weinberger, 1975; Vertes &

Miller, 1976).

Likely components of the engram for NMR conditioning are the neurons of the reticular formation in the core of the brain stem (Groves & Lynch, 1972; Ramon-Moliner & Nauta, 1966; Scheibel & Scheibel, 1958, 1967; Valverde, 1961, 1962). As described by the Scheibels in their Golgi studies, these cells have morphological features that make them attractive candidates for mediating conditioning. Of special interest are the large cells of nucleus reticularis pontis caudalis, which are strategically situated near the polysynaptic reflex arc of the unconditioned NMR—ventral to the sensory trigeminal nucleus and the abducens nucleus.

Like other large neurons in the reticular core, the dendritic arborizations of the cells are oriented in a transverse plane, perpendicular to the neural axis. Although polysensory, the dominant afferents may be from the more laterally situated small-celled or parvicellular reticular neurons (Groves & Lynch, 1972). The larger reticular units in the core are thought to be premotor in that their firing patterns in normal awake animals are closely correlated with motor activity (Siegel & McGinty, 1977).

The principal myelinated (fast) axonal projections of the large cells are either rostral or caudal, but rostral–caudal bifurcation of the main axon is common. The rostrally projecting axons can cover great distances—as far as the base of anterior neocortex (Scheibel & Scheibel, 1967). Finer (slow) axon collaterals project with "exuberant" bushy terminations onto cells in the cranial nerve nuclei. The most relevant of these for NMR conditioning would be axonic terminations directly onto the motoneurons in the abducens nucleus that innervate the retractor bulbi muscles of the eye. Alternatively, these fine collateral axons could project onto smaller interneurons within the abducens nucleus that drive the large motoneurons (Goldberg, Hull, & Buchwald, 1974).

The neuronal model for conditioning that emerges from these considerations portrays large neurons of reticular formation located ventral to the abducens nucleus as taking the converging inputs of the CS and US and somehow adjusting their output to drive abducens motoneurons either directly or indirectly through intervening synaptic relays. Precisely how these cells become conditioned remains to be determined.

An alternative theory has been proposed by Black-Cleworth, Woody, and Neimann (1975) and Young, Cegavske, and Thompson (1976; see also Thompson, 1976) in connection with evidence of conditioned increases in CS excitability based on direct suprathreshold stimulation of motoneurons. In brief, these authors suggest that the mechanisms for conditioning may be endogenous to the motor elements of the final common path of the CR. Black-Cleworth et al. (1975), for example, obtained eye-blink conditioning to a click CS in rats using antidromic firing of facial motoneurons as the US. Trigeminal rhizotomy ensured that the normal sensory drive to these cells was not involved in their experiments.

Young et al. (1976) tested the excitability of the NMR in rabbits to direct stimulation of abducens motoneurons. The ESB-elicited NMR was enhanced by prestimulation with a tone at an interstimulus interval favorable to conditioning. The degree of this facilitation increased as a function of paired tone–ESB presentations. Direct stimulation of motoneurons presumably bypassed the normal sensory drives to these cells via polysynaptic projections from the trigeminal complex.

From this evidence, it would appear that abducens motoneurons can selectively increase the weight of synaptic excitability at those postsynaptic receptor sites that are active concurrently with discharge firing. In this way previously subliminal inputs from a CS pathway would be accentuated as a direct result of coincidental input from the US. As is the case for the large neurons of the reticular formation, the precise mechanisms by which motoneurons would carry out this selective tuning and amplification of their receptive membrane, if indeed this is what they do, remains a matter of intriguing speculation.

Of the two hypotheses, the one stating that conditioning is mediated by the motoneurons themselves has less to recommend it than the other, which envisages the critical events occurring at a premotor stage. There is little doubt whether some large neurons of the reticular core can alter their firing patterns as a function of paired presentation of two stimuli. In a recent demonstration, Vertes and Miller (1976) recorded from reticular formation neurons in freely moving, unanesthetized, water-deprived rats. After some initial conditioning, a small but distinct population of cells was located in nucleus reticularis pontis caudalis that increased their activity to stimuli signaling grid shock but not to neutral stimuli or stimuli signaling water. The activity of these cells, unlike the majority of those encountered, could not be correlated with a particular pattern of motor activity. The authors suggested that these cells were a substrate of conditioned fear.

The reticular formation hypothesis of the engram also lends itself more easily to Mis's (1977) account of a midbrain–brain stem circuit for conditioned inhibition. Stimulation of the structures indicated by Mis as being essential components of conditioned inhibition of the NMR—namely, the nucleus of Darkschewitsch and the interstitial nucleus of Cajal—produces EPSPs in neurons described as lying ventral to the abducens nucleus (Graybiel & Hartwieg, 1974). Nor is there any evidence of direct projections from the oculomotor areas onto abducens motoneurons (Maciewicz, Kaneko, Highstein, & Baker, 1975). Space limitations prevent discussion of all the available evidence, but for purposes of elaborating the model, let us assume that the midbrain centers for conditioned inhibition ultimately excite reticular formation neurons lying ventrolateral to the large neurons at the core. These in turn make inhibitory connections onto the large cells. Activation of the midbrain–brain stem circuit by presentation of a conditioned inhibitor would then dampen or eliminate the output of the large reticular-formation neurons, and in this way reduce or eliminate the conditioned (but not unconditioned) NMR.

IX. THE ROLE OF THE HIPPOCAMPUS

In developing the model further, let us suppose that some of the large reticular-formation neurons that mediate conditioning project to the septohippocampal complex. There is considerable anatomical support for this assumption, and Lindsley and Wilson (1976) have shown that stimulation of nucleus reticularis pontis caudalis produces a desynchronized pattern of activity in the hippocampal EEG of freely moving cats. Desynchronization of EEG is usually taken as a sign of general excitabilty. If these observations can be extended to the rabbit, it becomes possible to account for the neural substrate of NMR conditioning in hippocampal units described previously (Berger et al., 1976; Thompson, 1976).

The puzzling aspect of the results of Berger et al. (1976) is that they clearly implicate the dorsal hippocampus in NMR conditioning; yet these structures are not essential for conditioning, because their destruction has virtually no effect on acquisition or performance. Recall also their observation of a phase lead of 25 to 35 msec between conditioned-unit activity from the hippocampus and the behavioral NMR.

The phase lead might be explained in terms of the morphology of the large reticular-formation neurons. The main rostrally projecting axon is basically a fast-conducting system consisting of a myelinated fiber of large diameter. An "efference copy" of activity originating in these cells would ascend to the forebrain relatively quickly and activate hippocampal units. By contrast collateral projections to the abducens nucleus consist of slower-conducting thin fibers that terminate in a bushy array. The proliferation of nodal points introduces an additional delay and cascading points of low conduction safety (Waxman, 1975). In brief, the hippocampus receives an image of the conditioned response in complete topographic detail before abducens motoneurons can transform this information into a response observed at the periphery.

What conceivable use is this premature image of the CR to the animal? Recalling Solomon and Moore's (1975) and Solomon's (1977) findings that dorsal hippocampectomy eliminates latent inhibition and blocking, while having no effect on acquisition and conditioned inhibition, one use for this premature image of the CR could be to assist the hippocampus in tuning out irrelevant or redundant signals. So long as the hippocampus receives an "efference copy" of the CR that is correlated with the reception of a particular stimulus, it somehow protects the input pathway to the site of CR initiation from interference or decay.

The hippocampus probably has nothing to do with tuning in the stimulus. In light of earlier discussion, that function could perhaps be assigned to the neocortex—possibly with a boost from the amygdala if the stimulus proves to be one of adaptive significance (see Douglas, 1972). A bit of reflection suggests further that the sequence of events in extinction may also begin among the reticular-formation neurons at the heart of the model. As the image of the CR generated by reticular-formation neurons begins to fade because the CS is no longer paired with a suitable reinforcer, the hippocampus tunes out the CS—that is, prevents it from intruding further into the critical zone.

One of the more intriguing possibilities is that the hippocampus can play an active role in preserving the engram within the intricacies of the reticular core. Perhaps reacquisition is so rapid because the hippocampus, on detecting a stronger image of the CR, lets open the floodgates for efficient and finely tuned channeling of CS-induced excitability to the site of CR initiation.

On relating this facet of how the hippocampus might be involved in retention of the CR, it is appropriate to quote Professor Estes (1973)—who in turn credits Konorski (1967, p. 267) — to the effect that "short-term memory is nonassociative and represents... persisting excitation or reexcitation of a stimulus and an organism's response to it." (See also Estes, Chapter 16.) The importance of the hippocampus for certain kinds of short-term memory is familiar and requires no elaboration, except to note the parallel between Estes's statement and the portrayal of hippocampal function in the present model. One is tempted to go so far as to suggest that the hippocampus uses its early copy of the CR to permit coordinated adjustive reactions to take place and perhaps to file away the events of conditioning in more highly organized and exotic memory stores.

Some hint as to how the hippocampus, perhaps acting in concert with the septal nuclei, tunes out potential CSs at the reticular core of the brain stem can be obtained from the anatomical literature (e.g., Nauta, 1958). For purposes of the model, let us assume that the septohippocampus complex sends projections to the reticular formation that terminate in presynaptic (axoaxonic) appositions onto the synaptic junctions between the CS pathway and the dendritic and somatic elements of the large reticular-formation neurons. Activation of selected units of the septohippocampal complex would then effectively prevent excitation by unimportant stimuli. Findings by Salafia et al. (1977) that posttrial stimulation of the hippocampus retards conditioning might be understood in these terms. Such stimulation could result in the indiscriminant tuning out of the reverberating neural trace of an important signal, such as the CS, and thus impose an obstacle to the acquisition process.

X.   THE MODEL

The bare bones of the neuronal model that summarizes these speculations are shown in Fig. 5.4. The large neuron on the right is a representative motoneuron in the abducens nucleus projecting to the retractor bulbi muscles. The large neuron to the left is a representative reticular-formation neuron of nucleus reticularis pontis caudalis. The model could be embellished in many ways to bring it more closely in line with the real nervous system. The synaptic organizations depicted in the diagram are from classical neurophysiology and do not indicate the subtle interactions among neurites that multiply the information-processing capabilities of this system by at least an order of magnitude. (See Scheibel & Scheibel, 1975; Schmitt, Dev, & Smith, 1976; Waxman, 1975.)

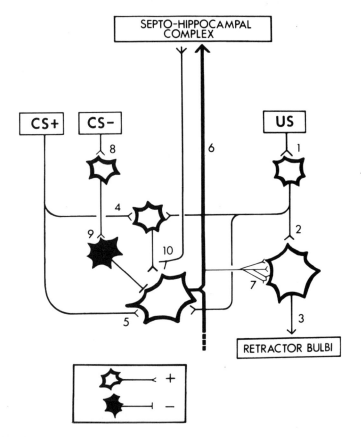

FIG. 5.4.   Proposed synaptic organization of midbrain–brain-stem neurons involved in rabbit NMR conditioning. Explanation in text.

Tracing through the diagram, 1 through 3 is the polysynaptic reflex arc of the NMR; 4 is one possible site of associative convergence between the CS and US; 5 is another; 6 is the mainline rostrally projecting axon, which transmits the efference copy of the CR; 7 shows the finer axon collateral system, which initiates the CR; 8 and 9 are the final links of the midbrain–brain-stem circuit for conditioned inhibition; and 10 is a representative axoaxonic apposition, which could tune out the CS en route to the large reticular-formation neuron. This is but one of the points where low-level tuning out might occur under higher control by the septo-hippocampal complex. In the diagram, the axoaxonic apposition (presynaptic inhibition) projecting from the septohippocampal complex could interpret information flow from the CS at points 4 and 5 instead of (or in addition to) point 10. It is also possible that the septohippocampal complex tunes out information flow serving conditioned inhibition at points close to the critical zone — points 8 or 9, for example. Although as yet undocumented for rabbit NMR conditioning, latent inhibition and blocking of conditioned inhibition have both been reported in CER situations with rats. Rescorla (1971) has shown that preexposure to a tone CS retards its later conversion to a conditioned inhibitor, and Suiter and Lolordo (1971) have reported that the development of conditioned inhibition to a stimulus is blocked in the presence of another stimulus that already possesses inhibitory properties. In light of evidence summarized here (e.g., Solomon, 1977; Solomon & Moore, 1975), it is possible that hippocampectomy might disrupt latent inhibition and blocking of conditioned inhibitors, just as it does for excitatory CSs.

For the uninitiated who might be unduly impressed by Fig. 5.4, I offer a very apt quotation from Scheibel and Scheibel. In characterizing the reticular core of the brain stem, they note (1967):

> The picture that emerges is one of continuous, intensive interaction between large numbers of conductors and the surrounding matrix of core neurons. The nature of these contacts must vary, depending on the presence of a myelin sheath and the frequency of axonal specialization. . . . However, the pattern of connectivity that can be traced out through the core suggests a circuit scheme so richly redundant that convincing arguments could be advanced for almost any conceivable loop or chain [p. 592].

Rich redundancy does not mean that the mystery of how conditioning occurs cannot be understood in the most fundamental molecular terms. But before research at the intracellular level can proceed, the locus of an engram must be established beyond question. Unhappily, I have little but circumstantial evidence to offer regarding the presumed pivotal role of neurons of the reticular core in rabbit NMR conditioning. Nevertheless, through the use of micropipette injections of metabolic poisons (Berthier, Spinelli, Solomon, & Moore, 1977), we hope to be able to produce the kind of selective and limited destruction of neurons that could perhaps bring the engram to bay.

ACKNOWLEDGMENTS

This chapter was prepared under a National Institute of Mental Health postdoctoral fellowship to the author during his tenure at the MRC Unit on Neural Mechanisms of Behaviour, I. Steele-Russell, Director, and as an Honorary Research Fellow in Psychology, University College, University of London. Research by the author and by his collaborators described in this chapter was supported by grants from the National Science Foundation to the author and from the National Institutes of Health (BMRS) to the University of Massachusetts–Amherst.

REFERENCES

Berger, T. W., Alger, B., & Thompson, R. F. Neuronal substrate of classical conditioning in the hippocampus. *Science,* 1976, *192,* 483–485.

Berthier, N., Spinelli, D. N., Solomon, P. R., & Moore, J. W. Fiber-sparing lesions of the central nervous system produced by cyanide. Presented at the European Brain and Behavior Society's Workshop on *The structure and function of the cerebral commissures.* Rotterdam, March 30–April 2, 1977.

Black-Cleworth, P., Woody, C. D., & Neimann, J. A. A conditioned eyeblink obtained by using electrical stimulation of the facial nerve as the unconditioned stimulus. *Brain Research,* 1975, *90,* 45–56.

Buchwald, J. S., & Brown, K. A. Subcortical plasticity. In J. D. Maser (Ed.), *Efferent organization and the integration of behavior.* New York Academic Press, 1973.

Carlton, P. L. Brain acetylcholine and inhibition. In J. Tapp (Ed.), *Reinforcement: Current theory and research.* New York: Academic Press, 1969.

Cegavske, C. F., Thompson, R. F., Patterson, M. M., & Gormezano, I. Mechanisms of efferent neural control of the reflex nictitating membrane response in rabbits (*Oryctolagus cunicilus*). *Journal of Comparative and Physiological Psychology,* 1976, *90,* 411–423.

Douglas, R. J. Pavlovian conditioning and the brain. In R. A. Boakes & M. S. Halliday (Eds.), *Inhibition and learning.* New York: Academic Press, 1972.

Eccles, J. C. *The neurophysiological basis of mind.* Oxford: Clarendon Press, 1953.

Enser, L. D. *A study of classical nictitating membrane conditioning in neodecorticate, hemidecorticate and thalamic rabbits.* Unpublished dissertation, University of Iowa, 1976.

Estes, W. K. Memory and conditioning. In F. J. McGuigan & D. B. Lumsden (Eds.), *Contemporary approaches to conditioning and learning.* Washington, D. C.: Winston, 1973.

Gabriel, M. Short-latency discriminative unit response: Engram or bias? *Physiological Psychology,* 1976, *4,* 275–280.

Gastaut, H. Some aspects of the neurophysiological basis of conditioned reflexes and behaviour. In G. E. W. Wolstenholme & C.M. O'Connor (Eds.), *Neurological basis of behaviour. Boston: Little, Brown, 1958.(a)*

Gastaut, H. *The role of the reticular formation in establishing conditioned reactions. In H. H. Jasper, L. D. Proctor, R. S. Knighton, W. C. Noshay, & R. T. Costello (Eds.), Reticular formation of the brain.* Boston: Little, Brown 1958.(b)

Goldberg, S. J., Hull, C. D., & Buchwald, N. A. Afferent projections in the abducens nerve: An intracellular study. *Brain Research,* 1974, *68,* 205–214.

Graybiel, A. M., & Hartweig, E. A. Some afferent connections of the oculomotor complex in the cat: An experimental study with tracer techniques. *Brain Research,* 1974, *81,* 543–551.

Grossberg, S. A neural model of attention, reinforcement and discrimination learning. In C. C. Pfeiffer & J. R. Smythies (Eds.), *International review of neurobiology* (Vol. 18). New York: Academic Press, 1975.

Groves, P. M., & Lynch, G. S. Mechanisms of habituation in the brain stem. *Psychological Review,* 1972, *79,* 237–244.

Halas, E. S., Beardsley, J. V., & Sandlie, M. E. Conditioned neuronal responses at various levels in conditioning paradigms. *Electroencephalography and Clinical Neurophysiology*, 1970, *28.*, 468–477.

Isaacson, R. L., & Pribram, K. H. (Eds.), *The hippocampus* (Vol. 2). *Neurophysiology and behavior*. New York: Plenum Press, 1976.

Iverson, S. D. Do hippocampal lesions produce amnesia in animals? In C. C. Pfeiffer & J. R. Smythies (Eds.), *International review of neurobiology* (Vol. 19). New York: Academic Press, 1976.

Kamin, L. J. Predictability, surprise, attention, and conditioning. In B. A. Campbell & R. M. Church (Eds.), *Punishment and aversive behavior*. New York: Appleton-Century-Crofts, 1969.

Konorski, J. *Conditioned reflexes and neuron organization*. Cambridge: Cambridge University Press, 1948.

Konorski, J. *Integrative activity of the brain: An interdisciplinary approach*. Chicago: University of Chicago Press, 1967.

Kuhar, M. J. Cholinergic neurons: Septal–hippocampal relationships. In R. L. Isaacson & K. H. Pribram (Eds.), *The hippocampus* (Vol. 1). *Structure and development*. New York: Plenum Press, 1976.

Lewis, P. R., & Shute, C. C. D. The cholinergic limbic system: Projections to the hippocampal formation, medial fornix, nuclei of the ascending cholinergic reticular system, and the sub-fornical organ and preoptic crest. *Brain*, 1967, *40*, 521–540.

Lindsley, D. B., & Wilson, C. L. Brain stem hypothalamic systems influencing hippocampal activity and behavior. In R. L. Isaacson & C. H. Pribram (Eds.), *The Hippocampus* (Vol. 2). *Neurophysiology and behavior*. New York: Plenum Press, 1976.

Lockhart, M., & Moore, J. W. Classical differential and operant conditioning in rabbits *(Oryctolagus cuniculus)* with septal lesions. *Journal of Comparative and Physiological Psychology*, 1975, *88*, 147–154.

Lorente de No, R. The interaction of the corneal reflex and vestibular nystagmus. *American Journal of Physiology*, 1933, *103*, 704–711.

Lubow, R. E. Latent inhibition. *Psychological Bulletin*, 1973, *79*, 398–407.

Maciewicz, R. J., Kaneko, C. R. S., Highstein, S. M., & Baker, R. Morphophysiological identification of interneurons in the oculomotor nucleus that project to the abducens nucleus in the cat. *Brain Research*, 1975, *96*, 60–65.

Mackintosh, N. J. A theory of attention: Variations in the associability of stimuli with reinforcement. *Psychological Review*, 1975, *82*, 276-298.

Maser, J. D., Dienst, F. T., & O'Neal, E. C. The acquisition of a Pavlovian conditioned response in septally damaged rabbits: Role of a competing response. *Physiological Psychology*, 1974, *2*, 133–136.

Marchant, H. G., III, Mis, F. W., & Moore, J. W. Conditioned inhibition of the rabbit's nictitating membrane response. *Journal of Experimental Psychology*, 1974, *95*, 408–411.

Marchant, H. R., III, & Moore, J. W. Blocking of the rabbits conditioned nictitating membrane response in Kamin's two-stage paradigm. *Journal of Experimental Psychology*, 1973, *101*, 155–158.

Mis, F. W. A midbrain-brain stem circuit for conditioned inhibition of the nictitating membrane response in the rabbit *(Oryctolagus cuniculus)*. *Journal of Comparative and Physiological Psychology*, 1977, *91*, 975–988.

Moore, J. W., Goodell, N. A., & Solomon, P. R. Central cholinergic blockade by scopolamine and habituation, classical conditioning, and latent inhibition of the rabbit's nictitating membrane response. *Physiological Psychology*, 1976, *4*, 395–399.

Nauta, W. J. H. Hippocampal projections and related neural pathways to the midbrain in the cat. *Brain*, 1958, *81*, 319–340.

Norman, R. J., Buchwald, J. S., & Villablanca, J. R. Classical conditioning with auditory discrimination of the eye blink in decerebrate cats. *Science,* 1977, *196,* 551-553.

Norman, R. J., Villablanca, R. R., Brown, K. A., Schwafel, J. A., & Buchwald, J. S. Classical eyeblink conditioning in the bilaterally hemispherectomized cat. *Experimental Neurology,* 1974, *44,* 363-380.

Oakley, D. A., & Russell, I. S. Neocortical lesions and Pavlovian conditioning. *Physiology and Behavior,* 1972, *8,* 915-926.

Oakley, D. A., & Russell, I. S. Differential and reversal conditioning in partially neodecorticate rabbits. *Physiology and Behavior,* 1974, *13,* 221-230.

Oakley, D. A., & Russell, I. S. Role of cortex in Pavlovian discrimination learning. *Physiology and Behavior,* 1975, *15,* 315-321.

Oakley, D. A., & Russell, I. S. Subcortical nature of Pavlovian differentiation in the rabbit. *Physiology and Behavior,* 1976, *17,* 947-954.

Oakley, D. A., & Russell, I. S. Subcortical storage of Pavlovian conditioning in the rabbit. *Physiology and Behavior,* 1977, *18,* 931-937.

Oleson, T. D., Ashe, J. H., & Weinberger, N. M. Modification of auditory and somatosensory systems activity during pupillary conditioning in the paralyzed cat. *Journal of Neurophysiology,* 1975, *38,* 1114-1139.

Patterson, M. M. Mechanisms of classical conditioning and fixation in spinal mammals. In A. H. Reisen & R.F. Thompson (Eds.), *Advances in psychobiology* (Vol. 3). New York: Wiley, 1976.

Patterson, M. M., Olah, J., & Clement, J. Classical nictitating membrane conditioning in the awake, normal, restrained cat. *Science,* 1977, *196,* 1124-1126.

Powell, D. A., Milligan, W. L., & Buchanan, S. L. Orienting and classical conditioning in the rabbit *(Oryctolagus cuniculus);* Effect of septal area lesions. *Physiology and Behavior,* 1976, *17,* 955-962.

Ramon-Molinar, E., & Nauta, W. J. H. The isodentritic core of the brain stem. *Journal of Comparative Neurology,* 1966, *126,* 311-335.

Reiss, S., & Wagner, A. R. CS habituation produces a "latent inhibition effect" but no active conditioned inhibition. *Learning and Motivation,* 1972, *3,* 237-245.

Rescorla, R. A. Summation and retardation tests of latent inhibition. *Journal of Comparative and Physiological Psychology,* 1971, *75,* 77-81.

Rosner, B. S. Recovery of function and localization of function in historical perspective. In D. G. Stein, J. J. Rosen, & N. Butters (Eds.), *Plasticity and recovery of function in the central nervous system.* New York: Academic Press, 1974.

Russell, I. S. Personal communication, 1977.

Salafia, W. R. Personal communication, 1977.

Salafia, W. R., Romano, A. G., Tynan, T., & Host, K. C. Disruption of rabbit *(Oryctolagus cuniculus)* nictitating membrane conditioning by posttrial electrical stimulation of hippocampus *Physiology and Behavior,* 1977, *18,* 207-212.

Scheibel, M. E., & Scheibel, A. B. Structural substrates for integrative patterns in the brain stem reticular core. In H. H. Jasper, L. D. Proctor, R. S. Knighton, W. C. Noshay, & R. T. Costello (Eds.), *Reticular formation of the brain.* Boston: Little, Brown, 1958.

Scheibel, M. E., & Scheibel, A. B. Anatomical basis of attention mechanisms in vertebrate brains. In G. C. Quarton, T. Melnechuk, & F. O. Schmitt (Eds.), *The neurosciences. A study program.* New York: Rockefeller University Press, 1967.

Scheibel, M. E., & Scheibel, A. B. Dendrites as neuronal couplers: The dendrite bundle. In M. Santini (Ed.), *Golgi centennial symposium: Perspectives in neurobiology.* New York: Raven Press, 1975.

Schmaltz, L. W., & Theios, J. Acquisition and extinction of a classically conditioned response in hippocampectomized rabbits *(Oryctolagus cuniculus). Journal of Comparative and Physiological Psychology,* 1972, *79,* 328-333.

Schmitt, F. O., Dev, P., & Smith, B. H. Electrotonic processing of information by brain cells. *Science,* 1976, *193,* 114-120.

Siegel, J. M., & McGinty, D. J. Pontine reticular formation neurons: Relationship of discharges to motor activity. *Science,* 1977, *196,* 678-680.

Solomon, P. R. Role of the hippocampus in blocking and conditioned inhibition of the rabbit's nictitating membrane response. *Journal of Comparative and Physiological Psychology,* 1977, *91,* 407-417.

Solomon, P. R., Kiney, C. A., & Scott, D. R. Disruption of latent inhibition following systemic administration of para-chlorophenylalanine. (PCPA). *Physiology and Behavior,* 1978, *20,* 265-272.

Solomon, P. R., & Moore, J. W. Latent inhibition and stimulus generalizaton of the classically conditioned nictitating membrane response in rabbits *(Oryctolagus cuniculus)* following dorsal hippocampal ablation. *Journal of Comparative and Physiological Psychology,* 1975, *89,* 1192-1203.

Suiter, R. D., & LoLordo, V. M. Blocking of inhibitory Pavlovian conditioning in the conditioned emotional response procedure. *Journal of Comparative and Physiological Psychology,* 1971, *76,* 137-144.

Thompson, R. F. The search for the engram. *American Psychologist,* 1976, *31,* 209-227.

Valverde, F. Reticular formation of the pons and medulla oblongata. A Golgi study. *Journal of Comparative Neurology,* 1961, *116,* 71-99.

Valverde, F. Reticular formation of the albino rats' brain stem. Cytoarchitecture and cortifugal connections. *Journal of Comparative Neurology,* 1962, *119,* 25-53.

Vertes, R. P., & Miller, N. E. Brain stem neurons that fire selectively to a conditioned stimulus for shock. *Brain Research,* 1976, *103,* 229-242.

Waxman, S. G. Integrative properties and design principles of axons. In C. C. Pfeiffer & J. R. Smythies (Eds.), *International review of neurobiology* (Vol. 18). New York: Academic Press, 1975.

Woody, C., Yarowsky, P., Owens, J., Black-Cleworth, P., & Crow, T. Effect of lesions of cortical motor areas on acquisition of conditioned eye blink in the cat. *Journal of Neurophysiology,* 1974, *37,* 385-394.

Young, R. A., Cegavske, C. F., & Thompson, R. F. Tone-induced changes in excitability of abducens motoneurons and of the reflex path of nictitating membrane response in rabbit *(Oryctolagus cuniculus). Journal of Comparative and Physiological Psychology,* 1976, *90,* 424-434.

6 Instrumental (Type II) Conditioning

N. J. Mackintosh
A. Dickinson
University of Sussex

I. INTRODUCTION

Conditioning experiments can be regarded as attempts to find out how animals learn about the relationships between different events in their environment. Konorski argued that such learning must involve the formation of associations or connections in the brain of the animal that reflect the associations that exist, or appear to the animal to exist, between the events in its environment. In the case of classical conditioning, therefore—where the experimenter arranges a relationship between events that occur independently of the animal's behavior, a CS and a reinforcer — Konorski (1948), more explicitly than Pavlov, assumed that connections were formed in the brain between internal representations of the CS and the reinforcer (see Hearst, Chapter 2). This remains the main theoretical account of classical conditioning to this day (Dickinson & Mackintosh, 1978; Mackintosh, 1974).

There is, of course, another important relationship that can operate in the animal's environment—namely, that between the animal's own behavior and some environmental event. The learning that results from exposure to such a relationship is typically studied in an instrumental conditioning experiment in which the experimenter arranges that the presentation of a reinforcer is contingent on the occurrence of a particular response.

Since the time of Thorndike, most Western research on conditioning has employed what are, by operational definitions at least, instrumental procedures. Mazes, runways, discrimination boxes, shuttle boxes, and operant chambers have been the preferred type of apparatus, and instrumental reward, punishment, and avoidance learning the preferred experimental paradigms. The last 10 to 15 years have witnessed a striking change of emphasis. New procedures for studying classical conditioning have been developed (Gormezano & Kehoe, 1975); responses long thought to provide paradigm examples of instrumental conditioning, such as key pecking in pigeons, are now recognized to be modifiable by classical contingencies alone (Hearst & Jenkins, 1974; Schwartz & Gamzu, 1977); theoretical analyses of the laws of associative learning have eschewed instrumental procedures because they are inadequately controlled (Wagner, 1969a) and are based almost exclusively on data from classical experiments (Rescorla & Wagner, 1972). Perhaps the single most important reason for this change of emphasis, however, is that we have no adequate theory of instrumental conditioning. In spite of considerable empirical and theoretical endeavor, there is still little general consensus about what associations are formed when an animal is exposed to an instrumental contingency, and how such learning is manifest in behavior. We believe that the early work of Konorski and Miller (Konorski, 1948, 1967; Konorski & Miller, 1937a, b; Miller & Konorski, 1928) contains some important clues to the nature of these associations.

Although we usually and rightly credit Thorndike with having been the first to undertake any systematic analysis of instrumental conditioning, Miller and Konorski (1928) must be acknowledged as the first investigators to recognize some of the important distinctions and similarities between instrumental and classical conditioning. Their experimental procedures were quite different from Thorndike's. Whereas Thorndike placed animals in the apparatus and left them free to discover the arbitrarily designated "correct" response by trial and error, Miller and Konorski exercised much stricter control over the relevant experimental contingencies by ensuring that the instrumental response (usually leg flexion) always occurred in the presence of a particular discriminative stimulus and that this compound was always followed by a reinforcer (which might be either appetitive or aversive). After a dog had been exposed to these contingencies, they discovered (1928), "a phenomenon will appear which is not predicted by Pavlov's theory (p. 1156)." If the reinforcer had been appetitive, the dog would start spontaneously lifting its leg as soon as the discriminative stimulus was presented; if the reinforcer, however, had been aversive, the presentation of the discriminative stimulus would inhibit the flexion response, and indeed appeared to elicit an antagonistic response.

The procedure satisfies the usually accepted operational definition of an instrumental experiment: The presentation of a reinforcer is contingent on the occurrence of a particular response (in the presence of a particular discriminative stimulus). The unusual feature of the experiment is that, initially at least, the experimenter controls the occurrence of the instrumental response — for example, by mechanically lifting the dog's leg. Although Skinner (1937) would disagree, we do not see any compelling reason why this might invalidate the procedure, and as we shall see, the ability of the experimenter to exercise some control over the performance of an instrumental response brings substantial advantages in experimental control.

II. THE DISTINCTION BETWEEN CLASSICAL AND INSTRUMENTAL CONDITIONING

Konorski and Miller distinguished between instrumental (Type II) and classical (Type I) conditioning, basically on the grounds that they could not see how the Pavlovian stimulus-substitution theory of conditioning could account for the changes in behavior seen in the instrumental experiment. There were two reasons for this. First, the conditioned response, or CR, (flexion in their experiments) was not related (in form) to the response elicited by the reinforcer, as it is in classical conditioning. Secondly, although appetitive and aversive reinforcers have similar effects in classical conditioning (both increasing the probability of a CR), in instrumental conditioning the probability of a CR is increased by appetitive reinforcers and decreased by aversive reinforcers. Skinner (1937) objected to this basis for the distinction on the grounds that it was tied to a particular theoretical position. According to Skinner (1937), the distinction between classical and instrumental conditioning "is solely in terms of the contingency of reinforcing stimuli—other properties of the types being deduced from the definition (p. 278)." There is much to be said for Skinner's view. As we shall see, however, although we might wish for a distinction that was not based on a particular theory of conditioning (which might well be wrong), in practice it is difficult to specify the effective contingency of reinforcement that actually controls behavior in any particular experiment.

A. Stimulus–Reinforcer and Response–Reinforcer Contingencies

We are all agreed on the operational distinction between classical and instrumental conditioning. In a classical conditioning experiment, the experimenter arranges a relationship between a stimulus (the CS) and a reinforcer, regardless of the subject's behavior; in an instrumental experiment, he arranges a relationship between the subject's behavior (usually in the presence of a particular stimulus — the discriminative stimulus) and a reinforcer. It seems clear to us that the question whether this operational distinction is of any theoretical importance must reduce to the question whether the subject's behavior can *in fact* sometimes be modified by stimulus–reinforcer relationships, and sometimes by response–reinforcer relationships. Unfortunately the question will never be answered simply by reference to the experimenter's operations, since we know that as soon as we arrange a stimulus–reinforcer relationship that develops a consistent response, a response–reinforcer relationship is created. Similarly, arranging a response–reinforcer relationship usually ensures that a particular stimulus is correlated with that reinforcer. For example, ever since Brown and Jenkins (1968) demonstrated that simply pairing a lighted key with food caused pigeons to peck the key, it has been impossible to assume automatically that the pecking observed in an instrumental experiment with pigeons was controlled by the response–reinforcer relationship.

In our judgment, however, there is now essentially incontrovertible evidence that the contingency between a CS and reinforcer alone is responsible for certain changes in behavior in certain conditioning experiments. No appeal to implicit response–reinforcer contingencies will explain why a pigeon will come to approach and direct pecks at a response key whose illumination signals the delivery of food, even though approaching and pecking cancels the food scheduled for delivery on that trial (Hearst, 1977; Schwartz & Williams, 1972); nor why a rabbit will learn a jaw movement CR to a CS signaling water, even though the delivery of water is cancelled on any trial on which a CR occurs (Gormezano & Hiller, 1972). Experiments employing aversive reinforcers yield similar conclusions (e.g., Coleman, 1975; Soltysik & Jaworska, 1962).

If these experiments establish that behavior is sometimes modified by classical stimulus–reinforcer contingencies, it might seem an equally easy matter to discover whether response–reinforcer contingencies can also affect behavior. In fact, however, the question is not so easily resolved. It is simple enough to show that the addition of a response–reinforcer contingency results in a particular change in behavior, but less easy to prove that the contingency is not working by producing a correlated change in the relationship between certain controlling *stimuli* and reinforcement. It is true, for example, that pigeons will come to peck a key whose illumination signals food, despite the omission contingency that cancels the delivery of food whenever a peck occurs. Nevertheless it is well established that the probability of pecking is in fact significantly less than it would have been in the absence of the omission contingency (Schwartz & Williams, 1972). We cannot, however, be certain that the relationship between the response of pecking and its consequences is responsible for this effect, because it is obvious that the addition of the omission contingency may have altered the stimulus–reinforcer relationship that we know to be capable of producing autoshaped key pecking. If the illumination of the key signals food only provided that the pigeon does not peck, then a relatively distant view of the key will be more highly correlated with food than will a closer view. As the pigeon approaches the key, therefore, the stimulus situation will change from one correlated with the presence of food, thus eliciting approach, to one correlated with the absence of food, thus eliciting withdrawal. If the difference between these sets of stimuli is sufficiently discriminable, the omission contingency will produce less approach and pecking than the strict classical contingency, but there would still be no reason to suppose that the contingency between pecking and food was directly responsible for this difference.

There is thus an asymmetry in the conclusions that may be drawn from an omission experiment. It will be easy to show that stimulus–reinforcer contingencies are important, if indeed they are; but it may be less easy to show that a subject's behavior is directly affected by a response–reinforcer contingency. The execution of most responses inevitably produces some correlated change in external stimuli: Pecking a key or pressing a lever is necessarily correlated with a close view of the key or lever. If a given response appears to be modified by its reinforcing consequences, how can we be certain that the effective contingency is not between these correlated changes in external stimuli and reinforcement? One answer, presumably, is to study a response whose execution produces little or no change in external stimuli. Thus it is certain that a pigeon's key peck can be directly affected by its consequences, since a contingency between pecking and food will maintain pecking in a totally dark box (Rudolph & Van Houten, 1977). Equally, Konorski and Miller's choice of leg flexion as an instrumental response has much to recommend it: It is difficult to see what changes in extroceptive stimuli are correlated with the flexion of one leg.

B. Principles of Reinforcement: law of effect and stimulus substitution

Skinner's analysis, therefore, of the distinction between classical and instrumental conditioning, although attractively simple in its freedom from theoretical preconceptions, may be hard to pin down. In the last analysis, we may have to follow Konorski and Miller in introducing more theoretical considerations. In fact, we have already surreptitiously done so, for we have implicitly allowed only certain changes in behavior to count as instances of instrumental conditioning. In effect we required that an animal respond so as to increase the probability of rewarding consequences or decrease the probability of punishing consequences (i.e., that behavior changes in accordance with the law of effect), before we would accept that instrumental conditioning had occurred. No restriction was placed on the ways in which classical conditioning could alter behavior. As soon as we impose such a restriction — that is, as soon as we propose a theory of how behavior is modified by stimulus-reinforcer contingencies — it becomes rather easier to show that certain changes in behavior are not produced by such stimulus contingencies because this *theory* of classical conditioning could not predict such changes.

In this paper we shall accept Pavlov's and Konorski's general theory of classical conditioning. If animals are exposed to a positive contingency between a CS and a reinforcer, they may establish an association between central representations of these events, such that activation of the CS representation (by presentation of the CS) also activates the representation of the reinforcer. Because activation of the latter normally elicits a certain set of responses, it follows that presentation of the CS will also elicit these responses. The theory is in two parts. It assumes first that an association is established between representations of events that mirrors the relationship between those events in the world; and secondly that this associative link is directly responsible for the observed change in the subject's behavior. Objections to the latter part of the theory, usually known as stimulus-substitution theory, have been well rehearsed by others (see Hearst, Chapter 2; Boakes, Chapter 9). Not all of these objections stand up to a close scrutiny (Mackintosh, 1974); most of those that do can, we believe, be reconciled with the theory if we allow that the nature of the response elicited both by a CS and by a reinforcer depends partly on the sensory properties of these events (Dickinson & Mackintosh, 1978).

We can now ask whether our theory of classical conditioning is really capable of explaining the results of all conditioning experiments, whether operationally classical or instrumental. We should not underestimate its success. In the hands of a skillful advocate, the theory does a remarkable job (Bindra, 1976; Moore, 1973). We might allow that pigeons approach and peck key lights signaling food because food elicits approach and pecking; that rats run down an alley to a goal box associated with food or with safety from shock because food elicits approach and danger withdrawal; even that rats press a lever for food because food may elicit not only approach but also manipulatory responses. We must certainly concede that relatively few supposedly instrumental responses can be unequivocally shown to lie outside the scope of a Pavlovian analysis. Nevertheless the arguments advanced by Konorski and Miller take on a new force. The Pavlovian analysis requires that the instrumental response either move the subject toward or away from stimuli associated with appetitive or aversive reinforcers, or resemble the responses elicited by those reinforcers. Flexion of one leg for food reward does not obviously satisfy either of these conditions, and the same claim can reasonably be made for a number of other responses: scratching or licking the body; running in a running wheel, and thus staying in the same place; making a right turn in a T-maze placed in a totally homogeneous environment; lever pressing with ex-afferent stimulation excluded (Mackintosh, 1974). Given the Pavlovian theory of classical conditioning, there is no way of explaining how certain changes in behavior could be produced by animals learning stimulus–reinforcer relationships alone. Unless we can propose a quite different theory of classical conditioning, the obvious implication is that these changes are a direct consequence of the relationship between the behavior and reinforcement. We need a theory of instrumental conditioning.

III. ASSOCIATIVE THEORIES OF
INSTRUMENTAL CONDITIONING

The dominant theory of instrumental conditioning among Western psychologists was first clearly enunciated by Thorndike and later taken over by Hull and Guthrie as an explanation of all conditioning, classical as well as instrumental. Stimulus–response effect theory asserts that if a response occurs in the presence of a particular stimulus and is followed by an appetitive reinforcer, the association between that stimulus and response will be strengthened. The theory resembles the Pavlovian theory of classical conditioning in one respect: The change in behavior observed in an instrumental experiment is assumed to be a direct consequence of the underlying associative structure. If conditioning is a matter of strengthening a stimulus–response association, it immediately follows that the presentation of that stimulus will, as a consequence of conditioning, be more likely to elicit the instrumental response. In other respects, however, the theory is radically different from the Pavlovian theory of classical conditioning, since it assigns a quite different role to the reinforcer. In the theory of classical conditioning, the reinforcer (or its immediate effects) is associated with an antecedent event; according to S–R theory, the reinforcer is a necessary condition for the formation of an association between two other events. The Pavlovian theory implies that the association established by the subject in a classical conditioning experiment reflects the relationships that actually exist in the environment; the experimenter arranges a contingency between a stimulus and a reinforcer, and the subject associates the two. In an instrumental experiment, however, although an environmental contingency exists between a response (in the context of a particular stimulus) and a reinforcer, S–R theory assumes that the subject associates the response with the antecedent stimulus. The important difference is that this association does not reflect the real contingencies in the world.

Although Konorski (1967) himself was a late convert, S–R effect theory has long fallen into disfavor among Western psychologists. Indeed a sustained critique of the theory is likely to be dismissed as a tedious exercise in beating a dead horse, even though the theory still provides the major context for the description and discussion of many instrumental experiments. A horse so thoroughly interred, one might imagine, would long since have been replaced by a fresh mount. In the West, however, the major opposition to S–R theory has come from those who interpret all conditioning as the learning of relationships between stimuli (Bindra, 1976; Deutsch, 1960; Mowrer, 1960; Tolman, 1932). Rather than an alternative theory of instrumental conditioning, these writers have been advocating what we should regard as a Pavlovian theory of all conditioning. Tolman and Deutsch, for example, argued that a rat rewarded for traversing the correct path through a maze learned the relationship between the stimuli of the true path and food encountered in the goal box, and that such learning established a series of subgoals, each of which attracted the rat in turn (elicited approach responses). We have no quarrel with this as an account of typical maze learning, but we should recognize, even if Tolman and Deutsch did not, that it assimilates such learning to a process of simple Pavlovian conditioning. It is not a theory of instrumental conditioning, since the rat's behavior is said to be modified, not because it learns that certain responses result in reinforcement and others do not, but because it learns that some stimuli encountered in the course of traversing the maze are correlated with reinforcement while others are not.

The major exception here is Skinner (1938). Although Skinner is an articulate opponent of S–R theory, his opposition was based on a quite different argument. For Skinner, of course, the defining feature of instrumental or operant behavior is that it is modified by its consequences. His objection was to S–R associationism. Operants for Skinner are precisely responses without eliciting stimuli. The relationship between an operant and the discriminative stimulus in whose presence it is reinforced is not one of association. Discriminative stimuli do not elicit operants; they "set the occasion" for them. We are not certain that we understand this theory. In any case, if we wish to analyze the associative structure of instrumental conditioning, there is little point in turning to an account that deliberately eschews associationism.

For serious consideration of an alternative *associative* theory of instrumental conditioning, we must return to Konorski and Miller's early work. As befits good followers of Pavlov, their analysis of instrumental or Type II conditioning followed, as far as possible, a Pavlovian line of reasoning. Consider a classical conditioning experiment, with two CSs, A and B, in which presentation of the AB compound signals food, while presentation of either A or B alone signals no food. We should expect that initially both A and B would be individually established as excitatory CSs, each eliciting salivary CRs, but that in due course only the AB compound would elicit a CR while the two components in isolation would become ineffective. The Type II experiment can be viewed in a similar light, with the discriminative stimulus (S^D) as CS A, and the flexion response as CS B. The response is elicited in the presence of S^D and is reinforced, but the S^D when presented alone is never followed by food; and if the subject responds during the intertrial interval, such flexion responses are also never reinforced. Because the compound is reinforced, both components initially become excitatory CSs. In due course, however, performance of the flexion response during the intertrial interval ceases, for it is no longer a signal for food, and presentation of S^D alone also fails to elicit a salivary CR; instead, according to Konorski (1967), the S^D elicits the flexion response "and thus completes the compound CS (p. 364)."

The associative parallel with a Pavlovian analysis of classical or Type I conditioning is exact, and in each case the associations established directly reflect the environmental relationships arranged by the experimenter. The difference lies in the way in which these associations produce changes in behavior. In the classical case, the association between CS and reinforcer ensures that the CS will elicit responses normally elicited by the reinforcer. But how does a set of associations between an S^D-response compound and a reinforcer cause the S^D to evoke the response? The traditions of reflex theory required Konorski and Miller to treat a response made in the presence of a discriminative stimulus as a conditioned reflex, with the stimulus serving as CS and the instrumental response as CR. In addition to the associations established with the reinforcer, therefore, which paralleled the CS–reinforcer associations established in a classical experiment, they assumed that Type II conditioning involved the formation of an additional association between the S^D and instrumental response. As Konorski (1967) later noted, their theory of instrumental conditioning now "consisted of two rather independent but mutually complementary ideas (p. 393)." They are certainly independent; whether they are mutually complementary seems to us open to question. We can accept that a complete account of instrumental conditioning will require more than the postulation of associative links between instrumental response and reinforcer; the problem, however, is one of translating such an association into performance, and it is not obvious that this should be solved by postulating a new associative link between S^D and response. This aspect of the theory takes us back to an S–R position, with its attendent implication that performance of the instrumental response is not directly controlled by the response–reinforcer association. In what follows we ignore for the time being the problem of response evocation and concentrate on what seems to us the novel and important feature of the Konorski–Miller theory, the assumption that associative links are established between the response and the reinforcer that reflect the actual contingencies arranged by the experimenter.

IV. THE ROLE OF RESPONSE–REINFORCER ASSOCIATIONS IN INSTRUMENTAL CONDITIONING

Several lines of evidence are relevant to the question whether instrumental conditioning involves the estabishment of associations between response and reinforcer. We examine this evidence in ascending order of importance.

A. Responses as stimuli

If animals are to associate their behavior with its consequences in the same way that they associate other events with their consequences in classical conditioning experiments, then it must be possible to show that responses share other properties of such events in their ability to enter into associations controlling behavior. There is substantial experimental evidence establishing this point (e.g., Buchman & Zeiler, 1975; Rilling, 1967). In Buchman and Zeiler's experiment, pigeons were trained on a chained schedule to peck a key that was alternately red and blue. When the key was red, the first response after 3 min. turned the key blue for 30 sec. Whether or not responses to the blue key were reinforced depended on the rate at which the pigeon had pecked in the preceding red component. Appropriate variations in the rate of responding to the blue key showed that subjects had learned the discriminative significance of their own rate of responding to the red key.

Beninger, Kendall, and Vanderwolf (1974) and Morgan and Nicholas (in preparation) have similarly shown that rats can use differences between such patterns of behavior as washing, rearing, or scratching as the cue to signal which of several alternative responses will be reinforced. In Morgan and Nicholas's experiment, for example, rats learned that pressing one of two levers would be reinforced if they had been washing at the moment when the two retractable levers were inserted into the chamber, whereas pressing the other would be reinforced if they had been rearing.

B. Response as classical CSs

If, as these experiments establish, an animal's own responses can serve as discriminative stimuli signaling which further responses will be reinforced, it seems probable that responses could also serve as CSs directly associated with reinforcers, and would thus come to elicit classical CRs. Konorski and Miller were the first to report just this effect: During the course of instrumental conditioning of the leg-flexion response for food reward, they observed that performance of the response was accompanied by salivation. The flexion response, therefore, came to act as a classical CS; there was, indeed, a close parallel between the initiation of active flexion responses (i.e., successful instrumental conditioning) and the first appearance of a classical CR correlated with the response, and a similar parallel between the decline of instrumental responding and the disappearance of the salivary CR during the course of extinction. A similar parallel was observed in instrumental-avoidance conditioning: If acid was injected into the dog's mouth in the presence of a CS but omitted if the CS was accompanied by flexion of the leg, the CS alone elicited a salivary CR; but the performance of the flexion response in the presence of the CS inhibited this classical CR.

These early results clearly suggest that associations (both excitatory and inhibitory) are formed between a response and a reinforcer during the course of instrumental conditioning. The alternative view of the parallel is that the classical responses represent motivational processes conditioned to environmental stimuli that serve to motivate the instrumental response

(Rescorla & Solomon, 1967). The idea that the classical CR reflects response–reinforcer associations suggests that, if anything, the classical CRs should follow the instrumental behavior whereas the motivational view supposes that they should precede the instrumental response. Most of the studies that have attempted to analyze this temporal relationship by using a long chain of instrumental responding (e.g., Deaux & Patten, 1964; Ellison & Konorski, 1964; Williams, 1965) have come out against the motivational viewpoint by showing that the classical CRs occur only late in the chain. Clearly this pattern is what would be expected if the instrumental chain itself acted as a prolonged classical CS.

C. Necessity of response–reinforcer associations for successful instrumental conditioning

Even if the parallel between instrumental responding and classical CRs were perfect, we could conclude only that an association between response and reinforcer was in fact formed during the course of instrumental conditioning. It would not follow that such an association was responsible for the emergence of instrumental responding. Response–reinforcer associations might be incidental rather than central to the development of instrumental conditioning. In order to show they were a necessary ingredient of successful instrumental conditioning, we should have to show that conditioning would fail to occur if we had prevented the establishment of the appropriate association between response and reinforcement.

1. Konorski and Miller's Experiment

Once again, the necessary experiments were first undertaken by Konorski and Miller, although in this instance the experiments were designed for a different purpose, and Konorski and Miller's own interpretation differed from the one we offer. In their standard instrumental paradigm, elicitation of the flexion response in the presence of S^D was followed by reinforcement, but neither S^D alone nor the response performed in the absence of S^D was reinforced. In order to test a particular theoretical analysis of their data proposed by Pavlov (Konorski, 1967, p. 361), Konorski and Miller asked what would happen if the nonreinforced trials to S^D alone were omitted. The answer was that, although the dog might occasionally perform the flexion response during intertrial intervals, such nonreinforced responses rapidly extinguished, and the animal never learned to respond in the presence of S^D. Successful instrumental conditioning, it appeared, required exposure not only to the contingency between reinforcement and the joint occurrence of S^D and responding, but also to the contingency between S^D alone and nonreinforcement.

Contemporary theories of conditioning suggest a ready explanation of this finding. Numerous studies of classical conditioning have established that conditioning to one element of a compound CS depends not only on its own salience, contiguity, and correlation with reinforcement, but also on the salience and relationship to reinforcement of other elements of the compound CS. The presence of a more salient or more valid signal for reinforcement may reduce or "overshadow" conditioning to a less salient or valid stimulus (e.g., Kamin, 1969; Wagner, 1969b). Konorski and Miller's results are readily interpreted as an instance of overshadowing. Reinforcement is predicted by the joint occurrence of the S^D and the flexion response. Either because the response is the less salient of the two, or because (since intertrial responses go unreinforced) it is a less valid predictor of reinforcement, it may be overshadowed by the S^D—unless the validity of the S^D is itself reduced by scheduling nonreinforced trials to S^D alone. Only when animals are exposed to the contingency between S^D alone and nonreinforcement, therefore, will they learn the association between the flexion response and reinforcement, and thus show evidence of successful instrumental conditioning. Parenthetically, it is worth noting that it was Konorski and Miller's procedure of eliciting the target instrumental response themselves that enabled them to make this discovery. In a normal instrumental experiment, the experimenter does not deliver reinforcement in the presence of S^D until the subject chooses to perform the instrumental response. The subject is thus automatically exposed to the two sides of the contingency: S^D alone signals nonreinforcement; S^D plus response signals reinforcement.

2. Subsequent Replications

Of all the discoveries made by Konorski and Miller in their pioneering work on instrumental conditioning, this surely must count as one of the most intriguing. Until very recently, however, apart from a related study by Konorski and Wyrwicka (1950), there have been very few attempts to pursue this question further. Williams (1975) has shown that pigeons will fail to learn to peck an illuminated key for food if the delivery of food is delayed several seconds after the required key peck and is immediately preceded by a second signal. The implication is that the occurrence of the second signal overshadowed learning of the contingency between pecking and food; but given the importance of classical, stimulus–reinforcer contingencies in the control of key pecking by pigeons (Hearst & Jenkins, 1974), it is possible that Williams's data are simply another instance of overshadowing effects in classical conditioning.

Two other sets of experiments have reported apparent overshadowing of the contingency between lever pressing and reinforcement in rats. St. Claire Smith (cited by Mackintosh, 1974) showed that rats exposed to a schedule of intermittent punishment for lever pressing would fail to suppress responding (i.e., fail to learn the punishment contingency) if the occurrence of each punished response was accompanied by a visual or auditory signal. Since the signal occurred only on each punished trial, it was a more valid predictor of punishment than was the response of pressing the lever, and should thus have overshadowed learning of the response–reinforcer contingency. More recently, Pearce and Hall (in press) have obtained results suggesting that the contingency between lever pressing and food may be overshadowed by the provision of an external stimulus predicting the occurrence of food. They trained rats to press a lever for food on a variable-interval schedule with a delay of reward of 0.5 sec. Control animals, for whom no event occurred in this 0.5-sec. interval between lever press and reward, learned to respond rapidly despite the delay. Experimental animals, however, learned to lever press only slowly, if a brief visual stimulus immediately followed each rewarded response, thus acting as a competing signal for the occurrence of reinforcement.[1]

3. Some New Data

Two features of Konorski and Miller's own procedure, it seemed to us, were of special relevance to any systematic attempt to pursue their original observations. First, they were surely studying an unequivocal instance of instrumental conditioning: No appeal to implicit stimulus–reinforcer contingencies will explain why dogs should learn to flex one leg in order to obtain food. Secondly, Konorski and Miller were able to control the occurrence of the response and thus manipulate its relationship to reinforcement with the same freedom and precision as is granted the experimenter in studies of classical conditioning. We have attempted to capture these features by training rats to run in a running wheel for food. First, because the response of running does not move the animal in space, it is difficult to argue that subjects will be simply learning, by a process of classical conditioning, to approach stimuli associated with the delivery of food. Moreover, food was delivered to a magazine opening in one wall at the bottom of the wheel, and the rat did not therefore have to run to approach the site of reinforcement. Secondly, since we could prevent animals from running by locking the wheel, or force them to run by driving the wheel from an external motor, we hoped to achieve a similar control over our target response as Konorski and Miller achieved over their dogs' flexion response. Another possibly important parallel was that just as Konorski and Miller's dogs could voluntarily perform a leg-flexion response at the same time as the experimenter was eliciting the response by passive flexion of the leg, so in our preparation the wheel, when driven by the external motor, revolved relatively slowly (6 r.p.m.), and the rat could (and did) override the motor by running faster as soon as the motor was engaged.

In all our experiments, therefore, the wheel could be in 1 of 3 states: locked, driven, and free. It remained locked throughout intertrial intervals; unlike Konorski and Miller, we chose not to permit our subjects to perform

---

[1]St. Claire Smith (personal communication) has independently obtained the same result.

the target response freely during the intertrial interval. On training trials, the wheel was driven by the motor for a fixed period of time, at the end of which 2 45-mg pellets were delivered into the food magazine. The rate at which the wheel actually rotated could be measured on both training and test trials, and the question of interest was whether the speed at which rats ran on test trials was affected by experimental procedures designed to overshadow the contingency between running and food on training trials.

The design of our first experiment is shown in Table 6.1. There were 3 groups — each containing 4 subjects — designated T, T+, and T−. All 3 groups were exposed to a positive contingency between food and running in the presence of a tone. On 9 trials each day, a 3000-Hz tone was turned on, and the wheel was driven for 15 sec.; as soon as the wheel had completed a further sixth of a revolution after this 15-sec. period had elapsed, the tone terminated, the wheel was locked, and food was delivered. On 1 trial each day, the procedure was exactly the same except that the wheel was unlocked and free, rather than driven by the external motor. For Group T, there were no other trials, but for Groups T+ and T−, there were 10 additional trials each day on which the wheel remained locked but the tone was sounded. For Group T+ these trials ended in reinforcement; for Group T− the tone terminated without food. After 25 days, a second test trial was added each

TABLE 6.1
Design of Experiment Replicating Konorski & Miller

Groups	Training Trials		Test Trials	
			Days 1–30	*Days 26–30*
T	⌈ Tone ⎰ ⟶ Food ⌊ Forced Run		⌈ Tone ⎰ ⟶ Food ⌊ Free Run	Free Run ⟶ Food
T+	⌈ Tone ⎰ ⟶ Food ⌊ Forced Run Tone ⟶ Food		⌈ Tone ⎰ ⟶ Food ⌊ Free Run	Free Run ⟶ Food
T−	⌈ Tone ⎰ ⟶ Food ⌊ Forced Run Tone ⟶ No Food		⌈ Tone ⎰ ⟶ Food ⌊ Free Run	Free Run ⟶ Food

day (in place of 1 of the forced training trials), on which the wheel was unlocked and food delivered after 15 sec, but no tone was presented.

The treatments given to the three groups were designed to vary the validity of the tone as a predictor of food. In Group T the tone was as valid as, but no more than, the running response; in Group T+ it was more valid; and in Group T− it was less valid. We thus expected that any tendency for the tone to overshadow instrumental conditioning might be enhanced in Group T+, but attenuated in Group T−. The design of our experiment is in fact exactly the same as that employed by Wagner (1969b) in a study of relative validity and the acquisition of stimulus control. Wagner's rats were all rewarded for pressing a lever in the presence of a tone–light compound, but differed in the treatment they received in the presence of the tone alone. For Group T+ the tone alone signaled that lever pressing would be reinforced; and for Group T− the tone signaled nonreinforcement; while Group T received no trials to the tone alone.

Figures 6.1 and 6.2 show that both Wagner and we obtained essentially the same results. In Wagner's experiment, the degree of control over responding acquired by the light, as measured on test trials to the light alone, varied inversely with the validity of the tone: Compared with Group T, Group T+ showed less and Group T− more control by the light. In our experiment, speed of running was measured both on test trials with the tone present and on trials with the tone absent, but this did not affect the outcome. In either case, there was a significant difference between the three groups, with Group T− running significantly faster than either Group T or Group T+ (p.<.05), and Group T+ running somewhat (but not significantly) slower than Group T. Given the modest number of subjects, the results seemed

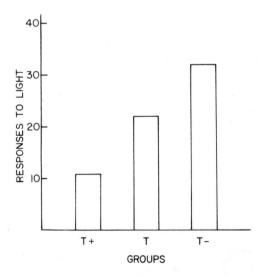

FIG. 6.1.   Control over responding acquired by a light, reinforced in conjunction with a tone as a function of the validity of the tone as a predictor of reinforcement. The number of responses to the light on test trials to the light alone. From Wagner, (1969b).

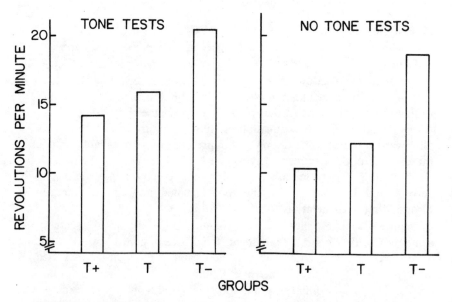

FIG. 6.2.   Speed of running on test trials when the wheel was unlocked. The left panel shows data from the last 10 test trials when the tone was presented, and the right panel the data from the 5 test trials without the tone.

reasonably encouraging. The extent to which rats ran on test trials, whether the tone was on or not, was inversely related to the relative validity of the tone and of running as predictors of food.

If we had given test trials only in the absence of the tone, one might have argued that the poor performance of Groups T and T+ in comparison with Group T− reflected a failure of performance rather than a failure to learn the response–reinforcer contingency. This in fact was the interpretation advanced by Konorski and Miller for their own results. They assumed that the contingency between the flexion response and food would normally be sufficient to establish an association between the two, but that the S^D would elicit the flexion response only if the compound of S^D and response were a better predictor of food than the S^D alone. In the absence of nonreinforced trials to S^D alone, it would be a fully adequate CS in its own right, and nothing would be gained by adding the response to the S^D. In the terminology of contemporary work on stimulus control (Mackintosh, 1977), they were invoking a process of masking rather than of overshadowing. Masking is said to occur if behavior is so strongly controlled by one stimulus that the control gained by a second cannot be detected on test trials in the presence of the first. In the case of overshadowing, the assumption is that the presence of the first stimulus has prevented the second from acquiring control in the first place, so that even if testing were conducted in the absence of the first stimulus, one would never detect control by the second.

There is, of course, nothing to prevent both masking and overshadowing from occurring in the same experiment (e.g., Farthing, 1972). We certainly do not wish to rule out the sort of process envisaged by Konorski and Miller. In fact, Konorski and Miller provided good independent evidence for such a process when they demonstrated that presenting a CS that had previously been paired with food, while the animals were performing an instrumental response for food, suppressed the instrumental response, a finding rediscovered in the West over 30 years later (Azrin & Hake, 1969; Boakes, Chapter 9; Morgan, Chapter 7). The conditions for translating a learned response–reinforcer association into performance may well be that the execution of the response in a particular context should increase the subject's spatial and/or temporal proximity to appetitive reinforcement. If, therefore, the context already includes a reliable predictor of food, it is possible that the response will not be executed. The differences between the groups in our experiment, however, cannot be simply due to such a masking effect: It was not the actual presence of the tone on test trials that prevented Groups T and T + from running as rapidly as Group T − ; since they also ran less rapidly in the absence of the tone, they must have learned the relationship between running and food less successfully.

The best single demonstration of the role of relative validity in overshadowing in classical conditioning remains the series of experiments by Wagner, Logan, Haberlandt, and Price (1968). In these experiments, animals were run under either "correlated" or "uncorrelated" conditions. In both cases, they received conditioning trials to a compound CS consisting of a light and one of two discriminably different tones, T_1 and T_2; and in both cases, 50% of trials were reinforced and 50% not reinforced. For the correlated group, reinforced trials were signaled by T_1 and nonreinforced trials by T_2; for the uncorrelated group, both T_1 and T_2 were equally often associated with reinforcement and nonreinforcement. Both groups, therefore, received exactly the same sequence of stimulus presentations and exactly the same frequency of reinforced and nonreinforced trials, and for both the light was associated with reinforcement on 50% of trials. The only difference was that the tones provided a better predictor of the outcome of each trial than the light for the correlated group but not for the uncorrelated group. When animals were tested with the light alone, the uncorrelated group showed reliable conditioning to the light, the correlated group little or none.

We followed this experimental design with forced running standing in the same relationship to reinforcement as did the light, and a tone and a clicker substituting for T_1 and T_2. Table 6.2 illustrates our procedure. During training there were 20 trials each day, 10 to each stimulus. On 9 of the 10 trials to each stimulus, the wheel was driven by the external motor, and on the remaining test trial the wheel was simply unlocked and the rat free to run.

TABLE 6.2
Design of Relative Validity Experiment
(The relationship of Tone and Clicker to Food
in the Correlated group was counterbalanced)

Groups	Training Trials	Test Trials	
		Days 1–30	Days 26–60
Correlated	Tone ⟶ Food, Forced Run	Tone ⟶ Food, Free Run	Free Run ⟶ 50% Food
	Clicker ⟶ No Food, Forced Run	Clicker ⟶ No Food, Free Run	
Uncorrelated	Tone ⟶ 50% Food, Forced Run	Tone ⟶ 50% Food, Free Run	Free Run ⟶ 50% Food
	Clicker ⟶ 50% Food, Forced Run	Clicker ⟶ 50% Food, Free Run	

After 25 days, the auditory stimuli were omitted on the 2 test trials of each day.

The results are shown in Fig. 6.3. Both on forced-training trials and on test trials with the tone and clicker present, the uncorrelated group ran faster than the correlated group ($p < .05$). Unfortunately, although the difference continued through the final series of test trials in the absence of tone and clicker, it fell short of significance on these trials. The results may not be definitive, therefore, but they suggest that just as Wagner et al.'s experimental subjects failed to attribute the occurrence of reinforcement to the light when the tones provided better information about the outcome of each trial, so here, our rats were less likely to attribute the occurrence of food to the fact that they had been running if the outcome of each trial was better predicted by the nature of the auditory stimulus on that trial. A second feature of these results is worth noting. Both during training and on test trials with the auditory stimuli present, the correlated group ran less rapidly on reinforced trials than on unreinforced trials ($p < .05$). Thus the presence of a reliable CS+ suppressed performance of the running response. This finding is, of course, reminiscent of the observation originally reported by Konorski and Miller and subsequently confirmed by Ellison and Konorski (1964), that the presentation of CS+ to a dog engaged in instrumental responding would normally suppress that response.

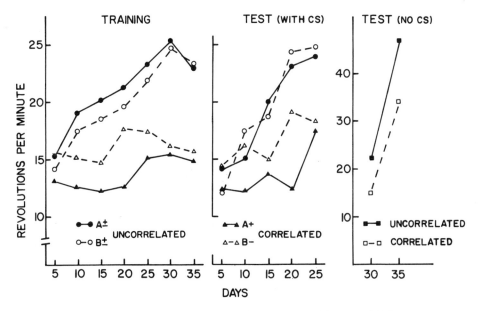

FIG. 6.3.   Speed of running on forced-training trials and free-test trials for correlated and uncorrelated groups throughout the experiment. Note the different scale for the ordinate of the right panel, which shows the final series of test trials without tone or clicker.

4. Conclusions

The evidence we have presented provides considerable encouragement for the view first seriously suggested by Konorski and Miller that successful instrumental conditioning depends on the establishment of an association between representations of the instrumental response and of the reinforcer. The evidence may not be particularly extensive, but it seems to us persuasive. The theory has the virtue of stressing the parallels between the associative learning underlying classical and instrumental conditioning. In both cases, organisms are assumed to associate those events that are, actually, related to one another in the real world, and there is every reason to believe that the associative laws that have been worked out for classical conditioning (e.g., the role of relative proximity and validity) will be found to apply equally to the establishment of response–reinforcer associations in instrumental conditioning. Moreover, as Konorski (1967) has noted, the theory provides an interesting and plausible explanation of differences in the conditionability of certain instrumental responses (see Shettleworth, Chapter 15). Just as the probability of successful classical conditioning to a particular CS depends on the salience of the CS relative to that of other potential signals for the reinforcer, so the probability of successful instrumental conditioning is likely to depend on the salience of the instrumental response relative to that of other potential signals. As Konorski wrote (1967):

> We thought that the inability of the organism to manipulate autonomic responses in an instrumental way depended on the fact that as a rule these responses were lacking feedback and in consequence their performance was not reported to the brain and could not become the type I CS [p. 367].

V.   INSTRUMENTAL PERFORMANCE

It is just as well that we can marshal good reasons for accepting that an association between response and reinforcer constitutes the knowledge on which instrumental performance depends, for there is no gainsaying the serious problem that confronts such a theory of instrumental conditioning. The problem is to see how such an association can produce a change in behavior.

To understand the nature of the problem, we must consider what it means to say that an association is established between the representations of any set of events. Associationism is, of course, a doctrine of venerable antiquity, but there has been little change in the central meaning assigned to the concept of an association. Thus Hartley, writing more than 200 years ago, could say (1749):

> Any sensations, A, B, C, etc., by being associated with one another a sufficient number of times, get such a power over the corresponding ideas, a, b, c, etc., that any one of the sensations, A, when impressed alone, shall be able to excite in the mind b, c, etc., the ideas of the rest [Proposition X].

This corresponds rather closely to the notion of an association inherent in Pavlov's and Konorski's theory of classical conditioning: An appropriate relationship between two events will ensure that the presentation of one of those events activates (excites in the mind) the representation of the other. The association is an excitatory link between two representations.

This seems to us adequate as a theory of classical conditioning: If CS and reinforcer are associated, the presentation of the CS activates the representation of the reinforcer and consequently elicits responses normally elicited by the actual presentation of the reinforcer. The association is sufficient by itself to explain the observed change in the organism's behavior. But no such association between response and reinforcer will produce the changes in behavior that we wish to explain in instrumental conditioning. Once the dog has flexed his leg, he will perhaps salivate if the flexion response has been associated with food, but what we need to explain is why he flexes his leg.

A. S-R theory

As we have noted, Konorski and Miller solved the problem by assuming that, in addition to an association between response and reinforcer, instrumental conditioning depended on the establishment of an association between the discriminative stimulus and the instrumental response. Our main objection to the postulation of this S-R link is that it seems to make the response–reinforcer association redundant (see, however, Morgan, Chapter 7). If presentation of the S^D is by itself sufficient to elicit the instrumental response reinforced in its presence, what need is there of any association between the response and the reinforcer? Konorski himself seems to have accepted the force of this point, for in his later theorizing he drops the emphasis on the response–reinforcer association, and adopts an account of instrumental conditioning remarkably similar to that of Hull (1943). Instrumental conditioning is assumed to involve S-R associations that are strengthened by a subsequent decrease in drive (Konorski, 1967; Morgan, Chapter 7).

The problem is that such a theory cannot easily explain those data that suggest that instrumental conditioning fails to develop if the response–reinforcer association is overshadowed. Konorski's later explanation of his own earlier experiments depends on the assumption that a well-trained, consistently reinforced CS+ inhibits the drive state, whose subsequent reduction is a necessary condition for the strengthening of S-R associations (Konorski, 1967, pp. 404–409; Morgan, Chapter 7). Instrumental conditioning will fail to develop if the instrumental response is always elicited and reinforced in the presence of such a CS+; it requires that the stimulus in whose presence the response is elicited should be only an imperfect predictor of reinforcement, which will therefore elicit the drive CR rather than the drive-inhibiting consummatory CR. This distinction seems to us a difficult one to sustain. One implication, for example, is that responding would initially develop and only later be suppressed if the elicitation of the response was always accompanied by the presentation of a CS; for early in training the CS will elicit preparatory, drive CRs rather than consummatory, drive-inhibiting CRs. A glance at the results shown in Fig. 6.2 shows that we obtained no evidence of such an effect.

B. The Bidirectional hypothesis

An alternative to S-R Theory, that still preserves the notion of an association as an excitatory link, is the bidirectional hypothesis of Pavlov (1932) and Asratyan (1974). The basic assumption is that when any two events are paired, the association established between their representations can transmit excitation in either direction. When an instrumental response is reinforced in the presence of a particular S^D, associations will be established between S^D and reinforcer and between response and reinforcer. The S^D will in due course come to elicit the response because its presentation will activate (via a forward association) a representation of the reinforcer, which will then (via a backward association) activate a representation of the response. The theory preserves not only the basic Pavlovian assumption that associated events elicit one another's representations, but also the view that the associations established in conditioning reflect the contingencies manipulated by the experimenter. The only new assumption is that associations are bidirectional.

The most obvious evidence for the bidirectionality of associations is the phenomenon of backward conditioning. Although the reliability of backward conditioning could reasonably be disputed less than five years ago (Mackintosh, 1974), we should now perhaps accept its reality (Dickinson & Mackintosh, 1978). But it would be a mistake to confuse the process we are interested in—namely, the possibility that a subject can retrieve a representation of one event from another that regularly follows it — with the experimental effect called backward conditioning. That effect, after all, is at best very fragile, and there would be a certain perversity in appealing to a phenomenon so insubstantial in order to explain one as robust and reliable as the occurrence of successful instrumental conditioning.

Better evidence for the bidirectional hypothesis is provided by the type of result first reported in the West by Trapold (1970), but already familiar to Konorski (1967, pp. 444–448). In Trapold's experiments, rats were required to press one lever in the presence of one stimulus, and another in the presence of a second. When both responses were reinforced with the same reinforcer, the problem was very difficult, but the provision of different reinforcers for the two different responses made the problem considerably easier. The bidirectional hypothesis assumes that if the two responses are reinforced with different reinforcers, each discriminative stimulus will be perfectly correlated with its particular reinforcer, and the presentation of one S^D will elicit a representation of its correlated reinforcer, which will then elicit a representation of the appropriate response. If the same event is used to reinforce both responses, then each S^D will elicit a representation of this single reinforcer, and there will be no way of retrieving the correct response. If the backward association between response and reinforcer were the only means of retrieving the correct response, the problem would have been insoluble (which it was not); in general it would follow that animals could never learn to perform more than one instrumental response for the same reinforcer. Although we must therefore allow animals other ways of retrieving instrumental responses, Trapold's results certainly suggest that a backward scan from the reinforcer associated with that response may be one important route (see Morgan, Chapter 7).

There remains one critical problem for the bidirectional hypothesis. It predicts that once an association has been established between a response and a reinforcer, any stimulus that elicits a representation of that reinforcer will also elicit that response. As we have already noted, the presentation of a CS+ for food to an animal engaged in responding instrumentally for food will typically suppress rather than facilitate responding. Konorski (1967, pp. 407–409, 450) reports certain exceptions to this generalization. In an experiment by Wyrwicka (1952), for example, once an instrumental response had been established to an S^D, a small number of classical pairings of a new CS with the reinforcer might initially enable that CS to elicit the instrumental response; and after the instrumental response had been partially extinguished, reinforced presentation of the S^D alone might again elicit the response. Although these exceptions surely warrant further experimental analysis, one must have grave doubts about a theory that predicts that the exception will be the rule. It seems to us quite certain that activation of the representation of a reinforcer previously used to establish a particular instrumental response, either by presenting the reinforcer itself or by presenting a CS associated with it, is not normally a sufficient condition for eliciting that instrumental response (cf. Morgan, Chapter 7). To the extent that the bidirectional hypothesis predicts that such activation will automatically result in the performance of the response, we must reject the hypothesis. If backward associations from the reinforcer are important, their operation must be constrained by some other factor.

C. Associations as Propositions

We know of no other viable account of instrumental conditioning that preserves both the view that response–reinforcer associations are of central importance and the traditional view of associations as excitatory links. Since we believe that the first view is correct, perhaps we should consider rejecting the second. A diligent search through the annals of associationism suggests that alternative conceptions have been held. Thus Hobbes distinguished between unguided and regulated trains of thought, one example of the latter being "when of an effect imagined we seek the causes, or means that produce it: and this is common to man and beast (Hobbes, 1651)." Regulated trains, according to Hobbes (1651), are:

> regulated by some desire, and design. . . . From desire, ariseth the thought of some means we have seen produce the like of that which we aim at; and from the thought of that, the thought of means to that mean; and so on continually, till we come to some beginning within our own power [Part 1, Chapter 3].

An association between two events is some form of proposition or premise representing the relationship between them; an association between response and reinforcer in instrumental conditioning is represented as the means to an end (Tolman, 1932) or as an act-outcome expectancy (Irwin, 1971).

The point to note about a propositional view of associations is that when a proposition about the relationship between a response and reinforcer is combined with some other premise, the appropriate rules of imperative inference (Rescher, 1966) will permit the derivation of instructions for the execution of that response. If exposure to a contingency between wheel running and food establishes the propositional association "The response of wheel running produces food," and if a state of food deprivation engages an imperative premise of the form "Perform any response that will produce food," the rules of imperative inference will permit derivation of the instruction "Perform the response of wheel running."

We do not wish to elaborate the details of such an account, although we should note that both the proposition and the imperative premise would need to be stated more carefully if such a theory is to explain both how instrumental responding can be brought under the control of discriminative stimuli, and how a discriminative stimulus can elicit an instrumental response while a classical CS+ may suppress such a response. We are concerned here only to stress the distinction we have drawn between the theory of associations as propositions and the more commonly held view of associations as the means whereby the activation of the representation of one event will activate the representations of other events associated with the first.

One problem that arises if we stress this distinction is to see how classical conditioning should be conceptualized. To preserve a unified theory of association, we should have to suppose that the stimulus–reinforcer associations underlying classical conditioning were also propositional in form. But (despite our earlier remarks) it is perhaps no accident that Tolman (1932) had so little to say about classical conditioning, for there is surely little to recommend a propositional view of classical associations. The fact that a dog has learned the proposition "The bell is followed by food" or a pigeon has learned the proposition "The keylight is followed by food" in no way explains why the dog should start salivating when the bell is rung, or why the pigeon should start pecking the key. Nor will any obvious combination of these premises with the rules of imperative inference permit the derivation of instructions to salivate or peck the key. What we are saying here is that an expectancy theory of classical conditioning simply will not explain why classical CRs take the form they do. The association between CS and reinforcer in classical conditioning should not be conceptualized as an expectation or a proposition that produces changes in behavior only by combination with an imperative premise. The relationship between association and performance in classical conditioning is the much simpler one of traditional associative theories: When two events are associated, presentation of one activates a representation of the other. If this is accepted, it follows that there is an important difference between the ways in which we should think of classical and instrumental associations, or at least between the ways in which an association between stimulus and reinforcer generates classically conditioned behavior and the ways in which an association between response and reinforcer generates instrumentally conditioned behavior.

REFERENCES

Asratyan, E. A. Conditioned reflex theory and motivational behavior. *Acta Neurobiologiae Experimentalis,* 1974, *34,* 15–31.

Azrin, N. H., & Hake, D. F. Positive conditioned suppression: Conditioned suppression using positive reinforcers as the unconditioned stimuli. *Journal of the Experimental Analysis of Behavior,* 1969, *12,* 167–173.

Beninger, R. J., Kendall, S. B., & Vanderwolf, C. H. The ability of rats to discriminate their own behaviors. *Canadian Journal of Psychology,* 1974, *28,* 79–91.

Bindra, D. *A theory of intelligent behavior.* New York: Wiley, 1976.

Brown, P. L., & Jenkins, H. M. Auto-shaping of the pigeon's key-peck. *Journal of the Experimental Analysis of Behavior,* 1968, *11,* 1–8.

Buchman, I. B., & Zeiler, M. D. Stimulus properties of fixed-interval responses. *Journal of the Experimental Analysis of Behavior,* 1974, *24,* 369–375.

Coleman, S. R. Consequences of response-contingent change in unconditioned stimulus intensity upon the rabbit (*Oryctolagus cuniculus)* nictitating membrane response. *Journal of Comparative and Physiological Psychology,* 1975, *88,* 591–595.

Deaux, E. B., & Patten, R. L. Measurement of the anticipatory goal response in instrumental runway conditioning. *Psychonomic Science,* 1964, *1,* 357–358.

Deutsch, J. A. *The structural basis of behavior.* Cambridge: Cambridge University Press, 1960.

Dickinson, A., & Mackintosh, N. J. Classical conditioning in animals, *Annual Review of Psychology,* 1978, *29,* 587–612.

Ellison, G. D., & Konorski, J. Separation of the salivary and motor responses in instrumental conditioning. *Science,* 1964, *146,* 1071–1072.

Farthing, G. W. Overshadowing in the discrimination of successive compound stimuli. *Psychonomic Science,* 1972, *28,* 29–32.

Gormezano, I., & Hiller, G. W. Omission training of the jaw-movement response of the rabbit to a water US. *Psychonomic Science,* 1972, *29,* 276–278.

Gormezano, I., & Kehoe, E. J. Classical conditioning: Some methodological–conceptual issues. In W. K. Estes (Ed.), *Handbook of learning and cognitive processes* (Vol. 2). Hillsdale, N. J.: Lawrence Erlbaum Associates, 1975.

Hartley, D. *Observations on man, his frame, his duty, and his expectations.* London: 1749.

Hearst, E., & Jenkins, H. M. Sign tracking: The stimulus–reinforcer relation and directed action. *Monograph of the Psychonomic Society,* Austin, Texas, 1974.

Hobbes, T. *Leviathan.* London, 1651.

Hull, C. L. *Principles of behavior.* New York: Appleton-Century-Crofts, 1943.

Irwin, F. W. *Intentional behavior and motivation: A cognitive theory.* Philadelphia: Lippincott, 1971.

Kamin, L. J. Predictability, surprise, attention and conditioning. In B. Campbell & R. Church (Eds.), *Punishment and aversive behavior.* New York: Appleton-Century-Crofts, 1969.

Konorski, J. *Conditioned reflexes and neuron organization.* Cambridge: Cambridge University Press, 1948.

Konorski, J. *Integrative activity of the brain.* Chicago: University of Chicago Press, 1967.

Konorski, J., & Miller, S. On two types of conditioned reflex. *Journal of General Psychology,* 1937, *16,* 264–272.(a)

Konorski, J., & Miller, S. Further remarks on two types of conditioned reflex. *Journal of General Psychology,* 1937, *17,* 405–407.(b)

Konorski, J., & Wyrwicka, W. Research into conditioned reflexes of the second type. I. Transformation of conditioned reflexes of the first type into conditioned reflexes of the second type. *Acta Biologiae Experimentalis,* 1950, *15,* 193–204.

Mackintosh, N. J. *The psychology of animal learning.* London: Academic Press, 1974.

Mackintosh, N. J. Stimulus control: Attentional factors. In W. K. Konig & J. E. R. Staddon (Eds.), *Handbook of operant behavior.* Englewood Cliffs, N. J.: Prentice-Hall, 1977.

Miller, S., & Konorski, J. Sur une forme particuliere des reflexes conditionnels. *Comptes Rendus des Séances de la Société de Biologie,* 1928, *99,* 1155–1157.

Moore, B. R. The role of directed Pavlovian reactions in simple instrumental learning in the pigeon. In R. A. Hinde & J. Stevenson-Hinde (Eds.), *Constraints on learning.* London: Academic Press, 1973.

Morgan, M. J., & Nichols, D. J. Behaviour discrimination and instrumental conditioning of naturally occurring action patterns in the laboratory rat. *In preparation*

Mowrer, O. H. *Learning theory and behavior.* New York: Wiley, 1960.

Pavlov, I. P. The reply of a physiologist to psychologists. *Psychological Review,* 1932, *39,* 91–127.

Pearce, J. M., & Hall, G. Overshadowing of instrumental conditioning of a lever press response by a more valid predictor of reinforcement. *Journal of Experimental Psychology, Animal Behavior Processes. In Press.*

Rescorla, R. A., & Solomon, R. L. Two-process learning theory: Relationships between Pavlovian conditioning and instrumental learning. *Psychological Review,* 1967, *74,* 151–182.

Rescorla, R. A., & Wagner, A. R. A theory of Pavlovian conditioning: Variations in the effectiveness of reinforcement and nonreinforcement. In A. H. Black & W. F. Prokasy (Eds.), *Classical conditioning II: Current research and theory.* New York: Appleton-Century-Crofts, 1972.

Rescher, N. *The logic of commands.* London: Routledge and Kegan Paul, 1966.

Rilling, M. Number of responses as a stimulus in fixed interval and fixed ratio schedules. *Journal of Comparative and Physiological Psychology,* 1967, *63,* 60–65.

Rudolph, R. L., & Van Houten, R. Auditory stimulus control in pigeons: Jenkins and Harrison (1960) revisited. *Journal of the Experimental Analysis of Behavior,* 1977, *27,* 327–330.

Schwartz, B., & Gamzu, E. Pavlovian control of operant behavior: An analysis of autoshaping and its implications for operant conditioning. In W. K. Honig and J. E. R. Staddon (Eds.), *Handbook of operant behavior.* Englewood Cliffs, N. J.: Prentice-Hall, 1977.

Schwartz, B., & Williams, D. R. The role of the response–reinforcer contingency in negative automaintenance. *Journal of the Experimental Analysis of Behavior,* 1972, *17,* 351–357.

Skinner, B. F. Two types of conditioned reflex: A reply to Konorski and Miller. *Journal of General Psychology,* 1937, *16,* 272–279.

Skinner, B. F. *The behavior of organisms.* New York: Appleton-Century, 1938.

Soltysik, S., & Jaworska, K. Studies on the aversive classical conditioning. 2. On the reinforcing role of shock in the classical leg flexion conditioning. *Acta Biologiae Experimentalis,* 1962, *22,* 181–191.

Tolman, E. C. *Purposive behavior in animals and men.* New York: Appleton-Century, 1932.

Wagner, A. R. Incidental stimuli and discrimination learning. In R. M. Gilbert & N. S. Sutherland (Eds.), *Animal discrimination learning.* London: Academic Press, 1969.(a)

Wagner, A. R. Stimulus validity and stimulus selection in associative learning. In N. J. Mackintosh & W. K. Honig (Eds.), *Fundamental issues in associative learning.* Halifax: Dalhousie University Press, 1969. (b)

Wagner, A. R., Logan, F. A., Haberlandt, K., & Price, T. Stimulus selection in animal discrimination learning. *Journal of Experimental Psychology,* 1968, *76,* 171–180.

Williams, B. A. The blocking of reinforcement control. *Journal of the Experimental Analysis of Behavior,* 1975, *24,* 215–225.

Williams, D. R. Classical conditioning and incentive motivation. In W. F. Prokasy (Ed.), *Classical conditioning: A symposium.* New York: Appleton-Century-Crofts, 1965.

Wyrwicka, W. Studies on motor conditioned reflexes. V. On the mechanism of the motor conditioned reaction. *Acta Biologiae Experimentalis,* 1952, *16,* 131–137.

7 Motivational Processes

Michael Morgan
University of Durham

I. THE REPRESENTATIONAL THEORY OF MOTIVES

Suppose a dog has been trained in a given experimental situation to raise one of its legs in response to the sound of bubbling, for a food reward. There are a variety of measures that the experimenter can take to alter the strength of the behavior. For example, the dog will probably respond less if made less hungry; or if taken into a different situation; or if the bubbling is changed to a slightly different stimulus; or if for several trials the bubbling is not followed by food. Despite the similarity between the effects of these different operations, a tradition has grown up in most theories of learning of classifying only some of them as "motivational." The effect of food deprivation would certainly be "motivational." That of changing the training stimulus would be much less so. Concerning extinction there would be room for much argument. The distinction between the "motivational" and the nonmotivational causes of a response has sometimes been expressed by distinguishing the "energizing" effects of variables like food deprivation, from the "associative" effects of factors such as specific external stimuli (e.g., Hebb, 1949). Associative factors determine which of a defined set of responses will occur—for example, leg flexion or barking; "motivational" factors, on the other hand, are necessary to provide a fiat for any behavior to occur at all.

It is hard to put this idea into anything like a logically coherent form. The term "motivation" was invented for propaganda purposes by the metaphysician Schopenhauer,[1] and has invaded ordinary speech and psychological textbooks with somewhat different meanings. Most psychologists readily admit that the term is little more than an attractive chapter heading. And yet there is a persistent feeling that the action of a "motivational" variable, such as food deprivation, differs in some fundamental way from the mode of action of associative factors. In Konorski's model of learning, this distinction is to a certain extent maintained, but in addition he described a number of ways in which ostensibly "motivational" variables, such as food deprivation, affect response instigation in much the same way as do external stimuli. The model is a highly eclectic one, in which most of the mechanisms that have been contrasted so energetically by other learning theorists seem to find their place from time to time and to coexist perfectly happily. It is far from clear which of the several factors affecting response instigation in Konorski's model would be called "motivational" by universal consent, and which would not. The view that is developed in this chapter is that such a question of terminology is of little importance in the long run, and that attention to specifics is more valuable than arguing about whether the effect of a given variable should be considered "motivational" or not.

In several places Konorski used the scholastic term "spiritus movens" for the collection of factors involved in the instigation of behavior. For example (1967), "the *spiritus movens* for instrumental responses is provided by emotive factors or drives which not only enable the formation of particular Type II CR's but are also indispensable for their elicitation (p. 424)." Against the Hullian tradition, which saw responses as being selected by one mechanism (learning) and then "motivated" by a different mechanism, Konorski tended to see the problem as a unitary one of response selection. It has become customary, though not perhaps very helpful, to group together theories of this general kind as "incentive theories of motivation." Bindra very clearly describes the core of this position as follows (1976): "The action performed is not first selected and then instigated by the motivational influence. Rather the selection of the action and its instigation go hand in hand; the motivational influence is an intrinsic part of the response selection procedure (p. 195)." For Konorski, however, the factors involved in

---

[1]It is a surprise to discover that "Motivation" was originally a German and not an English word (die Motivation). The *Oxford English Dictionary* records its first use in a review of Schopenhauer's *"The fourfold root of the principle of sufficient reason."* The translator of that work into English, Hildebrand, describes in a translator's preface her perplexity with the word, which she finally decided to reproduce in its German spelling. Schopenhauer probably coined the word because it sounded like *die Gravitation;* he wanted to stress the idea that the causes of human behavior were similar to physical forces (see his *"Essay on the freedom of the will"* published in the Library of Liberal Arts by the Bobbs-Merrill Company).

response selection are very heterogeneous, which may explain why he used the general terms "motivation" and "motivational" so seldom, and why he indicated skepticism concerning their usefulness by enclosing them typically in quotation marks. It may be helpful, therefore, to provide a preliminary classification of the various factors that, in the model, are responsible for the selection of learned behavior (see Section II). But first, it is necessary to give an account of what is selected according to Konorski, for his theory differs fundamentally on this point from other incentive theories, such as Bindra's. Briefly, because it is not the main purpose of this chapter to go into detail about the ideo-motor theory, Konorski shared with William James and with Pavlov (see Section III, subsection C) the view that the antecedent of a voluntary movement is the activation of a central kinaesthetic representation of that movement. This representation is activated by performance of the movement itself, but can also be activated by connections with representations of external stimuli, and with drive centers. When it is activated in the last two ways, it produces the movement by its direct connections (centrifugal) with motor neurons. Thus the strength of some motor movement depends on the degree of activation of its central representation, *and this in turn depends on the activation of associated representations.*

The ideo-motor theory allows an elegant and powerful way of stating the important problems of motivation. If the strength of some observed movement depends on the degree of activation of its central kinaesthetic representation, how do factors such as food deprivation or quality of reward affect strength of responding? Ex hypothesi, it would seem, they must do so by affecting the degree to which the motor representation is activated by its associated representations. The problem is to see how they do so, and whether their mode of action differs in principle from that of other factors, such as exteroceptive stimuli, which also affect response strength.

There is an implication that the problem of motivation is closely related to that of memory. Konorski is quite specific that motivational states involve the activation of images, such as that of the smell of a particular food or the sight of a particular food tray. These images depend on activation of central "gnostic" units, which may be activated through associative pathways rather than by their usual input. It is the activation of a response image through its associations that produces voluntary movement. Using mnemonic terminology, the question becomes one of how motivational variables such as food deprivation affect the retrieval of response memories or images underlying performance of the relevant behavior. In the following account, I speak interchangeably of the "activation" and the "retrieval" of such memories, with no special implications in the two cases.

The general hypothesis that motives consist in the activation of internal representations of goal objects and of behavior is one with a long history[2] and with a great deal of obvious support from introspection. Konorski made a powerful plea for the usefulness of evidence derived from introspection on this point (1967):

> To give a simple example, if from the fact that a subject salivates in response to the conditioned stimulus heralding the presentation of food we infer that the corresponding nervous structures have entered into the functional connection... so I can as well infer the existence of this connection in my own brain from the fact that hearing from another room the sounds of a dinner being served I clearly visualize it in my imagination [p. 3].

Observations if this kind lead Konorski to the view that a major effect of motivational variables such as food deprivation is to determine the kind and the strength of relevant representations in the imagination. A statement of this position is as follows (1967):

> Thus, if we are hungry, the first thing to appear is the image of a table set with a meal toward which our appetite is directed. Under the influence of strong hunger the visual or auditory images connected with food may be converted into illusions or even hallucinations, according to the mechanism discussed in the preceding section. The analogous facts are easily observed under a strong and unfulfilled sexual drive, when a person to whom our drive is directed, or an indefinite person of the opposite sex, appears vividly, and sometimes even obsessively, in our visual imagination. Such visual images stop completely when our drive is satiated and even *if we try we cannot voluntarily evoke them* [p. 184, italics added].

Among recent learning theorists, Bindra has consistently maintained the importance of motivational states in activating representations of goal objects. In his most recent statement of this position (1976) he suggests that "the influence of motivational states is directed primarily at the central perceptual representations of the relevant goal (incentive) objects. It may be supposed that a central motive state, once generated by certain motivational properties of the object, excites the gnostic assemblies representing the important features of that object (p. 191)." On this view, the priming of stimulus representations by internal states such as hunger leads the animal to respond selectively to certain kinds of afferent stimulation. Konorski argued for a similar point of view (1967): "There is no doubt that in the state of hunger the excitability of the gustatory analyser is strongly increased, and the same is true of the olafactory analyser; a subject, who in the state of satiation, does not pay any attention to weak odours, does so when he is in a state of hunger (p. 23)."

---

[2]In the *Nicomachian Ethics,* for example, Aristotle remarks that "a living creature would not have appetite without imagination (phantasia)."

The purpose of this Section has been to suggest a particular way of analyzing the problems of motivation within the context of Konorski's general approach. If the strength of a behavior is determined, as the ideo-motor theory has it, by the strength of internal representations of that behavior, and in turn by the strength of representations with which the kinaesthetic image is associated, our task becomes that of classifying the factors determining the retrieval of such representations. In a given case, such as that of hunger, we need to know whether the mode of action is directly upon the motor representation, or indirectly through associated gnostic assemblies, which are important for retrieval of the motor represen-tation. As William James said (1891), *"We reach the heart of our enquiry into volition when we ask by what process it is that the thought of a given subject comes to prevail stably in the mind* (Vol. II, p. 561, original italics)."

II.   GENERAL SURVEY OF MOTIVATIONAL
MECHANISMS

A. General principles of response activation

Figure 7.1 is an attempted synthesis of the various factors described by Konorski as important in response instigation; it also shows some of the postulated interconnections between these factors, such as the inhibitory effect of US representation on the drive (D) center. The nearest diagram given by Konorski himself is in Chapter 10 of *Integrative Activity of the Brain* (p. 447). Fig. 7.1 will show that Konorski considered four classes of motivational influence on an instrumental reflex. In the later version of the model (1974) only two of these are given prominence, but they are all surveyed in this section for completeness. Konorski emphasized that the development of his views on this subject was mainly influenced by ex-perimental results and ideas advanced in the Nencki Institute by Wyrwicka and by Soltysik. Wyrwicka (1960) also gives a useful diagram of the dif-ferent instigating factors in instrumental conditioning. For clarity and con-venience I confine the present account to responding maintained by food reinforcement. Konorski had little to say about other kinds of positive rein-forcement, and his account of negative reinforcement does not differ much in principle from the food case. We might bear in mind, however, that there are likely to be very important differences between motivational systems such as hunger and sex (Hinde, 1970, Ch. 15); in particular, Beach (1956) has raised the possibility that the relative importance of external and inter-nal eliciting agents may be quite different in the various cases.

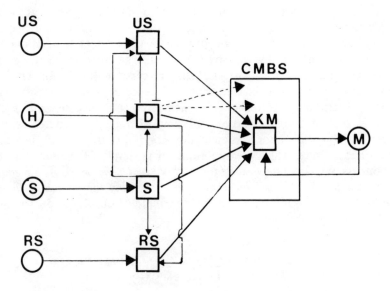

FIG. 7.1. Diagram of various factors affecting response instigation in Konorski's model. US, unconditional stimulus — e.g., food; H, hormonal factors; S, an external stimulus; D, drive center — e.g., of hunger; RS, releasing stimulus; CMBS, central motor behavioral system; KM, kinaesthetic-motor representation of a motor response, M.

We see that in Fig. 7.1 all the instigating influences exert their effect through activation of a motor representation (KM), which finds itself within a general "central motor behavioral system" (CMBS). Once a response representation has been sufficently activated, no further "motivational" factors intervene between it and the overt response: This is the logical core of any ideo-motor theory. Within this scheme there are the following four classes of factors affecting response (M) activation:

B. Direct effects of the Drive (D) center

A drive center is a neural mechanism promoting preparatory behavior relevant to a particular goal object, such as food. The behavior under its control is considered an unconditional reflex, and in the case of alimentary behavior, the unconditional drive reflex is termed *hunger*. From the physiological point of view, hunger may be considered a definite pattern of central-nervous processes manifested by particular efforts directed to procuring food and discontinued after food has been ingested ad libitum. The unconditional reflex is partly under the control of factors in the internal milieu, which are generally termed "hormonal," and these are in turn determined by food deprivation; but it is also affected by exteroceptive stimuli, of which the most important is the presence of food itself. Konorski considered that a small amount of food increases hunger after it has been ingested, and his observations in this point may be compared with Bruce's (1938) data on the facilitating efforts of a small quantity of water on rats running in a maze.

In addition to being elicited by internal (hormonal) factors and by unconditional stimuli connected with food, hunger can be conditioned and can thus be elicited in situations and by stimuli with which it has been previously associated. (See Bindra, 1976, pp. 197–205 for a critical discussion.) In Fig. 7.1, therefore, three distinct factors are shown as controlling the hunger drive: hormonal factors (H), food (the US), and external stimuli (S). In turn, the hunger drive in the model can activate the instrumental response in several ways, both direct and indirect. There are two direct effects. First, it has a nonspecific arousing influence on the whole motor system, which tends to potentiate any ongoing behavior. This influence is readily compared with the Hullian concept of "general drive." Second, there can be specific connections between the drive center and the motor response, formed conditionally because the response has been made at times when the drive center was active. The effect of such a connection is that whenever the animal is hungry it will tend to make the instrumental response. This effect of hunger is perhaps best compared with the hypothetical associative role of "drive stimuli" in Hullian theory.

The two direct effects of hunger are shown in a diagram by the D–M connection and by the dotted lines between D and the CMBS denoting general activation. The indirect effects of hunger are dealt with in Section II, subsection C.

C. Connections involving the US representation

The US center controls consummatory responding to food and is the center involved in classical conditioning of alimentary reflexes such as salivation.[3] The presentation of food after an instrumental response will activate the "food center," and as a consequence a connection will be formed between this center and the center for the instrumental response. On future occasions when the "food center" is activated, either directly by food or by a CS (in both cases with the help of the hunger drive), the associative connection

---

[3]Konorski identified the US center for food with the lateral hypothalamus, and the "off" center with the ventromedial hypothalamus. It has now become clear that this "two center" theory is much too simple to fit the physiological findings, for example, the loss of eating following lateral hypothalamic lesions is accompanied by a profound attentional loss that can impair many behaviors, such as behavioral thermoregulation (Satinoff & Shan, 1971).

with the instrumental response center will ensure that the relevant response is activated. This is a basic way in which "incentive stimuli" are held to activate instrumental responding, and it also provides an indirect mechanism for the activating effects of drive-deprivation states, through their facilitation of US centers (cf. Holland & Rescorla, 1975). The idea is similar to the one with which Pavlov (1932), in his "Reply of a Physiologist to Psychologists," attempted to rebut the criticism that his conditioned reflexes could not account for instrumental behavior. In some ways it also resembles the Hull–Spence r^g–s^g mechanism, in which a representation of the goal object becomes involved in elicitation of the instrumental response. However, a difference is that in the Hull–Spence mechanism, the relevant connection is not formed until the r^g comes to *precede* the instrumental response in time, whereas in Pavlov's theory it is formed despite the fact that the US representation is not activated until *after* the instrumental response has been made. It has seemed to some (see Mackintosh & Dickinson, Ch. 6) that a weakness of this scheme is that it relies on the un- proven notion of a "bidirectional connection" between centers, or as Astrayan (1974, p. 20) puts it, "a bidirectional conditioned connection with reciprocal conductivity." The idea of a bidirectional connection should not be confused with the logically distinct concept of "backward conditioning." The latter is a *procedure,* of at best limited potency in producing any kind of associative learning (review by Mackintosh, 1974, pp. 58–59), while a "bidirectional connection" is a hypothetical *construct* held to result from a normal forward conditioning procedure. The best evidence for bidirectional connections comes from human rote-learning experiments, where it has been shown that practice with A–B results in the association B–A, which can be a complication in transfer experiments (review by Ekstrand, 1966; Postman, 1971). The evidence from animal work, such as that of Asratyan, is weak (for criticisms see Soltysik, 1975); so the problem obviously requires more work. Konorski seems not to have had settled views on the question. In his last theoretical paper (1974), he stated that "in many cases connections be- tween the pairs of centers are bidirectional (p. 5); but in his 1967 account he did not evoke bidirectionality and presented a scheme for response activa- tion much more similar to the Hull–Spence $r^g s^g$ model, with additions from Guthrie's reinforcement theory (see Mackintosh & Dickinson, Ch. 6).

A complication in Konorski's model is that US centers, while having an associative role in response retrieval, are also held to have a generalized in- hibitory influence on responding (see below, p. 181). This is the reason that an inhibitory connection between US and D centers is shown in Fig. 7.1. Konorski proposed that a dog presented with food, or with a CS signal- ing food, becomes "quiescent" so that there is less likelihood of instrumen- tal responding. This rather unlikely hypothesis is considered in more detail in Section III, subsection A.

D. Direct sensorimotor connections

The mechanisms considered so far all involve the hunger drive in one way or another, for even the US representation cannot be fully activated in the absence of drive. Quite different in character are direct connections between exteroceptive stimuli and the response representation. When a dog, for example, is trained to lift its leg for food reward if, and only if, a tactile stimulus is applied to the appropriate leg, a direct connection between the tactile stimulus and the movement may be formed. Konorski used this mechanism to explain response differentiation with a common reinforcer, where evidently the US and D mechanisms cannot play a decisive role; and also to account for persistence of responding in the satiated animal. He called the latter state of affairs "resistance to satiation," and we consider the evidence for this idea more fully in Section IV. The idea of a direct connection is the logical equivalent of the S-R bond in American learning theory, except that the S-R bond is traditionally permitted to be efficacious only in the presence of a suitable activating "drive," which is not the case for Konorski. Also in Konorski's model the formation of direct S-R connections is highly constrained by biological factors, such that only certain privileged S-R connections are permitted (see Shettleworth, Ch. 15).

E. Surrogates of natural releasing stimuli

There is a fourth kind of response-elicitation mechanism to which Konorski gives little prominence, but which he suggests may be important in the instrumentalization of certain naturally occurring actions patterns (APs), such as face washing or scratching (see Shettleworth, Ch. 15). The idea is that the specific eliciting stimuli for such APs may come to be "hallucinated" or at least imagined by the animal as a result of conditioning. For evidence, Konorski quotes the informal observation that in the early stages of training to scratch with its hind leg, the animal also performs related but non-reinforced responses, such as rubbing with its ear against the wall or shaking its head.

Now that this preliminary survey is complete, we move on to a detailed consideration of the proposed mechanisms of response elicitation, beginning with the role of the hunger drive and US representation.

III. DRIVES AND US REPRESENTATIONS IN ACTIVATION

A. Nature of the evidence

The facts that led Konorski to include these "mediated" forms of activation in his model ("mediated" rather than "direct," as in straight sensorimotor connections) are substantially those that have traditionally led learning theorists to concede a role for "expectancies" in instrumental learning. Trapold and Overmier (1972) have reviewed recent evidence for what they call the "second learning system." Some of the facts consdered by Konorski himself are as follows:

1. Transfer of responding between discriminitive stimuli

In Experiment 23, Konorski (1967, p. 384) described a study in which an unrestrained dog was first taught to bark in the presence of a metronome. Then it was restrained on a stand in the same room and trained to lift its hind leg in the presence of a tone. Finally the metronome was presented on the stand: would this elicit the response trained to the metronome itself, or the response conditioned to the general experimental situation? In fact, the "stand response" (leg lifting) occurred, and this has considerable implications. Evidently the general experimental situation was far more important than the response selected. But in that case why does the metronome elicit any response at all on the stand? Konorski proposed that it does so through its connections to the D and US centers. Because these are common to the two responses, they can mediate the observed transfer of responding.

2. Response-independent reinforcement after extinction

After a dog has been extinguished by repeated presentations of a discriminative stimulus without food, it can be made to respond again by a few response-independent pairings of the stimulus with food (Wyrwicka, 1952). Konorski explained this by supposing that extinction involves breaking the S–US and S–D connections, while leaving the connections of US and D with the motor response intact. Thus as soon as S–US and S–D are restored by response-independent reinforcement, S will once more come to elicit R.

3. Effects of classical conditioning stimuli on instrumental responding.

It has sometimes been argued that the clearest demonstration of the activation of a response by an "expectancy" would be the elicitation of an instrumental response by a classical CS (Type I) with which it has never before been associated. This effect has indeed sometimes been observed (e.g., Estes, 1943, 1948), and Konorski (1967, p. 409) called it the "Wyrwicka phenomenon" after an experiment in which Wyrwicka (1952) (1) established an instrumental response to a discriminative stimulus; (2) introduced a second stimulus into the situation and gave it response-independent pairings with food; and (3) observed that in the initial stages of the classical training, the Type I CS has a tendency to evoke the instrumental response.

Konorski, however, considered that the facilitation effect is found only with a CS that has had a few response-independent reinforcements. Indeed, he considered that the effect of a well-established CS on instrumental performance is usually inhibitory. Three kinds of evidence are brought forward. First, if from the outset of training, leg flexion and a CS are paired with food, the classical conditioning to the CS will overshadow the association between the response and food. The response will occur in the intertrial interval (arguing against an associative interpretation, such as that proposed by Mackintosh & Dickinson, Ch. 6) but seems to be inhibited by the presence of the CS. Second, if a CS is given response-independent reinforcement and leg flexion subsequently is paired with food in the presence of the CS, the response does not seem to be learned (Konorski & Wyrwicka, 1950). This effect is discussed in Section III, subsection C. Finally, a well-established CS paired with a previously established discriminative stimulus for an instrumental response may inhibit that response (e.g., Soltysik, Konorski, Halownia, & Rentoul, 1976.)

There is not space here to review the often conflicting evidence on the inhibitory effect of CSs on instrumental responding (Boakes, Ch. 9; Mackintosh, 1974; Mackintosh & Dickinson, Ch. 6; Trapold & Overmier, 1970; Soltysik, 1975). It will suffice here to point to the evidence that the inhibitory effect may be specific to CS duration (Meltzer & Brahlek, 1970) as well as to CS quality (Soltysik et al., 1976.) The suppressive effect of a CS correlates poorly with its capacity to produce salivation (Soltysik, et al., 1976). The important point here is how Konorski attempted to explain the apparent contradiction between the suppressive effect of a CS and the theoretical proposal that arousal of a US representation can be motivating. (See especially the passage from Konorski, 1967, p. 184, quoted in Section I.) The answer lies in his elaborate distinction between drive and US centers, which is now reviewed briefly as a necessary preliminary to the experimental literature on response differentiation.

B. Distinction between drive and US centers in Konorski's model

For Konorski, unlike Pavlov, drive and US centers are distinct; and far from having the same effects on instrumental performance, they are often antagonistic. This is because activation of a US center, as well as providing one possible form of activation of the instrumental response by an associative connection, also inhibits the drive center (see Dabrowska, Ch. 12). The main evidence for this "antidrive" is the short-term satiation effect that follows ingestion of a quantity of food insufficient, in itself, to rectify the physiological defect that led to the drive in the first place. The inhibition in this sense is very similar to the short-term satiation following saccharine ingestion (Holman, 1975; Hsiao & Tuntland, 1971), and possibly also to Booth's (1976) idea of "conditioned satiation." The peculiarity of Konorski's account is that he goes on to suggest that, because food itself can decrease appetite, so must the expectancy of food. Common sense, as well as Konorski himself in several places (e.g., 1967, p. 184), suggests the reverse: that thinking about food when we are food-deprived can markedly increase appetite, just as the imagination can so markedly increase sexual arousal. Apart from the controversial data on the suppressive effects of a CS on instrumental behavior, Konorski provided no clear evidence for the alleged inhibitory effect of the US center on "drive." Nor, as a consequence of this, did he succeed in distinguishing experimentally between activation of a response by a drive-deprivation state per se, and its indirect activation through a US representation. This is obviously a very difficult problem. Activation of the US center certainly depends on the appropriate drive-deprivation state as shown by the effects of deprivation on classical conditioning (Bindra & Palfai, 1967; Konorski, 1967; Mitchell & Gormenzano, 1970); so that it will in principle be possible to explain an apparent effect of drive-deprivation as mediated by changes in the US retrieval system (Holland & Rescorla, 1975). We now turn to various experimental procedures that have been used to tackle this question, starting with response-differentiation studies.

C. Response-differentiation experiments with different drives or different USs

Suppose it is desired to train a dog to raise its left leg in the presence of one CS (S1) and its right leg in the presence of a different CS (S2). This involves a discrimination between the two stimuli, S1 and S2, and a differentiation between the two responses, R1 and R2. For traditional S–R theory there was no difficulty in explaining such learning: Two separate connections, S1–R1 and S2–R2, would be formed, and the ease of learning would depend simply on the extent of generalization between the two stimuli, and between the two responses. For Konorski, on the other hand, matters are more complicated, because part of the response-selection procedure depends on drive and US representations, so that a further factor affecting difficulty will be the extent to which the two responses share a single drive and/or US center. Because this center is common to the stimuli and the responses, either response can be facilitated by either stimulus. It is clear that this "second learning process" will hinder response differentiation. But if two responses are reinforced by different USs and have different drive centers, the second process will no longer cause interference. Konorski supported this hypothesis with several facts, mostly concerned with shock and food as USs. For example, lifting of one leg was established to S1 while a defensive response consisting of lifting the other leg was established in response to S2.

As Konorski (1967) wrote: "It was observed that the situation here was quite different from that encountered in the case of homogeneous CR's. The two instrumental responses... could be established quite independently of one another, and no interchange between them could be observed (p. 386)." No formal comparison between the one- and two-drive case seems to have been carried out. Konorski (1964) stated that it was possible for the dogs to lift the right leg to a buzzer for food reward and to lift the leg to metronome for shock avoidance. It is mentioned that such a differentiation is extremely difficult "where two instrumental acts are mediated by one and the same drive." But this difficulty applies only where the two CSs come from the same spatial location (Shettleworth, Ch. 15), and it is not specified in Konorski's (1964) account whether or not this was true in the two-drive case.

The experiments on food and shock do not speak directly to Konorski's drive-center hypothesis, for they can be equally well explained by the different US representations associated with different stimuli (e.g., Trapold, 1970). A way around this is to eliminate the discriminative exteroceptive stimuli S1 and S2 and to use the drive state as the only cue for response selection in the task. Zernicki and Ekel (quoted in Konorski, 1967, p. 445) in fact found it very difficult to train dogs to carry out one response for food on "hungry" days and a different response on "thirsty" days. This seems to confirm the finding of Hull (1933) that is indeed very difficult to train rats to go one way in a maze for food on hungry days and a different way on thirsty days. The finding is difficult for the drive-selection hypothesis but not fatal, because it can be argued that the reason for the difficulty is the lack of difference between "hunger" and "thirst" states, brought about because the animal voluntarily restricts its food intake when water deprived, and vice-versa. An experiment by Bolles and Petrinovitch (1954) has sometimes been considered to overcome this problem (e.g., Kimble, 1961). Rats simultaneously food- and water-deprived were made relatively hungry by giving them water for 1 hr before the session, and relatively thirsty by giving them food (cf. Van Hemel & Myer, 1970). Unfortunately, this means that they could have solved the problem not on the basis of drive states, but on the basis of the preexperimental treatment. To show that the animals were using drive stimuli, we would have to provide a control in which instead of being prefed or prewatered, the animals are placed in different distinctive goal boxes for 1 hr before being run.

This control was also lacking from an experiment in which Bailey (1955) examined the use of an irrelevant drive as a stimulus. This bears on a point that Konorski does not directly consider: Can associative connections be formed between a drive state and a response that is rewarded by a US not relevant to that drive state? In some ways this is the most crucial question to be answered by an associative theory of drive, but it has not been adequately resolved. Bailey found that rats could perform one response to turn off a bright light when hungry and a different response when thirsty. However, as in Bolles and Petrinovitch's experiment, drive was manipulated by prefeeding or prewatering, with no control for experimental treatment per se as a discriminative stimulus. There is thus very little experimental support for the attractive hypothesis that drive states control response selection, just as discriminative stimuli do, directly through an associative process (Gray, 1975; Mendelson, 1970; Morgan, 1969; 1974). Somewhat better results have been obtained using different levels of the same drive, hunger, as a discriminative stimulus (Jenkins & Hanratty, 1949); but to support a strictly associative view, we must carry out properly controlled versions of Bailey's experiment.

The evidence is much stronger that response selection can be aided by different US representations, as opposed to different drives. Trapold (1970) used different reinforcers for two responses with drive held constant. Rats were trained to press one lever for a food-pellet reward in the presence of S1 (a clicker) and to press the other lever for a sucrose-solution reward in the presence of S2 (tone). The discrimination turned out to be easier for these rats than for controls in which left- and right-lever presses, when correct, received the same reward (viz., either sucrose or food, in different groups). The Trapold result has been extended, with control for the effects of varied reward per se, to different magnitudes of the same reward (Carlson & Wielkewicz, 1976). Rats pressed one lever in the presence of S1 to get a 5-pellet reward and a different lever in the presence of S2 to get a 1-pellet reward. This group learned more quickly than controls for which there was no consistent association between the discriminative stimuli and the different magnitudes.

Both these experiments also showed that prepairing of the discriminative stimuli with the respective USs could influence subsequent acquisition of the discrimination. Similarly Bower and Grusec (1964) showed that a go:no-go discrimination was acquired more rapidly as a result of pairing S+ with reinforcement and S− with nonreinforcement before instrumental training began. Mellgren and Ost (1969) showed that a group in which CS+ and CS− were reversed in significance to become S− and S+ did even worse in the instrumental-discrimination phase than a group given random pairing of the stimuli with the US in the Pavlovian phase. There is a problem in relating these findings to Konorski's claim (see Section III, subsection A.3) that Type I conditioning of a CS makes it difficult to turn it later into an S+. Perhaps the locus of the "facilitation" effect is that a Cs− is even harder to turn into an S+ than is a CS+. The data of Mellgren and Ost are consistent with this interpretation, inasmuch as their absolute response rate data reveal that (1969) "the observed differences among groups in percentage of total responses are due largely though not exclusively to differences in S− response rate (p. 392).'' There is also some support for Konorski's effect in a study by Ost and Mellgren (1970), which showed that overtraining in the Pavlovian phase could somewhat ameliorate the difficulties that the rats had when the cues were reversed in significance in the instrumental phase. If extensive training of a stimulus as a CS+ makes it hard to turn it into an S+, this effect might be explained.

D. Drive and incentive shifts in instrumental responding

Results in the previous section supported the view that specific US representations are involved in the response-selection process, but provided no compelling evidence for the distinct role of drive-deprivation stimuli. If this is so, instrumental responding should be disrupted to a greater extent if the US is changed for a different US than if there is a change in drive deprivation. Following this logic, Capaldi, Smith, and Hovancik (1977) found that hungry rats running for food in a distinctive alley slowed down markedly when water was substituted for food, but slowed down little if at all when thirst was substituted for food deprivation. This occurred despite the fact that the change in deprivation considerably decreased consumption of the reward. Similar examples of persistence following a drive shift have been reported by Capaldi and Hovancik (1975), Zaretsky (1966), and by Mollenauer (1971), who found that the effect of a deprivation shift was influenced by the extent of training. Further examples of persistence of instrumental responding following a drive shift are reviewed in Section IV. A possible account of this persistence, which is a development of a view I have discussed elsewhere (Morgan, 1974), is that changes of deprivation state are often poor predictors per se to the animal of a changed reinforcement contingency. Suppose the animals have learned to expect reward in the food alley when hungry and have learned to run quickly on the basis of that expectancy. The implication of their persistence when shifted to water deprivation is that they continue to expect ''reward'' even though the attractiveness of the actual reward, food, as measured by the consummatory response, has been considerably reduced.

Another way of expressing this idea is that the nature of the US representation involved in the associative control of an instrumental response is not such that it is necessarily changed completely by a change in deprivation state. In other words, a US representation can survive a deprivation shift in the way that Capaldi et al. (1977) suggest. Some direct evidence for this comes from Jenkins and Moore's (1973) study of the form of the autoshaped key peck in pigeons under food or water deprivation. The topography of the peck of the food-deprived bird given a food US differs markedly from that of a water-deprived bird given water. However, if the drives were shifted and the birds given an extinction test, the topography of the peck remained appropriate to the previous US, and not to the new deprivation state.

The idea that the associative aspects of a US representation can survive a drive shift in some circumstances has also been argued by Holland and Rescorla (1975) on the basis of findings described in the following section.

E. Effects of changing the US representation on classical instrumental responding

In first-order classical conditioning, a variety of measures aimed at changing the exact nature of the US representation can have a decremental effect on a previously established CR. Holland and Rescorla (1975) paired a tone with food and looked at the increase in activity during the tone as a CR. A conditioned aversion to the food reward was then independently established by pairing it with high-speed rotation in some animals. In a subsequent test the tone produced a smaller CR in the experimental group than in controls for which the food and sickness had not been paired. This retroactive effect of a conditioned aversion is very strong evidence for the involvement of a US representation in first-order conditioning. However, the results with a second-order CR were quite different. First a light was paired with food to establish a first-order CR; then a tone was paired with the light to establish a second-order CR to the tone; finally the food was paired with sickness to change the representation of the US. This had the aforementioned effect on the first-order CR to the light; but it had a much smaller decremental effect on the second-order CR. The same was found with satiation, which had a large effect on the first-order CR, but very little effect on the second-order CR.

These results may be compared with an experiment by Holman (1975) on instrumental learning. Rats were trained to lever press for saccharine, and then a variety of treatments were carried out designed to alter the animal's representation of the reward. In one procedure the saccharine was paired with sickness produced by injection of lithium chloride. This had the expected effect of reducing saccharine consumption drastically in a simple drinking test, but in an extinction test it had no effect on lever pressing. Similarly, prefeeding with saccharine decreased consumption, but did not alter lever pressing in an extinction test. There is thus considerable evidence that treatments designed to alter a US representation by modification of the animal's internal state do not necessarily affect an instrumental response based on the US, or on a second-order Pavlovian reflex, or—if Jenkins and Moore's data are taken into account—on an autoshaped key peck. These findings, taken in conjunction with those that we have described earlier in Section III, support Konorski's view that there is no single mechanism of response activation, but rather a variety of mechanisms, which may sometimes work together and which may make different contributions in different situations. We saw earlier—from studies of response-independent reinforcement after extinction, for example, and from studies of response differentiation—that there is considerable evidence for "mediated" forms of activation, such as that involving US representations. But we now see, from the effects of conditioned aversion or satiation on responding, that this control cannot be the whole story, because the instrumental response may survive a manipulation that can be shown on independent grounds to exert a considerable effect on the US representation. These facts suggest that alternative "unmediated" forms of response activation exist. For Konorski, one such mechanism is the direct, unmediated, sensorimotor connection, and this is the idea that we turn to in the following section. The idea that habit can itself be a motive has received very little attention in animal learning, despite the strong evidence for its importance in people (Allport, 1937; James, 1891, pp. 549–599). We turn to this idea, which corresponds to Konorski's concept of a direct sensorimotor link, in Section IV and V.

IV. RESPONSE ACTIVATION BY EXTERNAL STIMULI AND RESISTANCE TO SATIATION

The ideo-motor theory holds that a response will be activated when its kinaesthetic representation is elicited by any of the inputs with which it is associated. For Konorski, one such association is a direct one with external stimuli. The simplest and most direct way of seeing whether such connections exist is by investigating the eliciting effects of an external stimulus when other inputs are prevented. If the animal is satiated, then an input from the drive center is ruled out. Equally, for Konorski an input from the US representation is prevented because activation of the latter is dependent on drive. The last point is contentious, as we have just seen, but let it for the moment be accepted that in the satiated animal, the direct sensorimotor connection is in complete possession of the field. The question becomes whether an evidence suggests that this association is adequate to elicit a response.

Wyrwicka (1950) is taken as the reference experiment. Instrumental leg flexion was established to a variety of different CSs, and then the dog was subjected to varying degrees of satiation by presession feeding. Satiation quickly abolished the instrumental response to certain of the CSs and the conditioned salivary response to all of them. But the surprising fact is that the response of certain "strong motogenic stimuli," such as bubbling and a whistle, persisted. Of particular interest was Wyrwicka's (1950) finding that "after being satiated Rex displayed no salivary conditioned reflex at all and did not take food (with the exception of the first trial) through the experiment and nevertheless raised his leg to all stronger stimuli." Similarly, Konorski (1967) quoted an experiment of Dobrzecka and Wyrwicka (1960) in which satiation abolished a leg-flexion response to an auditory stimulus, but not to a "specific tactile stimulus" consisting of a vibration applied to the instrumental limb (see Shettleworth, Ch. 15). Konorski referred to this as "resistance to satiation" and drew a parallel with the greater resistance to extinction also seen with the specific tactile stimulus. His account of these findings arose from notions developed in accounting for response differentiation. Certain stimulus-response connections have a privileged status and are formed in exceptional strength, with the result that they can produce a movement even in the absence of support from the drive and US inputs to the kinaesthetic representation.

The suggestion that "resistance to satiation" may be specific to certain S–R–S relationships is very clearly supported by some findings of Roper (personal communication) on satiation in mice. The animals were permitted to satiate progressively during a long session, obtaining the food by lever pressing in some sessions and by pushing against a key in others. In lever-pressing sessions, they continued to lever press for some time after ceasing to eat the pellets, so that the number of lever presses in the session exceeded the number of pellets consumed (see Fig. 7.2). This was never true in key-pushing sessions, however; in these sessions the mice stopped earning pellets just as soon as they ceased to eat them.

Two general points may be made from these experiments. When responding persists under satiation, a common reaction is to assume that the animal is not "really satiated." For example, in discussing some related findings, Kimble (1961) says that "inasmuch as some responding did occur in each experiment, it follows that the subjects were not completely satiated (p. 190)." As a definition of satiation this can perhaps be defended, but it begs the following questions. First, why does the response persist more to some stimuli than to others (Wyrwicka's result), and why do some responses show more persistence than others (Roper's result)? It is hard to see why the animal should be more satiated when exposed to "weak motogenic stimuli" than to those that continue to evoke a response. Second, why does the response sometimes persist when the consummatory response (salivation and eating) has disappeared? In a sense these points are significantly related, for in both cases we have to deal with the fact that the response to some stimuli (food, metronome) disappears, while that to others (bubbling, a specific tactile stimulus) is maintained. In other words, it is the dissociations that are interesting in these experiments, whether they be between instrumental responding to different CSs or between instrumental and consummatory responses. Attention to this aspect of dissociation will help to avoid

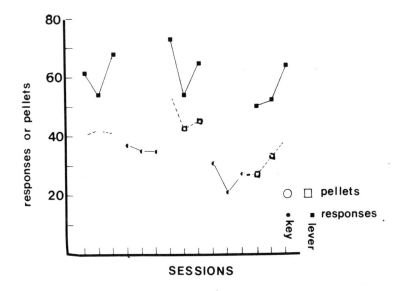

FIG. 7.2. Effects of satiation on two diferent operants (key pressing and lever pressing) in mice. Each session was long enough for the mice to cease consuming pellets as a result of satiation. In lever-pressing sessions, but not in key-pressing sessions, the mice continued to respond after satiation, resulting in an excess of presses over pellets consumed. Unpublished data of T. J. Roper, reproduced with permission.

academic disputes about the real meanings of "satiation," which is indeed revealed by the data to be not very meaningful as a unitary concept.

Dissociation between instrumental and consummatory responding in a single situation has been demonstrated by a number of physiological manipulations. Davis, Gallagher, Ladove, and Turausky (1969) found that blood transfused from a satiated rat to a deprived rat depresses food intake, but not lever pressing for food on a VI-30-sec schedule. The injected animals continued to work for pellets that they did not eat. Similarly, Margules and Stein (1969) found that a cholinergic blocking agent injected directly into the ventromedial hypothalamus led to a resumption of lever pressing after satiation, but not to any further consumption of the milk reward. A well-known phenomenon is that water deprivation in the rat can enhance food-rewarded instrumental responding in a food-satiated rat, even though thirst actually *depresses* food intake (Bolles, 1976, pp. 271–274). In the barbary dove, McFarland (1964) found that thirst increased a key-pecking operant, but the birds did not eat the food they obtained.

Dissociation between instrumental and consummatory responding is relevant to an assessment of the "conditioned-drive" explanation of resistance to satiation (Webb, 1952). If the animal continues with the instrumental response without eating, a "conditioned drive" cannot be invoked. However, in practice it is often very difficult to eliminate consummatory responding in a situation where the animal has learned to eat; and then matters become much more complicated (Capaldi & Myers, Haas, Shenel, Willner & Rescorla, 1975; Jackson & Walker, 1975; Zentall, Hogan, Compomizi, & Compomizi, 1976). As Konorski (1976) put it: "We have often observed that when a dog has been fully satiated before the experimental session and then placed on the experimental stand and given the Type II CS, he immediately performed the trained movement ... and often consumed the portion of food offered to him. Thus a strong hunger CS can produce hunger even in a state of satiation (p. 275)." For a critical account of "conditioned drive," Bindra (1976, pp. 200–205) may be consulted. It is not clear from Konorski's account whether the animal was satiated with the same "tasty food" (p. 275) as was used in the test of satiation; if not, the result may simply show the specificity of satiation to a particular kind of food. A further factor that needs to be assessed is the "disinhibition" of eating by novel stimuli (Drew, 1937). Finally, the recurrence of eating in a "satiated" animal may simply reflect the fact that the time for a further meal has arrived. An experiment by Morgan and Nicholas will make this point. Rats were trained to facewash and to rear on a concurrent VI-food-reinforcement schedule for food reward (Mellgren, Morgan, & Nicholas) and were then given several satiation tests in which the 30-min experimental sessions were preceded by a 1-hr access to the reward pellets. The mean instrumental and consummatory performance in these sessions is shown in Fig. 7.3, where it will be observed that levels of performance and of eating remained surprisingly unaffected. Because eating was so little altered by the satiation procedure, we decided to carry out a control test in which the rats were pre-fed for 1 hr as before, but then instead of being placed back in the experimental situation, were briefly lifted out of the prefeeding box and then replaced. Food consumption in the following 30-min period was then measured. It was found that levels of eating were comparable to those found in the satiation test in the experimental situation.

Our conclusion, in agreement with Bindra, is that the evidence for conditioned drive is not very strong. In any case, as pointed out earlier, it cannot be the explanation of continued instrumental performance when the consummatory response has been abolished. In Holman's study of conditioned aversion, for example (see Section III, subsection E), the consummatory response was affected, but not the instrumental response. Other examples of the dissociation are mentioned by Morgan (1974).

Persistence of earlier parts of a behavior sequence when later members have been reduced in strength is a commonplace in the ethological literature (Hinde, 1958; Hinde, 1970). Barras (1961), for example, observed successive courtships of male *Mormoniella vitipennis* (Hymenoptera) with nonreceptive females. He found that there was a rapid waning of responsiveness, but

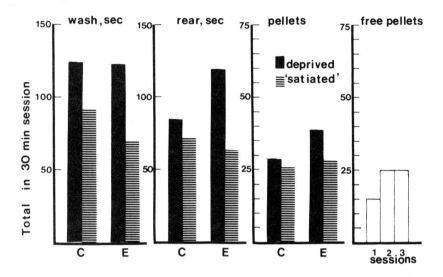

FIG. 7.3. Effects of prefeeding with reward pellets for 1 hr upon face washing and rearing instrumental activities in the rat. Experimental animals (Group E, N = 4) were prefed on "satiated" days and were not fed on "deprived" days. Control rats (C, N = 3) were not prefed on "satiated" days but were put on ad libitum home-cage feeding, as were the E animals. Panel 3 from the left shows the number of pellets consumed in the 30-min operant test. Panel 4 shows the pellets consumed in a 30-min period when the rats were simply removed from the prefeeding cage and then replaced.

that many instances occurred in which earlier parts of the courtship ritual were not followed up by later ones. Roper (1976) trained mice to bar press for nest material, which they were permitted to accumulate without the nest's being removed. The amount of material actually carried to the nest rapidly declined to zero, but the amount earned by lever pressing did not change.

Examples of resistance to satiation in the animal-learning literature have also been reviewed (Morgan, 1974). In a recent study Morgan, Einon, and Morris (1977) compared satiation and extinction procedures in a complex behavioral chain. Rats were trained to get from a start to a goal box through a cylindrical alley, which they first had to clear by removing an obstacle (a rubber ball) backwards into the start box. They were then either extinguished by removing pellets from the goal box or satiated by ad libitum home-cage feeding for 4 days, followed by 1 hr of prefeeding with reward pellets before each 12-trial session. The results (see Fig. 7.4) showed that both satiation and extinction had very different effects on different components of the chain. Little can be concluded concerning the tunnel entrance, where neither extinction nor satiation seems to have had any effect. The greatest increases in time to complete the component were found in the terminal component, which consisted of running to the goal box after clearing the alley. Satiation had a larger effect than extinction, although it is noteworthy that its effect was progressive despite the fact that the rats do not eat pellets even on the early trials. The effect of satiation on the ball-removal response was immediate, in contrast to the effect of extinction, but the response was by no means entirely abolished. In fact, the data show that even at the end of satiation, many of the rats were pulling the ball from the alley in less than 5 sec, which is quite impressive for a difficult response in an animal that is not eating its reward. Reductions in speed were least in the starting component and were about equal for extinction and satiation.

The fact that the terminal response of running to the food cup was much more seriously affected by satiation than the other components is further evidence that behaviors relatively remote from a goal may be activated somewhat differently from behaviors nearer to the goal. Evidence for this has also appeared from the brain stimulation reward literature, where it appears that performance in a two-lever chain has many differences from lever pressing reinforced directly by brain stimulation (review by Lenzer, 1972). An attractive interpretation in terms of Konorski's model is that the eliciting roles of US representations and drives may be relatively more important for

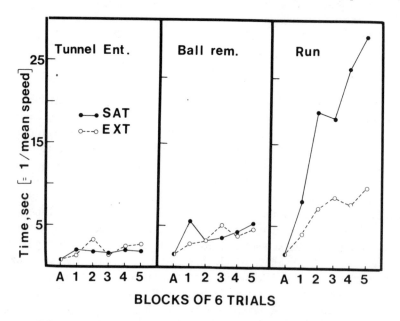

FIG. 7.4.   A comparison between the effects of extinction (EXT) and satiation (SAT) on three components (tunnel entrance, ball removal, and running) of a complex instrumental chain. Note the much greater effects on the terminal component (run). For details, see Section IV of the text.

nearer responses, whereas direct elicitation by external stimuli may be important for remote behaviors. This would be consistent with Holland and Rescorla's observations on differences between first- and second-order conditioning. Some evidence that may be relevant to this hypothesis has recently been obtained by Mellgren (see Fig. 7.5), using the aforementioned obstructed-alley task. Mellgren looked at the partial reinforcement extinction effect (PREE) in different components of the behavioral chain. The PREE has been very well established in straight-alley type situations and in certain operantdiscrete-trial studies, but there has been surprisingly little study of more complex chains. Mellgren found the typical PREE in the terminal-running response, consistent with straight-alley data; but he found that the effect for the ball-pulling component was, if anything, in the opposite direction. Rats continuously reinforced during training were numerically but not significantly more persistent in this component than a partially reinforced group. They were certainly not less persistent (see Fig. 7.5).

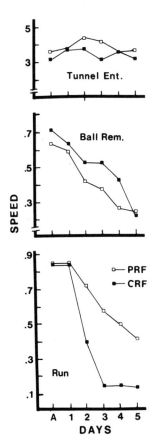

FIG. 7.5. Effects of reinforcement schedule (continuous reinforcement, CRF, or partial reinforcement, PRF) in training on resistance to extinction of the three components described in the legend to Fig. 7.4.

Note that partially reinforced animals are more resistant to extinction in the running component, but not in earlier components. Reproduced with permission of R. L. Mellgren (Mellgren, submitted for publication).

The following interpretation of the absence of a PREE in the earlier segments is highly speculative, because it relies on two independent hypotheses, either or both of which may be mistaken. First, it has already been argued that the role of US representations in response activation may be less for earlier than for later responses in a chain and may also be less for complex responses like pulling a ball than for running. Second, a large body of evidence implicates a role of US representations or reward expectancies in the PREE. Capaldi's sequential hypothesis holds, for example, that a memory of food-last-trial is a crucial part of the situation in which the continuously reinforced animal has learned to run; whereas the partially reinforced animal, on the other hand, has also learned to run when it remembers nonreward-last-trial. Put less formally, it is much more important for the continuously reinforced animal to have an active US representation before running than it is for the partially reinforced animal. But all this supposes that the US representation has some importance for activating the response in the situation. If we deal with a response in which the role of a US representation is much less than that of situational cues in eliciting the response, a PREE would not be expected. This may be the situation with the ball-pulling response, which, as the Morgan, Einon, and Morris (1977) result shows, is quite strongly resistant to satiation and therefore probably requires little support from US representation.

A prediction of this account is that partially reinforced animals will show enhanced resistance to satiation. This is because partial reinforcement lessens the dependence of a response on US representations, so that it will more readily survive a change in those representations. Pencer (1974) found that rats partially reinforced for running in a straight alley were more persistent in running when satiated than a continuously reinforced control group. Haas et al. (1970) reported a similar finding, but their animals were kept on partial reinforcement during the satiation, which complicates the issue. Pencer's animals were all shifted to CR in the satiation phase; so the training condition must have been responsible for the differences in resistance.

The fact that partial reinforcement increases resistance both to satiation and to extinction argues for a similarity between the mechanisms of extinction and satiation, as suggested by Morgan (1974). In both cases the animal ceases to be rewarded for carrying out an instrumental response that has been rewarded in the past — in the one case because eating is prevented by the absence of the reward pellets, and in the other because it is prevented by absence of the relevant drive-deprivation state. The analogy between the two processes has been further stressed in a recent study of Capaldi and Myers in which resistance to satiation was found to be enhanced not only by partial reinforcement in acquisition, but also by the use of small rewards in acquisition (cf. Wagner, 1961). However, a difference between extinction and satiation was also apparent, in that resistance to satiation was enhanced by small reward irrespective of percentage of reward in training. This suggests an important difference between mechanisms of extinction and satiation based, Capaldi and Myers suggest, on the presence of frustration in extinction but not in the satiation case.

A final question is whether resistance to satiation is basically a persistence phenomenon, or whether animals could learn to perform under satiated conditions, provided they also received reinforcement when hungry. An experiment by Morris, Einon, and Morgan (1976) suggests that performance under satiation will not develop if the animals meet it early in training. Rats were given one trial a day in a runway, with a random mixture of days on which they were hungry and days on which they were satiated by prefeeding. We were looking to see whether performance would develop on satiated trials, as a result of the reinforcement provided on deprived trials. In the event, running gradually developed on the deprived days, but showed no increases over the experiment on satiated trials (see Fig. 7.6). There was thus no evidence that the reinforcement received by the animals on deprived trials increased their speed when they were satiated. The deprivation state of the animal thus had excellent discriminative control over performance, if the differing states were met at the outset of training.

An experiment by Mendelson (1966) gives a rather different picture. Mendelson showed that even satiated rats would run in a T maze to a goal

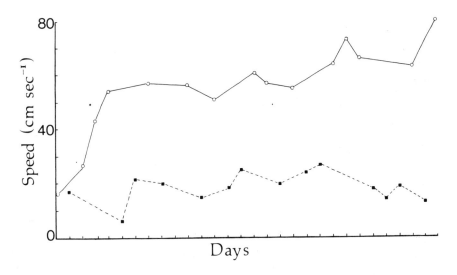

FIG. 7.6   Effects of satiation on the acquisition of a straight-alley running response in rats. The rats were given one trial a day; on some days (continuous curve) they were food-deprived, on other days (broken curve) they were prefed before being run. There is no tendency for speed to increase on satiated trials. Data replotted from group HRSR in Morris, Einon, and Morgan, 1976. See the original paper for the performance of other groups and for running speeds in extinction.

box where they ate in response to the presence of food and lateral hypothalamic stimulation, in preference to a goal box where they received lateral hypothalamic stimulation alone. This study has been quoted (Gray, 1975, pp. 188–189; Morgan, 1969, 1974); as showing that the rat need not be hungry to select a response that will be reinforced by eating. If this is so, it is surprising that the animals in the study by Morris et al. (1976) did not develop an initial tendency while satiated to run to a goal box where they ate on hungry trials. After all, if Mendelson is right, the rats should have anticipated eating on the basis of their experience on hungry trials, even when they were satiated, and should thus have run. It might be pointed out that they have a cue—namely, satiation—correlated with nonreinforcement, but this is something they should have learned gradually, and there should thus have been an acquisition followed by extinction on satiated trials.

It is not very clear, however, that even Mendelson's rats learned to run while satiated. The only animals that were tested for acquisition of running in the maze under satiation seem to have been a group of three naive animals that were given an acquisition followed by two reversals. (Two of the rats failed to accomplish the second running because of illness.) The finding of an original acquisition in these conditions conflicts with the result of an experiment by Wyrwicka, Dobrzecka, and Tarnecki (1960), who, using the same logic as Mendelson, failed to condition an instrumental leg flexion in goats using eating as a reinforcer in the satiated animal. It therefore appears that the question is still open.

Mendelson (personal communication) points out that his animals could be said to have shown acquisition in the sense that they accomplished a number of position-reversals under the satiated-running condition. The reason why some animals may have failed in later reversals is perhaps that on odd-numbered reversals they were being reinforced against their initial position bias. One possibility is that reversal of a previously established habit differs from acquisition of a habit *de novo* as in the Wyrwicka et al (1960) experiment. Another reason why Mendelson obtained learning and Wyrwicka et al did not, may have been that in the latter study the brain stimulation was clearly not reinforcing in its own right. This, if true, suggests that the Mendelson hypothesis applies only in the special case where brain stimulation provides additional reward.

Mendelson also makes the interesting point that the results of the Morris et al experiment contrast with a report by Teel & Webb of T-maze learning on satiated trials by rats. Rats were given 6 trials a day, four before their daily 1 hr meal and 2 afterwards. They showed just as strong a tendency to choose the "correct" side on satiated trials as on deprived trials. The difficulty here, however, is this is a much weaker way of producing "satiation" than the one used by Morris et al. Quite possibly, the rats in the Teel & Webb study were ready to begin another meal at the time they were run (cf. the experiment by Morgan & Nicholas, Section IV and Fig. 7.3.). Unfortunately, Teel & Webb did not put food in the maze on satiated trials, so it is really impossible to conclude anything about the rats' actual motivational state.

V. CONCLUSIONS

This chapter has tried to show that the mechanisms of response activation may be very different, depending on such factors as the nature of the response, the nature of the CS, and the animal's previous history. Konorski's model makes provision for a number of activating principles, which interact in complex ways. The following interpretation of the effects of deprivation and satiation on an instrumental response has a certain amount of evidence. Satiation appears to act primarily to affect the nature of the animal's representation of the US in the situation. Either retrieval of the US representation is completely abolished (the animal simply does not remember the food) or its representation is sufficiently altered in some way, perhaps with respect to its hedonic aspects. As Holland and Rescorla put it: (1975) "A major consequence of changes in deprivation may be to influence the organism's evaluation of the UCS. Indeed, in the present situation, that view would provide a complete description of the mode of action of drives (p. 362)." The effect of changing the US representation, however, will depend on many factors. If the US representation is importantly involved in activation of the response, then satiation will have an immediate effect in decreasing the strength of the behavior. This is often the case, particularly in a simple situation like a straight alley. If, on the other hand, the US representation is relatively less important, the role of external stimuli is more important — as with Konorski's "specific tactile stimulus" — and a shift to satiation will have much less of an effect. There will be considerable differences in the extent of resistance to satiation, depending on such factors as the nature of the exteroceptive eliciting stimuli (Konorski, 1967; Wyrwicka, 1950, Wyrwicka, 1960); the nature of the response (Roper, personal communication) and its position in a chain (Morgan, Einon & Morris, 1977); and possibly also the nature of the US and its sensitivity to a deprivation shift, although there seems to be no definite evidence on the last point.

REFERENCES

Allport, G. W. The functional autonomy of motives. *American Journal of Psychology,* 1937, *50,* 141–156.

Asratyan, E. A. Conditional reflex theory and motivational behaviour. *Acta Neurobiologiae Experimentalis,* 1974, *34,* 15–32.

Bailey, C. J. The effectiveness of drives as cues. *Journal of Comparative and Physiological Psychology,* 1955, *48,* 183–187.

Beach, F. A. Characteristics of the masculine "sex drive." In M. R. Jones (Ed.), *Nebraska symposium on motivation.* Lincoln: University of Nebraska Press, 1956.

Barras, R. A quantitative study of the behaviour of the male *Mormoniella vitripennis* (Walker) (Hymenoptera, Pteromalidae) towards two constant stimulus-situations. *Behavior,* 1961, *18,* 288–312.

Bindra, D. *A theory of intelligent behavior.* New York: Wiley, 1976.

Bindra, D., & Palfai, T. Nature of positive and negative incentive-motivational effects on general activity. *Journal of Comparative and Physiological Psychology,* 1967, *63,* 288–297.

Bolles, R. *Theory of motivation.* New York: Harper International, 1967.

Bolles, R., & Petrinovitch, L. A technique for obtaining rapid drive discrimination in the rat. *Journal of Comparative and Physiological Psychology,* 1954, *47,* 378–380.

Booth, D. A. Approaches to feeding control. In T. Silverstone (Ed.), *Appetite and food intake.* Berlin: Abakon, 1976.

Bower, G., & Grusec, T. Effect of prior Pavlovian discrimination training upon learning an operant discrimination. *Journal of the Experimental Analysis of Behavior,* 1964, *7,* 404–404.

Bruce, R. H. The effect of lessening the drive upon performance by white rats in a maze. *Journal of Comparative Psychology,* 1938, *25,* 225–248.

Capaldi, E. D., Smith, N. S., & Hovancik, J. R. Reversal learning as a function of changed reward location or changed drive. *Learning and Motivation,* 1977, *8,* 98–112.

Capaldi, E. D., & Hovancik, J. R. Transfer of discrimination under hunger to discrimination under thirst: The role of expectancy and habit. *Learning and Motivation,* 1975, *6,* 230–240.

Capaldi, E. D., & Myers, D. E. *Resistance to satiation of consummatory and instrumental performance.* Manuscript submitted for publication, 1977.

Carlson, J. G., & Wielkewicz, R. M. Mediators of the effects of magnitude of reinforcement. *Learning and Motivation,* 1976, *7,* 184–196.

Davis, J. D., Gallagher, R. J., Ladove, R. F., & Turausky, A. J. Inhibition of food intake by a humoral factor. *Journal of Comparative and Physiological Psychology,* 1969, *76,* 407–414.

Dobrzecka, C., & Wyrwicka, W. On the direct intracentral connections in the alimentary conditional reflex type II. *Bulletin of the Academy of Polish Science, Cl. 11, Ser. Sci. Biol.,* 1960, *8,* 373–375.

Drew, G. C. The recurrence of eating in rats after apparent satiation. *Proceedings of the Zoological Society of London,* Series A, 1937, *107,* 95–106.

Ekstrand, B. R. Backward associations. *Psychological Bulletin,* 1966, *65,* 60–64.

Estes, W. K. Discriminative conditioning. I. A discriminative property of conditioned anticipation. *Journal of Experimental Psychology,* 1943, *32,* 150–155.

Estes, W. K. Discriminative conditioning. II. Effects of a Pavlovian conditioned stimulus upon a subsequently established operant response. *Journal of Experimental Psychology,* 1948, *38,* 173–177.

Gray, J. A. *Elements of a two-process theory of learning.* London: Academic Press, 1975.

Haas, R. B., Shessel, F. M., Willner, H. S., & Rescorla, R. A. The effect of satiation following partial reinforcement. *Psychonomic Science,* 1970, *18,* 296–297.

Hebb, D. O. *The organization of behavior:* A neuropsychological theory. New York: Wiley, 1949.

Hinde, R. A. The nest-building behaviour of domesticated canaries. *Proceedings of the Zoological Society of London,* 1958, *131,* 1–48.

Hinde, R. A. *Animal behavior. A synthesis of ethology and comparative psychology* (2nd ed.). New York: McGraw-Hill, 1970.

Holland, P. C., & Rescorla, R. A. The effect of two ways of devaluing the unconditioned stimulus after first- and second-order appetitive conditioning. *Journal of Experimental Psychology: Animal Behavior Processes,* 1975, *1,* 355-363.

Holman, E. W. Some conditions for the dissociation of consummatory and instrumental behavior in rats. *Learning and Motivation,* 1975, *6,* 358-366.

Hsiao, S., & Tuntland, P. Short term satiety generated by saccharine and glucose solutions. *Physiology and Behavior,* 1971, *7,* 287-289.

Hull, C. L. Differential habituation to internal stimuli in the albino rat, *Journal of Comparative Psychology,* 1933, *16,* 255-273.

Jackson, D. E., & Walker, J. S. Rats respond for food on a variable-interval schedule following satiation. *The Psychological Record,* 1975, *25,* 415-418.

James, W. *Principles of Psychology* (Vol. 2). London: Macmillan, 1891.

Jenkins, J. J., & Hanratty, J. A. Drive intensity discrimination in the white rat. *Journal of Comparative and Physiological Psychology,* 1949, *42,* 228-232.

Jenkins, H. M., & Moore, B. R. The form of the auto-shaped response with food or water reinforcers. *Journal of the Experimental Analysis of Behavior,* 1973, *20,* 163-181.

Kimble, G. A. *Hilgard and Marquis' conditioning and learning* (2nd ed.). London: Methuen, 1961.

Konorski, J. Some problems concerning the mechanism of instrumental conditioning. *Acta Biologiae Experimentalis,* 1964, *24,* 59-72.

Konorski, J. *Integrative activity of the brain: An interdisciplinary approach.* Chicago: University of Chicago Press, 1967.

Konorski, J. Classical and instrumental conditioning: The general laws of connections between "centers." *Acta Neurobiologiae Experimentalis,* 1974, *1,* 5-14.

Konorski, J., & Wyrwicka, W. Research into conditioned reflexes of the second type. I. Transformation of conditioned reflexes of the first type into conditioned reflexes of the second type. *Acta Biologiae Experimentalis,* 1950, *15,* 193-204.

McFarland, D. J. Interaction of hunger and thirst in the Barbary dove. *Journal of Comparative and Physiological Psychology,* 1964, *58,* 174-179.

Lenzer, I. I. Differences between behavior reinforced by electrical stimulation of the brain and conventionally reinforced behavior: An associative analysis. *Psychological Bulletin,* 1972, *78,* 103-118.

Mackintosh, N. J. *The psychology of animal learning.* London: Academic Press, 1974.

Margules, D. L., & Stein, L. Cholinergic synapses in the ventromedial hypothalamus for the suppression of operant behavior by punishment and satiety. *Journal of Comparative and Physiological Psychology,* 1969, *67,* 327-335.

Mellgren, R. L., Morgan, M. J., & Nicholas, D. J. *Instrumental behaviors as discriminative stimuli.* In preparation.

Mellgren, R. L., & Ost, J. W. P. Transfer of a Pavlovian differential conditioning to an operant discrimination. *Journal of Comparative and Physiological Psychology,* 1969, *67,* 390-394.

Meltzer, D., & Braklek, J. A. Conditioned suppression and conditioned enhancement with the same positive UCS: An effect of CS duration. *Journal of the Experimental Analysis of Behavior,* 1970, *13,* 67-73.

Mendelson, J. Role of hunger in T-maze learning for food by rats. *Journal of Comparative and Physiological Psychology,* 1966, *62,* 341-349.

Mendelson, J. Self-induced drinking in rats: The qualitative identity of drive and reward systems in the lateral hypothalamus. *Physiology and Behavior,* 1970, *5,* 925-930.

Mitchell, D. S., & Gormezano, I. Effect of water deprivation on classical appetitive conditioning of the rabbit's jaw movement response. *Learning and Motivation,* 1970, *1,* 199-206.

Mollenauer, S. J. Repeated variations in deprivation level: Different effects depending on amount of training. *Journal of Comparative and Physiological Psychology,* 1971, *77,* 318-322.

Morgan, M. J. Motivation. *Cambridge Research,* 1969, *5,* 11–13.

Morgan, M. J. Resistance to satiation. *Animal Behavior,* 1974, *22,* 449–466.

Morgan, M. J., Einon, D. F., & Morris, R. G. M. Inhibition and isolation rearing in the rat: Extinction and satiation. *Physiology of Animal Behavior,* 1977, *18,* 1–5.

Morris, R. G. M., Einon, D. F., & Morgan, M. J. Persistent behavior in extinction after partial deprivation in training. *Quarterly Journal of Experimental Psychology,* 1976, *28,* 633–642.

Ost, J. W. P., & Mellgren, R. L. Transfer from Pavlovian differential to operant discriminative conditioning: Effect of amount of Pavlovian conditioning. *Journal of Comparative and Physiological Psychology,* 1970, *71,* 487–491.

Pavlov, I. P. Reply of a physiologist to psychologists. *Psychological Review,* 1932, *39,* 91–127.

Pencer, E. L. Persistence in the absence of primary motivation in the albino rat. *Journal of Comparative and Physiological Psychology,* 1974, *87,* 787–792.

Postman, L. Transfer, interference and forgetting. In J. W. Kling & L. A. Riggs (Eds.), *Woodworth and Schlosberg's experimental psychology.* New York: Holt, Rinehart and Winston, 1971.

Roper, T. J. Self-sustaining activities and reinforcement in the nest building behavior of mice. *Behaviour,* 1976, *59,* 40–58.

Satinoff, E., & Shan, S. Y. Y. Loss of behavioral thermoregulation after lateral hypothalamic lesions in rats. *Journal of Comparative and Physiological Psychology,* 1971, *77,* 302–312.

Soltysik, S. S. Post-consummatory arousal of drive as a mechanism of incentive motivation. *Acta Neurobiologiae Experimentalis,* 1975, *35,* 447–474.

Soltysik, S. S., Konorski, J., Halownia, A., & Rentoul, T. The effect of conditioned stimuli signaling food upon the autochthonous instrumental responses in dogs. *Acta Neurobiologiae Experimentalis,* 1976, *36,* 277–310.

Teel, K. & Webb, W. B. Response evocation on satiated trials in the T-maze. *Journal of Experimental Psychology,* 1951, *41,* 148–152.

Trapold, M. A. Are expectancies based upon different positive reinforcing events discriminably different? *Learning and Motivation,* 1970, *1,* 129–140.

Trapold, M. A., & Overmier, J. B. The second learning process in instrumental learning. In A. A. Black & W. F. Prokasy (Eds.), *Classical conditioning.* II. Current research and theory. New York: Appleton-Century-Crofts, 1972.

Van Hemel, P. E., & Myer, J. S. Control of food motivated instrumental behavior in water-deprived rats by prior water and saline drinking. *Learning and Motivation,* 1970, *1,* 86–94.

Wagner, A. R. Effects of amount and percentage of reinforcement and number of acquisition trials on conditioning and extinction. *Journal of Experimental Psychology,* 1961, *62,* 234–242.

Webb, W. B. Responses in absence of the acquisition motive. *Psychological Review,* 1952, *59,* 54–61.

Wyrwicka, W. The effect of diminished alimentary excitability upon conditioned reflexes of the second type. *Acta Biologiae Experimentalis,* 1950, *15,* 205–214.

Wyrwicka, W. Studies on motor conditioned reflexes. V. On the mechanism of the motor conditioned reaction. *Acta Biologiae Experimentalis,* 1952, *16,* 131–137.

Wyrwicka, W. An experimental approach to the problem of mechanism of alimentary conditioned reflexes, Type II. *Acta Biologiae Experimentalis,* 1960, *20,* 137–146.

Wyrwicka, W., Dobrzecka, C., & Tarnecki, R. Elaboration of alimentary conditioned reflex Type II with the use of electrical stimulation of the hypothalamus. *Bulletin de l'Académie Polonaise des Sciences,* 1960, *8,* 108–111.

Zaretsky, H. H. Learning and performance in the runway as a function of the shift in drive and incentive. *Journal of Comparative and Physiological Psychology,* 1966, *62,* 218–221.

Zentall, T. R., Hogan, T. E., Compomizi, K., & Compomizi, C. Responding to a positive stimulus by "satiated" pigeons. *Learning and Motivation,* 1976, *7,* 141–159.

8 Appetitive-Aversive Interactions and Inhibitory Processes

Anthony Dickinson
University of Cambridge

Michael F. Dearing
University of Sussex

I. INTRODUCTION

One of the greatest compliments that can be paid to a theorist is to recognize that his ideas provide a general framework for one's own research. We discovered during the initial development of our interest in the behavioral interactions between stimuli of contrasted affective or hedonic value that Konorski's general view of conditioning and motivational processes as outlined in the *Integrative activity of the Brain* provided just such a framework. Because we have employed Konorski's ideas as a general perspective within which to work, this chapter in no way represents a detailed exposition of his theory; rather we hope to exploit the flexibility of his ideas by bending and modifying them to meet our own needs.

Like many students of behavior, Konorski took as his starting point the classification of the basic activities of organisms. One of the main distinctions he drew relates to the biological role of an activity and led him to distinguish between *preservative* and *protective* reflexes. Preservative reflexes are those that are "absolutely indispensable to the preservation of the organisms (1967, p.9)," such as the ingestion of nutrients, whereas protective reflexes "are thrown into action only in states of emergency when the organism is confronted with, or endangered by, a harmful agent (1967, p.9)." Although this classification is ostensibly in terms of biological role, Konorski related this to another distinction based on the affective value of the stimulus controlling an activity and the relationship between this stimulus and the elicited behavior. Many preservative reflexes involve *appetitive* behavior directed toward and elicited by some *attractive* stimulus

(Konorski, 1967, p.10). By contrast, *defensive* behavior directed away from some noxious or *aversive* stimulus is characteristic of many protective reflexes.

Although recognition of the biological role of an activity is important for certain enterprises, Konorski mainly concentrated on the distinction in terms of the affective value of the eliciting stimulus, and emphasized the importance of understanding the behavioral interactions between aversive and attractive stimuli. He had two main reasons for such an emphasis. First, such a concern has certain ethological validity in that many situations and events in real life occur in an emotionally ambiguous context, involving both attractive and aversive stimuli. Konorski (1967, p. 340) took as his example the dilemma of a wife with a loving, but wayward spouse, and analyzed the emotional tension and inconsistency she might well experience. Second he argued that investigations of affective interactions are of general importance for understanding the functional organization of motivational and reinforcement mechanisms, a point that is especially valid in the analysis of conditioned inhibition. We follow Konorski in considering the significance of appetitive–aversive interactions for the mechanism of conditioned inhibition at a later stage. Our starting point, however, is with a description of the basic effects of scheduling both attractive and aversive stimuli in the same experimental situation.

II. BASIC APPETITIVE–AVERSIVE INTERACTIONS

Conditioned and unconditioned stimuli can be classified according to two orthogonal dimensions (Dickinson & Mackintosh, 1978; see Table 8.1). The first, the excitatory –inhibitory dimension, distinguishes between an excitor that is capable of eliciting a certain target response of interest and an inhibitor that is capable of opposing the elicitation of that response. An unconditioned stimulus is assumed to exert its excitatory effect through preestablished associations or connections. However, a neutral stimulus can acquire excitatory properties through excitatory Pavlovian conditioning in which this CS is paired with the US in the context of an overall positive contingency or correlation between the CS and US. By contrast, inhibitory properties appear to result from pairing a CS with the omission of the same US, or, in other words, arranging a negative correlation between the CS and US. The second, the affective or hedonistic dimension, characterizes a stimulus on the basis or whether it is itself an attractive or aversive US, or, in the case of a CS, was associated with an attractive or aversive US during conditioning. As we have mentioned, Konorski defined attractive and aversive stimuli in terms of the approach and withdrawal behavior elicited by them. An alternative conventional definition relates to their capacity to act as reinforcers in instrumental conditioning. An attractive US is a stimulus that could potentially act as a positive reinforcer for an instrumental response, whereas an aversive US is a stimulus whose omission or termination would provide a source of negative reinforcement and whose presentation would serve to suppress responding in a punishment procedure.

The arrows in Table 8.1 specify the interactions open to investigation. These interactions can be studied by looking at the effect of a stimulus on responses and reinforcement processes controlled by a target excitor. The interaction between excitors and inhibitors based on the same US is determined, of course, by definition—at least in respect to the responses elicited by the excitor. Although there is a dearth of empirical information, we assume that an inhibitor established with a given US can inhibit the action of an excitor based on a different US but with the same affective polarity. The effect of stimuli of one affective class on responses elicited by an excitor of the opposite affective value is, however, an empirical question, and we can ask whether an excitor or inhibitor of one hedonic value has an excitatory or inhibitory effect on the action of an excitor of the opposite affective value. This chapter is an attempt to specify the nature of these interactions.

One way of showing that a CS, negatively correlated with a US, has an inhibitory effect is by subsequently pairing that CS with the US and observing whether development of the conditioned response is retarded (Hearst, 1972; Rescorla, 1969a). Similarly, we could take an excitor of one affective value, pair it with a US of the opposite affective value, and then observe the development of the conditioned response normally reinforced by that US. The excitor would be said to have excitatory properties on the response of opposite affective significance if conditioning is enhanced, and inhibitory properties if conditioning is retarded.

TABLE 8.1
Classification of Stimuli

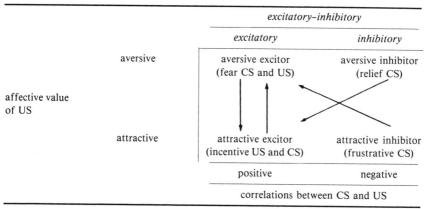

A. Attractive stimuli and defensive responses

The classic experiment in this area is by Konorski and Szwejkowska (1956). The experiment was carried out on one dog in a Pavlovian conditioning stand in which the amount of salivation (the appetitive response) conditioned by a food US and the magnitude of a leg-flexion response (the defensive or aversive response) conditioned by a shock US could be measured. The general design of the experiment is outlined in Table 8.2. In the first stage, a salivary response was established to two CSs, a bell and a metronome, by pairing them with food. Interspersed with these trials were nonreinforced presentations of a whistle. In a second stage, appetitive reinforcement of the bell continued, and a new CS, a propeller, was introduced and paired with shock to establish a flexion response. In the third stage, the metronome, whistle, and propeller were all presented in the absence of food and paired with shock.

The critical result concerns conditioning to the metronome — the CS previously paired with food—in stage 3. Konorski and Szwejkowska found that the development of the defensive leg flexion to the metronome was severely retarded in comparison with the level of aversive conditioning to the control CS, the whistle, which was not paired with food during the first stage. Although this result has yet to be replicated (Dickinson & Pearce, 1977), it does suggest that prior appetitive conditioning to a CS has an inhibitory or antagonistic effect on the subsequent development of a defensive response to the same CS.

A second procedure commonly used to assess the inhibitory properties of a stimulus is the summation test (Hearst, 1972; Rescorla, 1969a). This involves showing that when the putative inhibitor is combined with an excitor for the target response of interest — in this case a defensive response—the compound elicits a weaker response than both the excitor alone and a compound of the excitor and a control stimulus. Although there are no direct assessments of the inhibitory effects of an attractive CS on a purely Pavlovian defensive response, a number of studies (Bull, 1970; Grossen, Kostansek, & Bolles, 1969; Overmier, Bull, & Pack, 1971) have shown that

TABLE 8.2
Stages of the Konorski & Szwejkowska (1956) experiment

Stimuli	Stages				
	1	2	3	4	5
Bell (B)	B⟶food	B⟶food			B⟶food
Metronome (M)	M⟶food		M⟶shock		
Whistle (W)	W		W⟶shock		
Propeller (P)		P⟶shock	P⟶shock		
Rattle (R)				R⟶shock	R⟶food
Disc (D)			D		D⟶food

an attractive CS will suppress aversively motivated instrumental responding. Because Konorski (1967) and others (e.g., Rescorla & Solomon, 1967) have argued that performance of an instrumental response involves a central motivational state, conditioned by Pavlovian contingencies, at least one interpretation of this suppressive effect is in terms of the inhibitory action of an attractive CS on some central aversive response.

B. Aversive stimuli and appetitive responses

In a second phase of their experiment Konorski and Szwejkowska studied the transformation of an aversive CS into an attractive one (see Table 8.2). In the fourth stage, two new stimuli, a rattle and a disc, were introduced; the rattle was paired with shock, and the disc was nonreinforced. Both the rattle and the disc were then paired with food alone, along with the original attractive CS, the bell. A salivary response of a magnitude comparable to that elicited by the bell rapidly developed to the disc, the CS that had not been paired with shock during stage 4. By contrast, the aversive CS, the rattle, elicited only a small salivary response in spite of numerous pairings with food. This basic retardation effect has now been replicated in an elegant experiment by Scavio (1974) using rabbits. Establishing a tone as an aversive CS for a nictitating membrane response by pairing it with shock retarded the subsequent development of an appetitive response — the jaw-movement component of the swallowing reflex — when the tone was subsequently paired with the delivery of water. Again there are no direct tests of the inhibitory effects of an aversive stimulus using a purely Pavlovian summation procedure. However, the basic conditioned-suppression phenomenon (Estes & Skinner, 1941) provides a demonstration that an aversive CS will suppress appetitively motivated responding.

So it appears that appetitive–aversive interactions are of an inhibitory or antagonistic nature (see Dickinson & Pearce, 1977, for full review), with a conditioned excitor of one affective value having an inhibitory effect in both summation and retardation tests on a response established with a reinforcer of oppositive affective polarity.

III. THE NATURE OF THE INHIBITORY INTERACTION

Konorski assumed that the interaction seen in his experiment reflected the organization of central mechanisms. There is, however, no immediately obvious reason why the summation and retardation effects should not arise simply from peripheral response competition between opposing behavioral tendencies. We can start to determine the locus of the interaction by examining the way in which conventional inhibitors work. Although potential conditioned inhibitors can, in certain circumstances, control responses opposed to those elicited by an excitor (e.g., Daly, 1974; Hearst & Franklin, 1977; Wasserman, Franklin, & Hearst, 1974), the properties of inhibitors cannot be entirely reduced to those of peripheral-response interaction. This is most clearly revealed by a property of inhibitors that goes beyond the simple attenuation of a response manifest in the summation and retardation tests. The property is seen in the *superconditioning* phenomenon, first reported by Rescorla (1971) using a conditioned-suppression procedure with rats.

The basic experimental design for the groups of interest is outlined in Panel A of Table 8.3. The unpaired group received stimulus A, a tone, explicity unpaired with a shock US in stage 1 to establish it as a potential aversive inhibitor, whereas a control group received A and the shock in a random relationship. In the second stage, A was compounded with a new stimulus X, a light, for both groups, and the compound was paired with shock.[1] Finally, the level of aversive conditioning to the added stimulus X was assessed by presenting it alone. The level of conditioning or supression was expressed in terms of the conventional suppression ratio. This suppression ratio equals $A/(A+B)$, where A is the rate of responding in the presence of the stimulus and B the rate during a pre-CS period immediately preceding the stimulus. The ratio has the value of zero when conditioning or suppresssion is maximum and a value of 0.5 in the absence of suppression. The test results, illustrated in Panel A of Fig. 8.1, show that more conditioning had accrued to X when it had been reinforced in compound with a conditioned inhibitor rather than with the control stimulus. This superconditioning effect has subsequently been replicated both in experiments on taste aversion (Taukulis & Revusky, 1975) and on autoshaping (Blanchard & Honig, 1976). It is very difficult to see how a conditioned inhibitor could enhance the reinforcing effects of shock in this way if its only effect was to control peripheral responses competing with the defensive response.

TABLE 8.3
Design of aversive superconditioning experiments

Groups	Stage 1	Stage 2	test
	Panel A - Superconditioning with aversive inhibitor (Rescorla, 1971)		
Unpaired	A⟶ shock omission	AX⟶ shock	X
Random	A/shock	AX⟶ shock	X
	Panel B - Superconditioning with attractive excitor (Dickinson, 1977)		
Paired	A⟶ food	AX⟶ shock	X
Random	A/food	AX⟶ shock	X

---

[1]Other control groups were included in Rescorla's (1971) experiment, but these have been omitted for clarity. They are presented later in the chapter.

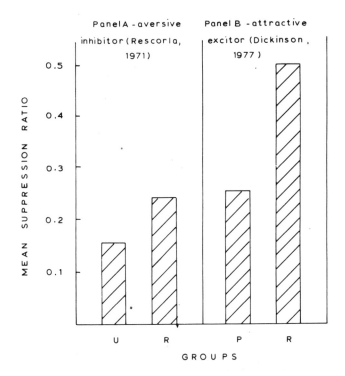

FIG.8.1. *Aversive Superconditioning:* Mean suppression ratio to the added element X on test. *Panel A:* Values estimated from the graphic data presented by Rescorla (1971) and average over all the test trials. U: unpaired group; R: random group. *Panel B:* Values taken from Dickinson (1977). P: paired group; R; random group.

We set out to see whether an attractive excitor would have the same effect as an aversive inhibitor on the superconditioning of suppression. The basic design of the experiment is shown in Panel B of Table 8.3, although the contingencies and control groups employed were somewhat more complex (see Dickinson, 1977). In the experimental or paired group, stimulus A —in this case a light—was established as an attractive excitor by pairing it with food, rather than as an aversive inhibitor as in Rescorla's experiment. The control or random group received both A and food semirandomly related. Thereafter the procedure basically followed that of Rescorla's experiment in stage 2, except that stimulus X in this case was a tone. The test results, shown in Panel B of Fig. 8.1, clearly demonstrated a higher level of suppression conditioned to X in the paired group. So it appears that an attractive excitor can also enhance aversive conditioning. This parallel between the superconditioning properties of an explicit aversive inhibitor and an attractive excitor strongly suggests that the inhibitory effect of attractive CSs on defensive response and reinforcement processes should also be viewed in terms of some central interaction.

Although there are, as yet, no direct tests of the ability of an aversive excitor to produce superconditioning of an appetitive response established by classical procedures, some recent work by Fowler and his colleagues (Fowler, in press; Fowler, Fago, Domber, & Hochhauser, 1973) suggests the existence of such an effect. Rats were exposed to a brief tone after making the correct choice in a food-reward situation involving a brightness discrimination in a T-maze. Surprisingly, they found that animals made fewer errors during acquisition if the tone had been previously paired with a mild shock. After considering a number of alternative accounts, Fowler (in press) favored an explanation in which the presence of the aversive tone led to the superconditioning of the appetitive properties to the positive discriminative stimulus.

Thus, conditioned excitors not only have similar inhibitory properties to an explicit conditioned inhibitor of contrasted affective value, but also show a comparable capacity to mediate superconditioning by a reinforcer of opposite emotional significance. The implication of this latter finding is that these stimuli exert their common effects through some central mechanism. To specify the nature of this mechanism, we need to look moe closely at Konorski's general view of conditioning.

IV. CENTRAL MOTIVATIONAL MECHANISMS AND COUNTERCONDITIONING

A. Epicritic versus protopathic properties

Konorski (1967, ch. 1) argued that the important properties of an effective reinforcer or US are mediated by two separate but interrelated systems. Presenting an aversive US, such as a shock, first activates some form of internal sensory representation of that stimulus. This is turn arouses a response mechanism that determines the actual form of the "consummatory" defensive reflex, such as leg flexion or eyeblink, seen in response to that particular US. This system mediates what Konorski (1967, p.43) called the "epicritic," or sensory, properties of the stimulus. Activation of an internal representation of a stimulus with emotional significance will also lead to the arousal of a central motivational or drive system related the affective quality of that stimulus. In the case of aversive USs, Konorski referred to this system as a fear system and to the state brought about by its arousal as a fear reflex. This system mediates the "protopathic" (1967, p.43), or general emotive and reinforcing, properties of the stimulus. Arousal of this motivational system has two main effects on the behavior elicited by the stimulus. First, it feeds back on the sensory representation of the stimulus to potentiate the consummatory response (1967, p.48). Second, it leads to the emission of "preparatory" behavior, such as withdrawal, that reflects the general affective character of the stimulus rather than its specific sensory properties. A schematic representation of such system is shown in Fig, 8.2.

When a neutral CS is paired with a US in a classical conditioning experiment, the internal representation of the CS becomes connected by independent excitatory associations with both the representation of the US and the motivational system, with the reinforcement process depending on the arousal of the motivational system by the US (Konorski, 1967, Chapter 6; Hearst, Chapter 2). Phasic stimuli presented with a relatively short CS–US interval become connected to both these systems, whereas more tonic and

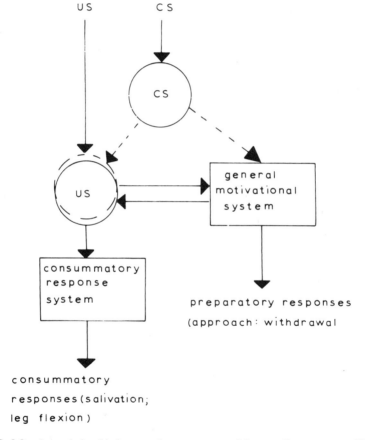

FIG. 8.2.   Interrelationship between the components of the overall system controlling responses and motivational states elicited by both unconditioned and conditioned stimuli. ——▶ : preestablished excitatory connections; ·· --▶ : excitatory connections established by conditioning; circle: internal sensory representation of a stimulus excited by either an unconditioned (solid circles) or conditioned (hatched circles) connection.

long-lasting CSs are primarily associated with motivational systems. Joint activation of both the motivational system and the US representation by a CS results in a consummatory response reflecting both the affective quality of the US and its specific sensory properties. By contrast, activation of the motivational system alone produces preparatory responses related only to the motivational properties of the reinforcer.

Konorski identified these conditioned excitatory associations with neural connections between systems in the brain. There is, however, no need for such an assumption, and on a functional level these conditioned connections might be embodied in a number of different ways. For instance, they might be viewed in terms of Wagner's (in press; Chapter 3) recent model of associative conditioning, which assumes that a conditioned excitor is capable of retrieving a representation of the US from some long-term store into a limited-capacity rehearsal mechanism. The presence of this representation in the rehearsal mechanism controls the conditioned response. In addition, there is no reason why the retrieved representation should contain all the information encoded in a representation of the same US that originates from the actual presentation of that US. For this reason we have distinguished, in Fig. 8.2, between representations of the US elicited by the actual US itself and by a CS.

In the case of appetitive conditioning with an attractive US, such as food, the underlying mechanism is essentially analogous except for one major factor. Konorski assumed that activation of the sensory representation of the US leads to inhibition rather than arousal of the associated appetitive motivational system (1967,pp. 44–49; Morgan, Chapter 7). Although the presentation of food to a hungry animal can arouse both systems, different aspects of this event are responsible in each case. Whereas the actual presence of food in the mouth stimulates the sensory representation, the sight of food and the termination of a limited eating bout activates the motivational system. One reason for this distinction is that Konorski wished to identify the motivational system activated by a CS paired with food presentation with that engaged by food deprivation, and therefore referred to it as a hunger system. Although not solely for the sake of our argument, we part company with Konorski here and assume that organization of the appetitive mechanism is essentially identical with that of the aversive system, with the sensory representation of an attractive US having an excitatory effect on the appetitive motivational system. This is more in line with the standard concept of incentive motivation (e.g., Bindra, 1972; Mowrer, 1960; Rescorla & Solomon, 1967) that distinguishes between the motivational effects of incentive and deprivational manipulations. We assume that the effect of deprivation is to open some form of gate that allows the US presentation to arouse the appetitive motivation system.

In talking about aversive and appetitive motivational systems, we are implicitly assuming that at some stage a wide range of different USs of the same affective value activate a common system. Although unclear about the unitary nature of the aversive-drive system, Konorski distinctly identified different drive systems related to different classes of attractive stimuli. For the time being, we continue to treat the appetitive and aversive motivational systems each as a unitary mechanism activated by a wide range of USs with different sensory properties but similar affective values. This assumption has little empirical support in the present context, especially because the experiments considered in this chapter are limited to using shock as an aversive US and food and water as attractive USs. The implication of this assumption is taken up again in the concluding section.

Konorski (1967, pp. 335–340) saw the difficulty in transforming a CS of one affective value into a CS of the opposite value as arising from the presence of reciprocal inhibitory connections between the appetitive and aversive motivational systems. Here Konorski is in good company, for the presence of such an opponent-process mechanism has been argued for in one form or another by a number of authors (e.g., Bindra, 1974; Estes, 1969; Gray, 1975; Millenson & de Villiers, 1972; Miller, 1963; Mowrer 1960; Olds & Olds, 1964; Stein, 1964).[2] According to this type of model (see Fig. 8.3), the transformation of, say, an aversive CS into an attractive CS is difficult for two reasons. First, the actual reinforcement process involved in establishing the association between the aversive CS and the appetitive centers depends on the arousal of the appetitive motivational system at the time of, or shortly after, the CS presentation. However, the very presentation of this stimulus opposes arousal of the appetitive system. Second, even if these associations are formed, performance of many consummatory CRs depends on arousal of the appetitive system that will again be inhibited by the aversive properties of the CS. The operation of such a performance factor was demonstrated by Konorski and Szwejkowska (1956). If an originally aversive CS, which had been successfully transformed into an attractive CS, was presented in the context of other reinforced aversive CSs, the salivary CR disappeared and the original defensive CR was again elicited.

B. Counterconditioning

An important feature of this opponent-process model rests in the fact that it

---

[2]This opponent-process model is to be distinguished from that suggested by Solomon and Corbit (1974). In their model both processes were engaged by presenting the same stimulus, whereas the present model assumes that the opponent processes are activated by stimulil of contrasted affective value.

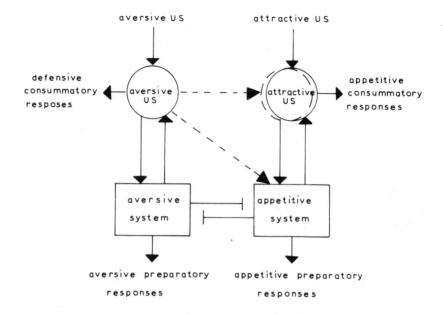

Fig. 8.3. Illustration of the reciprocal inhibitory interactions between the aversive and appetitive motivational systems and of the excitatory connections formed during the counterconditioning of an aversive US. ———▸: preestablished excitatory connections; —|: preestablished inhibitory connections; ---▸: excitatory connections established by conditioning; circle: internal sensory representation of stimulus excited by either unconditioned (solid circles) or conditioned (hatched circles) connections.

is the protopathic, or motivational, properties of stimuli with contrasted affective value that are incompatible, not their sensory, or epicritic, characteristics. This means not only that it should be difficult to transform consummatory responses elicited by an aversive stimulus from defensive to appetitive ones, but also that once successfully transformed the stimulus should have lost its original aversive character — or, in other words, have been counterconditioned. Many years ago, Erofeeva (1916; 1921) reported from Pavlov's laboratory that a strong salivary response, reinforced by food, could be conditioned to a painful electric shock as a CS in dogs. More interestingly, it appeared that as the alimentary response developed, the shock appeared to loose its noxious character. This basic counterconditioning effect has been reported by a number of other Russian investigators (e.g., Marukhanyan, 1954; Varga, 1955), although quantitive comparisons with the appropriate control groups have not been made. It should be noted that Erofeeva's study attempted to countercondition an aversive US rather than CS. Although a large number of counterconditioning studies have been run with conditioned stimuli (Dickinson & Pearce, 1977), there are distinct

advantages in using an unconditioned stimulus, because the original properties of the stimulus are not subject to extinction during the counterconditioning procedure.

Erofeeva esssentially reported a change in the defensive responses elicited
by the shock as a result of counterconditioning. However, as we have
pointed out, the opponent-process model maintains that the central effect of
counterconditioning is to change the motivational and reinforcing properties
of the shock. A number of experiments in our laboratory have looked for
such changes. In one such experiment (Pearce & Dickinson, 1975) rats were
exposed to an initial classical counterconditioning stage in which the experimental or paired group received pairings of footshock and freely
delivered food, with the onset of the shock preceding food delivery. Other
control groups received shock and food in an unpaired sequence, shock
alone, and food alone. Subsequently, the capacity of the shock to act as an
aversive reinforcer was assessed in a conditioned-suppression procedure. In
line with the opponent-process model, the rats in the paired condition
developed suppression more slowly than those in the control groups. Pairing
shock with food has also been shown to reduce the extent to which a rat will
avoid a location in which the shock is delivered and the extent to which it
will suppress instrumental responding in a punishment procedure (e.g.,
Dickinson & Pearce, 1976).

A crucial feature of the counterconditioning explanation of these results
is that the reduction in the aversiveness of the shock should depend on successfully establishing the shock as an attractive CS. A defect in the foregoing experiments was the absence of any direct measure of appetitive conditioning during the classical counterconditioning procedure. In the absence
of such a measure, we cannot be sure that the shock was successfully
established as an attractive CS. To see whether a reduction in the aversiveness of a shock would occur after successfully establishing it as an attractive CS, Dearing (1978) employed a rabbit-conditioning preparation
similar to that reported by Scavio (1974). Thirsty rabbits were placed in a
restraining stock and mounted with conventional head gear. This allowed
concurrent measurement of both a defensive outer-eyelid closure in response
to a paraorbital shock and the jaw-movement component of an appetitive
swallowing reflex in response to an interoral injection of water through a
cannula in the cheek. In addition, an instrumental response could be made
available to the animals by placing an inverted lever above their snouts. This
lever could be operated by a slight upward movement of the head. In the
first stage, operation of the lever was reinforced by water injections on a
variable interval schedule with a mean of 1 min. After the response rates
had stabilized, a discrete-trial punishment procedure was superimposed on
this baseline responding. Every so often, 70-sec tones were presented to the
rabbits during which lever pushes were punished on fixed-interval 15-sec

schedule by a paraorbital shock. Whenever a shock or water reinforcer was "set up", the clocks controlling both schedules were stopped to prevent fortuitous pairings of shock and water as the animals suppressed responding. Suppression of lever pressing during the tone was measured in terms of the previously defined suppression ratio. The shock level was progressively increased until it produced a mean suppression ratio of approximately 0.3 across all animals. This value was used throughout the remainder of the experiment. Finally, suppression to the tone was extinguished.

The next stage involved Pavlovian counterconditioning during which the lever was removed and the animals divided into two groups, a paired and an unpaired group. The paired group received trials in which presentation of the shock was immediately followed by a water injection, whereas the unpaired group received the same number of shock and water presentations in a random sequence. Appetitive conditioning to the shock was successful in the paired group; the percentage of jaw-movement responses on shock-alone test trials at the end of conditioning was 70% for the paired group and 6.6% for the unpaired group ($p < 0.05$).

To find out whether the appetitive conditioning of the shock had altered its general aversive properties, Dearing returned the rabbits to the instrumental punishment procedure. Fig. 8.4 illustrates clearly that the aversive properties of the shock had been attenuated by counterconditioning; the paired group showed significantly less suppression of the lever-press

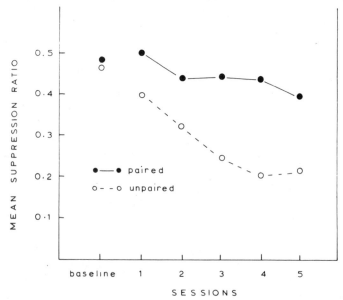

FIG. 8.4.   Mean suppression ratio to the tone during signaled punishment following counterconditioning.

response than the unpaired group. It is interesting to note that this change in the aversive properties of the shock occurred in the absence of a reduction in the defensive response actually elicited by the shock; Dearing could detect no change in the amplitude or latency of the eyeblink in the paired group during classical counterconditioning. Tait (1974) also failed to find any attenuation of the defensive response in similar rabbit-counterconditioning procedure, and we suspect that this insensitivity reflects the fact that even motivationally neutral stimuli applied directly to the orbital region can elicit eyelid closure and bulbar retraction. The unconditioned response to a stimulus should only be changed by counterconditioning if the elicitation of that response involves motivational processes. Because the occurrence of conditioned, as opposed to unconditioned, orbital defensive reflexes in the rabbit are critically dependent on motivational properties of the US (Bruner, 1966; Frey, Maisiak, & Dugue, 1976), the conditioned components of this reflex should be susceptible to counterconditioning, and in fact are so (Scavio, 1974).

In conclusion, these counterconditioning experiments provide good evidence for the general concept of reciprocal inhibitory interactions between central appetitive and aversive motivational systems, and it is clear that the interactions between the behavioral properties of stimuli of contrasted affective value should be viewed within the context of such a system.

V.   EFFECT OF AN INHIBITOR OF CONTRASTED AFFECTIVE VALUE

So far, we have looked at the action of an excitor of one affective value on behavioral processes controlled by an excitor of opposite affective significance, and have found that it is essentially an inhibitory one, at least at the level of motivational and reinforcement processes. Reference to Table 8.1 shows that the remaining interaction to consider is that between an inhibitor and excitor of opposite affective polarity. The opponent-process model makes some specific predictions about the nature of this interaction. To illustrate these predictions, however, we must first look at the sort of associative processes assumed to underly the action of a conventional inhibitor.

Konorski's ideas on the mechanism of inhibition are discussed in detail by Rescorla (Chapter 4). Briefly, Konorski assumed that inhibitory effects were due to the capacity of the inhibitor to alter arousal of the central mechanism mediating the excitatory response. He had two alternative views about the associative structure underlying such an action. According to the first (Konorski, 1948), a direct inhibitory association is formed during conditioning between the internal representation of the inhibitory CS and the central system activated by an excitor. The alternative view (Konorski, 1967) is that an excitatory association is formed between the CS representation and some other central mechanism — an "anti-drive" center or "no-US" unit—whose arousal in turn leads to the inhibition of the excitatory system. This is no place to go into Konorski's reasons for finally preferring the second view, and for the present purposes it is only important to note that, according to both models, the effect of an inhibitory stimulus is the presence of an inhibitory influence on the relevant motivational system. Fig. 8.5 illustrates the path of such an influence for an attractive inhibitor, with the question mark leaving open the actual associative structure of path.

In the absence of any excitatory influence on the appetitive and aversive systems, presentation of an attractive inhibitor, for example, will be without an effect. However, if both systems are under an excitatory influence, their potential levels of arousal will be reduced by mutual inhibition. If we now present an attractive inhibitor, the level of activity in the appetitive system will be reduced and correspondingly the level of activity in the aversive system increased. As far as the excitatory functions of the aversive system are concerned, the presentation of an attractive inhibitor should be functionally equivalent to that of an aversive excitor (see Fig. 8.5).

What we now need is a procedure with which to test this prediction. Kamin (1969) demonstrated that if a stimulus, A, was paired with shock

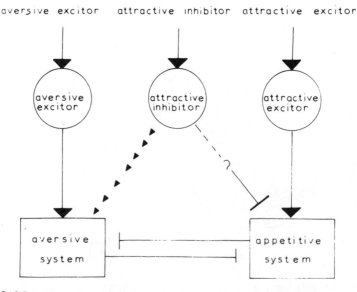

FIG. 8.5.   Illustration of the effect of an attractive inhibitor on the opponent-process system under the concurrent influence of attractive and aversive excitors. ———▶: excitatory connction; —|: inhibitory connection; ?: unspecified paths (see text); ▶▶▶: functionally equivalent excitatory influence (see text).

before aversive conditioning to a compound stimulus, AX, in a conditioned-suppression procedure, the amount of conditioning accruing to X was reduced or blocked. (See conditions 1 and 2 of Table 8.4.) This phenomenon of blocking can be illustrated by considering two further groups run by Rescorla (1971). In addition to groups receiving A, a tone, and shock unpaired (Group U) and randomly related (Group R) in stage 1, Rescorla also ran groups that received either A and shock paired to establish A an an aversive excitor (Group P), or no preexposure to A (Group no-CS) before aversive conditioning to the AX compound in stage 2. Panel A of Fig. 8.6 illustrates the suppression maintained by X alone during the first test session. Less aversive conditioning accrued to X in Group P (condition 1 of Table 8.4) than in Group R and Group no-CS (condition 2 of Table 8.4). Blocking can be explained if it is assumed that the amount of conditioning to a CS is positively related to the difference between the level of arousal of the aversive system when the shock is presented and during the CS (Konorski, 1948; Rescorla, 1973). Presentation of the pretrained aversive excitor, A, during conditioning to AX decreases this discrepancy, and hence the amount of conditioning to X.

TABLE 8.4
Blocking of aversive conditioning

Conditions	Stage 1	Stage 2	test
1	A⟶ shock	AX⟶ shock	X
2	control treatments	AX⟶ shock	X
3	A⟶food ommission	AX⟶ shock	X

In many ways this procedure provides an ideal way of testing the functional equivalence of an aversive excitor and attractive inhibitor. If such an equivalence exists, an attractive inhibitor should also be capable of blocking aversive conditioning by a shock. The presentation of both shock and food during the conditioned-suppression procedure will ensure the concurrent arousal of both the appetitive and aversive systems during presentation of the inhibitor. To test this prediction, Dickinson (1976) initially trained rats to lever press for food on a variable-interval schedule with a mean of 2 min. The lever was then withdrawn and classical appetitive conditioning administered. Stimulus A, a 30-sec overhead light, was established as a potential attractive inhibitor for Group U by associating it with the omission of food, as in stage 1 of condition 3 in Table 8.4. This was done by intermixing presentation of a 30-sec clicker, during which free food was delivered on a variable-time schedule with a mean of 7.5 sec, with nonreinforced presentations of a clicker–light compound. After 8 sessions, each containing 5 reinforced clicker and 5 nonreinforced clicker–light presentations, the lever was

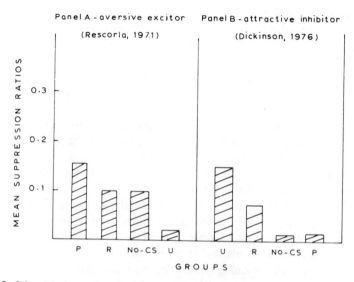

FIG. 8.6. *Blocking of aversive conditioning:* Mean suppression ratios to added element X on test. *Panel A:* Blocking when A established as aversive excitor in Group P. Values estimated from graphic data presented by Rescorla (1971) for first session of testing. *Panel B:* Blocking when A established as attractive inhibitor in Group U. P: paired groups; U: unpaired groups; R; random groups; no-CS: no-CS groups.

returned and pressing reestablished. In stage 2 the light, A, was compounded with a novel 3000-Hz tone, X, and the compound paired with a 0.5-ma 0.5-sec shock for 6 trials. Finally the amount of suppression conditioned to the tone was measured by presenting it alone on 2 test trials. The control groups received exactly the same training except during the appetitive conditioning stage. All control animals experienced the same clicker–food pairings as Group U. The differences were that the light was semirandomly associated with food for Group R, paired with food for Group P, and not presented at all during this stage for Group no-CS.

Panel B of Fig. 8.6 illustrates the suppression maintained by the tone, X, on the test trials. The tone maintained less aversive conditioning or suppression in Group U, for whom the light, A, was established as a potential attractive inhibitor than in the control groups.[3] An overall analysis of test suppression ratios showed that there was a significant difference between the groups ($p < 0.05$); and individual comparisons by the Newman–Keuls procedure revealed that Group U was significantly less suppressed than both Groups P and no-CS ($p < 0.05$ in both cases). Fowler (in press) has recently

---

[3]In view of the previous demonstration that an attractive excitor could produce superconditioning, one might have expected the paired group to show more suppression to X than the other groups in the test. Any superconditioning, however, was probably obscured by a "floor" effect in the present experiment.

reported a similar demonstration of the blocking of aversive conditioning by an attractive inhibitor.

The implication of the transreinforcer blocking experiment is that an attractive inhibitor is functionally similar to an aversive excitor in its capacity to modulate the aversive-reinforcement process. This parallel would, of course, be strengthened if the similarity could be extended to other properties of aversive excitors. Wagner (1971) reported an experiment in which it was found, using a rabbit eyelid-conditioning preparation, that when two excitors are presented in compound and nonreinforced, more extinction occurred to one stimulus if the other was a strong rather than a weak excitor. This enhancement of extinction can also be demonstrated in conditioned suppression (Dickinson, 1976). One group of rats, Group P, received the light, A, paired with shock in the first stage and the tone, X, also paired with shock in second stage, as in condition 1 of Table 8.5. These two aversive excitors were then presented in compound and nonreinforced in a third stage. Finally, the residual amount of conditioning to X was measured on tone-alone test trials. Control groups received exactly the same training except during the first stage. In this stage, the light, A, was associated with shock omission for the Group U, was semirandomly associated with shock for Group R, and was not presented for Group no-CS. The enhancement of aversive extinction is illustrated in Panel A of Fig. 8.7 by the fact that Group P showed less suppression to the tone, X, than the control groups on the test trials. Enhancement of extinction can be explained in the same terms as blocking, by assuming that the amount of extinction occurring to a CS is a positive function of the level of arousal of the aversive system at the time the CS is presented (Konorski, 1948; Rescorla, 1973). Simultaneous presentation of another aversive excitor just increases this level.

TABLE 8.5
Enhancement of aversive extinction

Conditions	Stage 1	Stage 2	Stage 3	test
1	A ⟶ shock	X ⟶ shock	AX	X
2	control treatments	X ⟶ shock	AX	X
3	A ⟶ food ommission	X ⟶ shock	AX	X

In the next experiment, Dickinson (1976) attempted to find out whether an attractive inhibitor would similarly enhance aversive extinction by comparing conditions 2 and 3 of Table 8.5. Again after lever pressing had been initially established, paired, unpaired, random, and no-CS groups were run, using exactly the same classical appetitive conditioning schedules during stage 1 as employed in the transreinforcer blocking experiment. As a result, the light, A, should have become an attractive inhibitor in Group U. Thereafter, the procedure, followed that for the enhancement of the

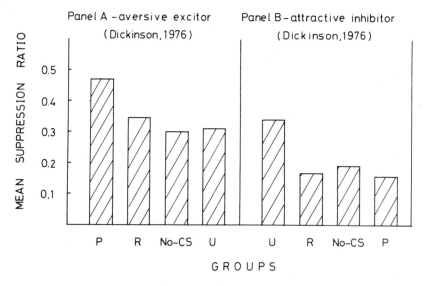

FIG. 8.7. *Enhancement of aversive extinction.* Mean suppression ratios to added element X on test. *Panel A:* Enhancement when A established as aversive excitor in Group P. *Panel B:* Enhancement when A established as attractive inhibitor in Group U. P: paired groups; U: unpaired groups; R: random groups; no-CS; no-CS groups.

aversive-extinction experiment during stages 2 and 3 (see Table 8.5). For all groups the tone, X, was established as an aversive excitor in stage 2 by giving 10 pairings with a 0.5-ma 0.5-sec shock in a single session. After the lever had been returned and pressing reestablished, the rats were give 8 nonreinforced presentations of the tone–light compound in stage 3, during which suppression was extinguished. Finally the residual suppression to the tone was measured in 2 tone-alone test trials, as illustrated in Panel B of Fig. 8.7. A comparison of Panels A and B shows that an attractive inhibitor enhances aversive extinction in the same manner as an explicit aversive excitor. An overall analysis of the test suppression ratios revealed a significant difference between the groups ($p < 0.02$), while individual comparisons showed that Group U was significantly less suppressed than each of the control groups ($p < 0.05$ in all cases).

Thus, it appears that an attractive inhibitor can modulate aversive conditioning and extinction in a manner that parallels an explicit aversive excitor — a parallel that is captured by an opponent-process model of appetitive–aversive interactions. At present there is little evidence about the remaining interactions — namely, the effect of an aversive inhibitor on processes controlled by an attractive excitor (see Table 8.l). Fowler et al.(1973), however, did find that presenting an aversive inhibitor just after a rat had made the correct choice in a food-reinforced brightness discrimination in a T maze increased the number of errors during acquisition. Fowler (in press) himself has recently interpreted this finding in terms of the blocking of appetitive conditioning to the positive discriminative stimulus of the maze by the aversive inhibitor.

VI. ATTRACTIVE–AVERSIVE AND EXCITATORY–INHIBITORY DISTINCTIONS

At this point it is worth summarizing the conclusions we have reached so far. Table 8.1 specifies two classes of appetitive–aversive interactions: first, that between excitors of opposite affective value and, second, that between inhibitors and excitors of opposite value. The evidence reviewed points to the following conclusions.

1. On the basis of the summation and retardation tests, an excitor appears to have an inhibitory effect on processes controlled by another excitor of opposite affective value.

2. The transreinforcer superconditioning effect argues that the locus of the interaction is central, whereas successful counterconditioning of the reinforcing properties of a stimulus suggests that the interaction occurs between systems mediating the motivating and reinforcing properties of stimuli.

3. The second interaction, that between excitor and inhibitor, appears to be excitatory in nature in that an inhibitor can function like an explicit excitor of the opposite value in modulating reinforcement and extinction processes controlled by that excitor. This interaction was revealed by the presence of transreinforcer blocking and enhancement of extinction.

4. The basic opponent-process model does a fairly good job of accounting for these interactions.

We cannot, however, let the analysis rest there, for once we realize that an implication of the model is that excitors and inhibitors both have excitatory and inhibitory actions, it raises doubts about the basic distinction between these two classes of stimuli.

A stimulus is regarded as an excitor for a particular response or motivational state because of its ability to elicit that response or state. We have seen, however, that it also has the capacity to inhibit other responses and states; why, therefore, should we not regard the stimulus as primarily an inhibitor? The reason is that, in terms of the present model, the internal association directly involving the stimulus representation is an excitatory one. This association is either preestablished in the case of a US or formed by the excitatory conditioning precedure in the case of a CS. If we wished to be perverse, however, the algebra of associations would allow us to argue that pairing a CS with, for example, food establishes a direct inhibitory connection between the CS representation and the aversive system. The appetitive properties of this CS are manifest because its presentation inhibits the aversive system, which in turn removes an inhibitory influence from the appetitive system. If this were so, we might wish to characterize a CS paired with food primarily as an aversive inhibitor. Besides its sheer perversity, there are a number of reasons for arguing against this position, the most obvious being that it implies that appetitive responses should only be elicited when the aversive system is aroused, so that its inhibition can in turn lead to the removal of an inhibitory influence from the appetitive system. Because appetitive responses can clearly be elicited in the absence of aversive stimulation, it is obvious that we should regard the CS as an attractive excitor, and treat its capacity to inhibit aversive behavior as a secondary property.

The appropriate characterization of an inhibitor, however, is less clear. The common reason for regarding the acquired properties of a stimulus that has been negatively correlated with food as primarily inhibitory is again because in the model the internal association directly involving the CS representation is an inhibitory one (see Fig. 8.5). Its capacity to act like an aversive excitor is then a secondary property brought about by the inhibitory connections between the appetitive and aversive systems. Alternatively, we could assume that negatively correlating the CS and food establishes an excitatory connection between the CS representation and the aversive system, so that its aversive properties are a primary effect and its inhibitory ones only secondary. This functional connection is illustrated in Fig. 8.5. Konorski (1967), in fact, assumed that inhibitory conditioning led to the formation of an excitatory association directly involving the CS representation. He argued, however, that the target of this association was an "anti-drive" center for the particular system aroused by the US, rather than the motivational system activated by stimuli of opposite affective value.

It is not so obvious that treating an "inhibitor" as primarily an excitor is nearly as perverse as regarding an "excitor" as primarily an inhibitor. In fact, frustration theory (e.g. Amsel, 1958) has long treated a CS paired with food omission as a direct excitor of specific responses and motivational states, and a number of authors (e.g., Gray, 1975; Wagner, 1969) have argued that there are common systems involved in mediating the behavioral effects of fear and frustration CSs. We argued earlier that a stimulus should be regarded as a true "excitor" for a particular response or state if it could elicit that response in the absence of stimulation of the opposite affective value. This is because a double-inhibitory path can only produce an excitatory effect if the first link serves to suppress a system with an antagonistic influence on the mechanism mediating the elicited response. Disinhibition of this mechanism requires that the opponent system is actually exerting an inhibitory effect at that time. It is equally true, however, that a double-inhibitory path can only produce an excitatory effect if the mechanism directly mediating the response is itself also under an arousing influence, for only then can disinhibition produce an increased level of activity. This means that a CS paired with the omission of food should only have aversive properties if there is some other independent source of aversive stimulation.

There is good evidence, however, that the omission of an attractive US and the presentation of CSs associated with such omission have aversive properties in their own right, and in the absence of explicit aversive stimulation (Coughlin, 1972; Leitenberg, 1965; Wagner, 1969). Animals will withdraw from potentially attractive inhibitors negatively correlated with an attractive US (e.g. Hearst & Franklin, 1977; Jenkins & Boakes, 1973; Wasserman, et al., 1974). The omission of an attractive US can provide a source of negative reinforcement for instrumental responding; animals will learn to perform an instrumental response to avoid (e.g., D'Andrea, 1971; Thomas, 1965) and to escape from (e.g., Daly, 1974) such omission and associated stimuli. Finally, the omission of an attractive US can suppress responding in both punishment (e.g., Baron & Kaufman, 1969; Hyde, 1976) and conditioned-suppression procedures (e.g., Holmes, 1972; Leitenberg, Bertsch, & Coughlin, 1968).

These findings are compatible with the idea that the associative path activated by an attractive inhibitor is a direct excitatory connection with the aversive system, and that the inhibitory properties of such a stimulus are mediated by the opponent-process system. A similar argument can be mounted for regarding an aversive inhibitor as primarily an excitor of the appetitive system. There is some evidence that aversive inhibitors are attractive stimuli in their own right and can provide a source of positive reinforcement for instrumental responding (e.g., Denny, 1971; Morris, 1975; Rescorla, 1969b; Weisman & Litner, 1969) in the absence of any source of explicit attractive stimulation.

VII. CONCLUSIONS

We have seen that an excitor of one affective class has an inhibitory effect on the general motivating and reinforcing properties of an excitor of the opposite affective value. And we have argued, along with Konorski and other authors, that such an effect can be explained in terms of reciprocal inhibitory connections between central appetitive and aversive motivational and reinforcement systems. Furthermore, it appears that an inhibitor has functional similarities to an excitor of the contrasted affective class in its capacity to modulate reinforcement and extinction. Both the inhibitory and excitatory properties of inhibitors are compatible with the opponent-process model and can be embodied within such a model by two alternative associative structures. The first assumes that the prime association formed by a negative correlation between a CS and US is an inhibitory connection between the CS representation and the system activated by the US during conditioning. In such a model, the excitatory effects of an inhibitor are a secondary consequence of the inhibitory relationship between the appetitive and aversive systems. In contrast, the alternative view argues that the prime association is an excitatory connection between the CS representation and the motivational system opposed to the one activated by the US during conditioning. In this case the excitatory properties of the "inhibitor" represent its direct action, with the inhibitory effects being mediated by the opponent-process system. This latter view is similar to Konorski's later theory (1967) in assuming that both excitatory and inhibitory conditioning procedures result in the formation of excitatory association with different target systems, but differs in suggesting that this target system is a motivational mechanism of opposite affective polarity rather than a specific "anti-drive" system.

We have argued for the second of these alternatives largely on the basis that inhibitors exhibit motivational and reinforcement effects characteristic of the opposite system in the absence of any explicit excitor of that system. A representation of this type of model with the framework of Konorski's theory of classical conditioning is illustrated in Fig. 8.8.

Throughout this paper, we have tended to treat the appetitive and aversive systems as though they are unitary mechanisms activated by a wide range of different USs. As we have mentioned, Konorski (1967) distinctly identified different drive systems related to different classes of attractive USs. In fact, he assumed that the interaction between different appetitive motivational states can also be explained in terms of reciprocal inhibitory connections between their respective systems. Contemporary research on the pattern of time sharing between different appetitive activities, however, suggests that such a model is not appropriate (McFarland, 1974); and we must assume that such time sharing is controlled by mechanisms outside the scope of the appetitive system. As far as the properties of excitors go, it matters little whether we treat the appetitive and aversive systems as some final, common pathway activated by a wide range of different stimuli within each affective category or as a collection of interrelated systems, each aroused by the appropriate type of attractive or aversive stimulus. Even within a general system, effects specific to a particular US (e.g., Overmier, Bull, & Trapold, 1970; Trapold, 1970) can still be mediated by the internal representation of the actual US employed. The nature of the motivational systems, however, becomes critical when we consider the properties of inhibitors.

If we assume that the motivational systems are unitary mechanisms, and that an inhibitor of one affective class directly activates the system of the opposite class, the problem that such a model faces is that the associative structure underlying the action of an inhibitor does not retain any information about the particular US employed during conditioning. This means that a conditioned inhibitor, established with a US of one particular affective category, should be capable of inhibiting responses established or elicited by a wide range of USs of the same affective polarity. At present we know of no critical evidence that allows us to decide whether conditioned inhibitors are US specific within a given affective category. Such specificity would point to the alternative conception of these general motivational systems as a collection of interrelated systems. So the general aversive mechanism would then be composed of a collection of both separate drive systems for different types of aversive USs and separate antidrive systems, each with a selective inhibitory influence on a particular appetitive drive system within the general appetitive mechanism. Even so we might want to

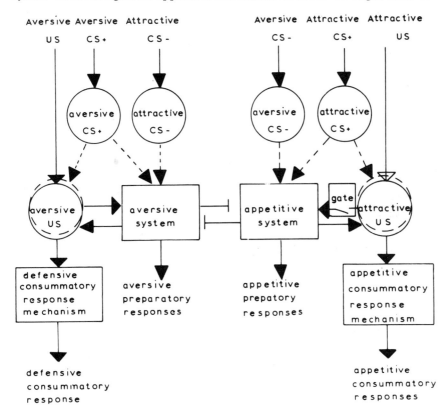

FIG. 8.8.   *Illustration of the opponent-process model:* System mediating appetitive-aversive interactions and conditioned inhibition. CS +: conditioned excitor; CS −: conditioned inhibitor; ——▶: preestablished excitatory connections; —|: preestablished inhibitory connections; - - -▶: excitatory connections established by conditioning; circle: internal representations of a stimulus excited by either unconditioned (solid circles) or conditioned (hatched circles) connections.

treat the collection of separate aversive drive and appetitive antidrive systems as some form of composite, general aversive mechanism because of the similarity of their excitatory and reinforcing properties.

Two final points should be made in connection with the present notion of conditioned inhibition. First, conditioned inhibition is purely motivational in character. An inhibitor does not signal that a US will not occur but simply elicits an affective state antagonistic to that necessary for the performance of the response. As a result, inhibitory mechanisms should not be engaged during associative conditioning using neutral stimuli. Although it remains an open question as to whether conditioned inhibitors can be established by, for example, sensory preconditioning, it is clear that any explanation within the present terms would have to resort to something like the "no-US" units entertained by Konorski. These would oppose the arousal of the sensory representation of the US. Second, as it stands, the model assumes that an attractive inhibitor, for example, can activate the aversive system independently of the corresponding appetitive state of the animal. This may well not be so, for it makes sense to suppose that an animal will only find an attractive inhibitor aversive it if expects food and is hungry—or, in other words, only if the appetitive system is also aroused. To accommodate this fact, we should have to assume the input to one motivational system from the sensory representation of an inhibitor is in some way gated by the state of arousal of the opposite system. It will be appreciated that this immediately produces complications in working out the interaction between the two systems. At present, we know of little evidence that bears directly on this question, although Rescorla (1976) has recently demonstrated that an aversive inhibitor has the capacity to reinforce higher-order inhibitory conditioning in the absence of an aversive excitor (see also Fowler, in press).

Whatever the particular nature of the appetitive and aversive mechanisms, the major implications of this type of model is that the excitatory–inhibitory distinction is, at least at the level of general motivational processes, secondary to the affective distinction. All motivationally significant stimuli have both excitatory and inhibitory effects, but on different classes of behavior.

REFERENCES

Amsel, A. The role of frustrative nonreward in noncontinuous reward situations. *Psychological Bulletin*, 1958, *55*, 102–119.
Baron, A., & Kaufman, A. Time-out punishment: Pre-exposure to time-out and opportunity to respond during time-out. *Journal of Comparative and Physiological Psychology*, 1969, *67*, 479–485
Bindra, D. A unified account of classical conditioning and operant training. In A. M. Black

& W. F. Prokasy (Eds.), *Classical conditioning II: Current research and theory.* New York: Appleton-Century-Crofts, 1972.

Bindra, D. A motivational view of learning, performance, and behavior modification. *Psychological Review,* 1974, *81* 199–213.

Blanchard, R., & Honig, W. K. Surprise value of food determines its effectiveness as a reinforcer. *Journal of Experimental Psychology: Animal Behavior Processes,* 1976, *2,* 67–74.

Bruner, A. Facilitation of classical conditioning in rabbits by reinforcing brain stimulation. *Psychonomic Science,* 1966, *6,* 211–212.

Bull, J. A., III. An interaction between appetitive Pavlovian CSs and instrumental avoidance responding. *Learning and Motivation,* 1970, *1,* 18–26.

Coughlin, R. C. The aversive properties of withdrawing positive reinforcement: A review of the recent literature. *Psychological Record,* 1972,*22,* 333–354.

Daly, H. B. Reinforcing properties of escape from frustration aroused in various learning situations. In G. H. Bower (Ed.), *The psychology of learning and motivation* (Vol. 8). New York: Academic Press, 1974.

D'Andrea, T. Avoidance of time-out from response-independent reinforcement. *Journal of the Experimental Analysis of Behavior,* 1971, *15,* 319–325.

Dearing, M. F. *Counterconditioning in rabbits.* Unpublished Masters dissertation, University of Sussex, 1978.

Denny, M. R. Relaxation theory and experiments. In F. R. Brush (Ed.), *Aversive conditioning and learning.* New York: Academic Press, 1971.

Dickinson, A. *Modulation of fear conditioning and extinction by frustration.* Unpublished manuscript, 1976.

Dickinson, A. Appetitive–aversive interactions: Superconditioning of fear by an appetitive CS. *Quarterly Journal of Experimental Psychology,* 1977, *24,* 71–83.

Dickinson, A., & Mackintosh, N. J. Classical conditioning in animals. *Annual Review of Psychology,* 1978, *29,* 587–612.

Dickinson, A., & Pearce, J. M. Preference and response suppression under different correlations between shock and a positive reinforcer in rats. *Learning and Motivation,* 1976, *7,* 66–85.

Dickinson, A., & Pearce, J. M. Inhibitory interactions between appetitive and aversive stimuli. *Psychological Bulletin* 1977, *84,* 690–711.

Estes, W. K. Outline of a theory of punishment. In B. A. Campbell & R. M. Church (Eds.), *Punishment and aversive behavior.* New York: Appleton-Century-Crofts, 1969.

Estes, W. K., & Skinner, B. F. Some quantitative properties of anxiety. *Journal of Experimental Psychology,* 1941, *29,* 390–400.

Erofeeva, M. N. Contributions a l'étude des reflexes conditionnels destructifs. *Compte rendu de la societé de biologie Paris,* 1916, *79,* 239–240.

Erofeeva, M. N. Additional data on nocuous conditioned reflexes. *Izvestiga Petrogradskogo Nauchnago Instituta im P. F. Lesgafta,* 1921, *3,* 69–73. (Russian)

Fowler, H., Fago, G. C., Domber, E. A., & Hochhauser, M. Signaling and affective functions in Pavlovian conditioning. *Animal Learning & Behavior,* 1973. *1,* 81–89.

Fowler, H. Cognitive associations as evident in the blocking effects of response-contingent CSs. In S. H. Hulse, H. Fowler, & W. K. Honig (Eds.), *Cognitive processes in animal behavior.* Hillsdale, N. J.: Lawrence Erlbaum Associates, in press.

Frey, P. W., Maisiak, R., & Dugue, G. Unconditional stimulus characteristics in rabbit eyelid conditioning. *Journal of Experimental Psychology: Animal Behavior Processes,* 1976, *2,* 175-190.

Gray, J. A. *Elements of a two-process theory of learning.* London: Academic Press, 1975.

Grossen, N. E., Kostansek, D. J., & Bolles, R. C. Effects of appetitive discriminative stimuli on avoidance behavior. *Journal of Experimental Psychology,* 1969, *81,* 340–343.

Hearst, E. Some persistent problems in the analysis of conditioned inhibition. In R. A. Boakes & M. S. Halliday (Eds.), *Inhibition & learning.* London: Academic Press, 1972.

Hearst, E., & Franklin, S. R. Positive and negative relations between a signal and food: Approach-withdrawal behavior. *Journal of Experimental Psychology: Animal Behavior Processes,* 1977, *3,* 37-52.

Holmes, P. W. Conditioned suppression with extinction as the signalled stimulus. *Journal of the Experimental Analysis of Behavior,* 1972, *18,* 129-132.

Hyde, T. S. The effect of Pavlovian stimuli on the acquisition of a new response. *Learning and Motivation,* 1976, *7,* 223-239.

Jenkins, H. M., & Boakes, R. A. Observing stimulus sources that signal food or no food. *Journal of the Experimental Analysis of Behavior.* 1973, *20,* 197-207.

Kamin, L. J. Predictability, surprise, attention and conditioning. In B. A. Campbell and R. M. Church (Eds.) *Punishment and aversive behavior.* New York: Appleton-Century-Crofts, 1969.

Konorski, J. *Conditioned reflex and neuron organisation.* Cambridge: Cambridge University Press, 1948.

Konorski, J. *Integrative activity of the brain: An interdisciplinary approach.* Chicago: University of Chicago Press, 1967.

Konorski, J., & Szwejkowska, G. Reciprocal transformations of heterogeneous conditioned reflexes. *Acta Biologiae Experimentalis,* 1956, *16,* 95-113.

Leitenberg, M. Is time out from positive reinforcements an aversive event? A review of the experimental evidence. *Psychological Bulletin,* 1965, *64,* 428-441.

Leitenberg, M., Bertsch, G. J., & Coughlin, R. C. "Time-out from positive reinforcement" as the UCS in a CER paradigm with rats. *Psychonomic Science,* 1968, *13,* 3-4.

Marukhanyan, E. V. The effect of the duration and intensity of a conditioned electroshock stimulus upon the magnitude of the conditioned food and acid reflexes. *Zhurnal Vysshei Nervio Deyatel'nosti,* 1954, *4,* 684-691. (Russian)

McFarland, D. J. Time-sharing as a behavioral phenomenon. In D. S. Lehrman, J. S. Rosenblatt, R. A. Hinde, & E. Shaw (Eds.), *Advances in the study of behavior (Vol. 5).* New York: Academic Press, 1974.

Millenson, J. R., & de Villiers, P. A. Motivational properties of conditioned anxiety. In R. M. Gilbert & J. R. Millenson (Eds.), *Reinforcement: Behavioral analyses.* New York: Academic Press, 1972.

Miller, N. E. Some reflections on the law of effect produce a new alternative to drive reduction. In M. R. Jones (Ed.), *Nebraska Symposium on Motivation (Vol. 11)* Lincoln: University of Nebraska Press, 1963.

Morris, R. G. M. Preconditioning of reinforcing properties to an exteroceptive feedback stimulus. *Learning and Motivation,* 1975, *6,* 289-298.

Mowrer, O. H. *Learning theory and behavior.* New York: Wiley, 1960.

Olds, J., & Olds, M. E. The mechanisms of voluntary behavior. In R. G. Heath (Ed.), *The role of pleasure in behavior.* New York: Harper & Row, 1964.

Overmier, J. B., Bull, J. A., III, & Pack, K. On instrumental response interaction as explaining the influences of Pavlovian CSs upon avoidance behavior. *Learning and Motivation,* 1971, *2,* 103-112.

Overmier, J. B., Bull, J. A., III, & Trapold, M. A. Discriminative cue properties of different fears and their role in response selection in dogs. *Journal of Comparative and Physiological Psychology,* 1971, *76,* 478-482.

Pearce, J. M., & Dickinson, A. Pavlovian counterconditioning: Changing the suppressive properties of shock by association with food. *Journal of Experimental Psychology: Animal Behavior Processes,* 1975, *1,* 170-177.

Rescorla, R. A. Pavlovian conditioned inhibition. *Psychological Bulletin,* 1969, *72,* 77-94.(a)

Rescorla, R. A. Establishment of a positive reinforcer through contrast with shock. *Journal of Comparative and Physiological Psychology,* 1969, *67,* 260–263. (b)

Rescorla, R. A. Variations in the effectiveness of reinforcement and nonreinforcement following prior inhibitory conditioning. *Learning and Motivation,* 1971, *2,* 113–123.

Rescorla, R. A. A model of Pavlovian conditioning. In V. S. Rusinov (Ed.), *Mechanisms of formation and inhibition of conditional reflex.* Moscow: "Nauka" Academy of Sciences of the USSR, 1973. (Russian)

Rescorla, R. A. Second-order conditioning of Pavlovian conditioned inhibition. *Learning and Motivation,* 1976, *7,* 161–172.

Rescorla, R. A., & Solomon, R. L. Two-process learning theory: Relationships between Pavlovian conditioning and instrumental learning. *Psychological Review,* 1967, *74,* 151–182.

Scavio, M. J., Jr. Classical–classical transfer: Effects of prior aversive conditioning upon appetitive conditioning in rabbits. *Journal of Comparative and Physiological Psychology,* 1974, *86,* 107–115.

Solomon, R. L., & Corbit, J. D. An opponent-process theory of motivation: I. Temporal dynamics of affect. *Psychological Review,* 1974, *81,* 119–145.

Stein, L. Reciprocal notion of reward and punishment mechanisms. In R. G. Heath (Ed.), *The role of pleasure in behavior.* New York: Harper & Row, 1964.

Tait, R. W. *Assessment of the bidirectional conditioning hypothesis through the UCS_1–UCS_2 conditioning paradigm.* Unpublished doctoral dissertation, University of Iowa, 1974.

Taukulis, H. K., & Revusky, S. H. Odor as a conditioned inhibitor: Applicability of the Rescorla–Wagner model to feeding behavior. *Learning and Motivation,* 1975, *6,* 11–27.

Thomas, J. R. Time-out avoidance from a behavior-independent contingency. *Psychonomic Science,* 1965, *3,* 217–218.

Trapold, M. A. Are expectancies based upon different positive reinforcing events discriminably different? *Learning and Motivation,* 1970, *1,* 129–140.

Varga, M. N. On the role of power ratios in the functioning of the bidirectional conditioned reflex bond. *Zhyrnal Vysshei Nervhoi Deyatel'nosti,* 1955, *5,* 723–731. (Russian)

Wagner, A. R. Frustrative nonreward: A variety of punishment? In B. A. Campbell & R. M. Church (Eds.), *Punishment and aversive behavior.* New York: Appleton-Century-Crofts, 1969.

Wagner, A. R. Elementary associations. In H. H. Kendler & J. T. Spence (Eds.) *Essays in neobehaviorism: A memorial volume to Kenneth W. Spence.* New York: Appleton-Century-Crofts, 1971

Wagner, A. R. Expectancies and the priming of STM. In S. H. Hulse, H. Fowler, & W. K. Honig (Eds.), *Cognitive processes in animal behavior.* Hillsdale, N. J.: Lawrence Erlbaum Associates, in press.

Wasserman, E. A., Franklin, S. R., & Hearst, E. Pavlovian appetitive contingencies and approach versus withdrawal to conditioned stimuli in pigeons. *Journal of Comparative and Physiological Psychology,* 1974, *86,* 616–627.

Weisman, R. G. & Litner, J. S. Positive conditioned reinforcement of Sidman avoidance behavior in rats. *Journal of Comparative and Physiological Psychology,* 1969, *68,* 597–603.

9 Interactions Between Type I and Type II Processes Involving Positive Reinforcement

R.A. Boakes
University of Sussex

I. INTRODUCTION

"Classical-instrumental interactions" and similar phrases have been used to refer to a specific set of issues that arise from attempting to understand the behavior of an animal exposed to both classical and instrumental conditioning procedures. The first study of this kind was carried out by Konorski and Miller (1930). An instrumental procedure was used to train a dog to lift a limb for food and, independently, a bell was established as a CS for food by means of a classical procedure. They then tested the effect of sounding the bell during performance of the instrumental response and found that instrumental responding decreased. Somewhat later a study by Estes and Skinner (1941), different in many respects from that of Konorski and Miller but similar in general principles, showed that a CS for response-independent shock suppressed bar pressing by rats for food. In the years that have followed, a very large number of experiments of this general form have been undertaken—both in the Nencki Institute, where the tradition of working with constrained subjects have been maintained, and in Western laboratories, where, in general, Estes and Skinner's lead in studying freely moving subjects has been followed.

Phrases such as classical-instrumental interactions have also been used in a much more general way, which is not confined to the consideration of situations in which both classical and instrumental procedures are deliberately introduced. This possibility arises because a situation that may appear to involve only classical contingencies can also contain *implicit* instrumental contingencies and, similarly, because a situation that may appear

233

to involve only instrumental contingencies may contain implicit classical contingencies. In this more general sense such labels may be applied to a variety of phenomena that have in common only that their analysis seems to demand involvement of two separate conditioning processes, which can compete, complement, or otherwise interact with each other. This chapter includes discussion of such phenomena, as well as issues arising from the specific experimental paradigm first used by Konorski and Miller.

Basic assumptions

It was clearly no accident that these two experimenters were the first to study this kind of problem. Their studies were prompted by their early adoption of two basic assumptions about conditioned behavior: (1) that there exist at least two distinctive forms of conditioning process and (2) that these forms can make contact in determining behavior. It seems difficult to discuss fruitfully the present topic unless these assumptions are accepted; consequently they require some comment.

Discussion of the first assumption can become confusing as a result of the ambiguity of some of the terms normally used. For example, "classical conditioning" is used to refer both to a set of experimental procedures and to some of the theories produced to explain the effects of employing such procedures. In an attempt to reduce the possible confusion caused by such ambiguity, the terms "instrumental" and "classical" are used in this chapter to refer to *procedures,* while the terms introduced by Konorski and Miller, "Type I" and "Type II", are used to refer to presumed conditioning *processes.* Thus, Type I refers to processes that involve the formation of associations between stimulus events or aspects of the animal's world ("S-S learning"), but not to associations involving a response as an element; whereas Type II refers to processes involving associations that do contain responses as elements ("S-R" or "R-Rft" learning).

An important reason for this notational distinction is that adoption by an experimenter of a given kind of procedure does not guarantee that over a range of situations the same kind of conditioning process is involved. It has been recognized for some decades that a change in behavior obtained by employing a classical procedure, in which reinforcement is independent of an animal's behavior, may involve a Type II process. Skinner (1948) applied the term "superstitious conditioning" to such cases, and Konorski used the term "parasitic conditioning (1967, p.268)."

The converse possibility is one in which an instrumental procedure is in effect, whereby reinforcement depends on an animal's responses, but the resulting change in behavior is based on Type I processes alone. Thus, in a situation in which an animal starting from A finds food at B, the change that occurs in its behavior may simply be based on learning that food is associated with place B and on using previously acquired skills to get from A to B. Although obtaining food is dependent on making the response "go from A to B," it is not necessary to assume that any kind of associative learning involving such a response is involved. This second possibility has often been discussed by theorists of animal learning. It was first given detailed attention by Morgan (1894), some 50 years before Skinner applied the term "superstitious conditioning" to its converse. Yet it has not acquired an equivalent label and, perhaps because of this, is a possibility that has often been overlooked outside of the Pavlovian tradition.

The second assumption behind Konorski and Miller's early study of interactions also requires some comment. Although the idea of two forms of conditioning became widely held in Western psychology, it was generally accompanied by the belief that the two forms act independently. This was particularly true within the Skinnerian tradition, despite the early study of conditioned suppression by Estes and Skinner (1941). A major exception was provided by Mowrer (1950), who, together with other two-factor theorists, assumed both that there are two forms of conditioning process and that the two make contact, at least in avoidance learning. In more recent years the increased interest in interactions among Western psychologists stems from two main sources, both of which stressed in different ways the effect of Type I processes on behavior maintained by instrumental contingencies. One was the review paper by Rescorla and Solomon (1967), in which concepts developed in the context of avoidance learning were applied to a wider range of issues. The second was the discovery of the autoshaping phenomenon by Brown and Jenkins (1968); subsequent research on this phenomenon has made it clear that direct effects on behavior can occur as a result of implicit stimulus–reinforcer contingencies, which may engage Type I processes, in situations used very widely for the study of instrumental conditioning. Detailed examples of this are discussed in Section IV.

Levels of interaction

Assumption B, that the two processes make contact, prompts the question, Where does contact occur? Reasonably explicit in the work of Konorski and of other theorists is the idea that interactions can take a number of different forms. It seems worth spelling this out.

Without making very much commitment to any particular theory of conditioning, one can distinguish between three components of both Type I and Type II processes: an associative, a motivational, and a response component, as illustrated schematically in Fig. 9.1. When, for example, a Type I process appears to be affecting the outcome of a Type II process, we can ask whether the interaction is occurring at an associative, motivational, or response level, or at more than one level. Although detailed distinctions between such components may turn out to be difficult (see Morgan, Chapter 7), this kind of categorization appears helpful in considering interactions, as the following examples are intended to suggest.

TYPE I PROCESS   TYPE II PROCESS

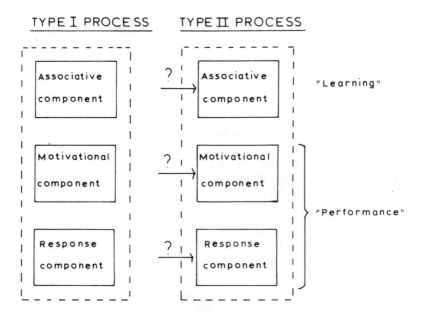

FIG. 9.1.   Schema illustrating possible forms of Type I-Type II interactions.

The earliest example of an interaction at an *associative* level is the experiment by Miller and Konorski (1928), discussed by Mackintosh and Dickinson (Chapter 6). Passive flexion of a dog's leg was consistently followed by the presentation of food, but in addition a stimulus — the sound of bubbling water—always accompaned leg flexion and did not occur at any other time. Bubbling water did not come to elicit active flexion responses under these particular conditions. The set of experiments to which this example belongs, together with those described in Chapter 6, suggest that the instrumental response fails to appear because acquisition of the Type II association, on which performance depends, has been overshadowed by the development of a Type I association. Learning that bubbling water predicts food prevents learning that leg flexion predicts food.

The phenomenon described by Estes and Skinner (1941) provides the most obvious example of a motivational interaction. In a typical experiment producing a "conditioned emotional response" (CER), a subject is independently exposed to both a classical procedure for establishing (say) a tone as a CS for shock and an instrumental procedure to establish some arbitrary response that is maintained by food reinforcement. If the tone is sounded during performance of the Type II response, the latter decreases in frequency. The procedural sequence and independent evidence for the development of both Type I and Type II associations rule out the possibility that the effect is due to interference at an associative level. Instead there are good reasons for believing that response suppression in CER occurs at least partly because the CS for shock tends to inhibit a system whose activation is necessary for maintained performance of the instrumental response. For example, Leslie (1977) has shown that the effects of such a CS closely resemble those of changing an animal's body weight. A full discussion of the issue appears in a recent review by Dickinson and Pearce (1977).

One of the reasons for concluding that CER involves a motivational interaction is that one can rule out an alternative possibility that it results solely from *response* competition between behavior conditioned to the tone as part of the Type I process engaged during classical training and the instrumental response. Similar competition would be expected whether the instrumental behavior is maintained by a schedule of attractive or aversive events. However, a CS for shock can produce an *increase* in instrumental responding when the latter is maintained by an avoidance schedule (Herrnstein & Sidman, 1958; Scobie, 1972).

Nevertheless recent evidence suggests that response interactions may well be involved in CER effects. Changes in the location of the CS, which are known to affect the Type I behavior elicited by the CS, can affect the degree of response suppression obtained, whereas such changes are not generally assumed to affect the motivational function of the CS. In an experiment by Karpicke, Christoph, Peterson, and Hearst (1977), a light was used as the CS for shock; this suppressed chain pulling for food by rats more when it was close to the chain (less than 1 cm away) than when at a distance (over 14 cm away). Since in general pairing a localized stimulus with shock results in withdrawal from the stimulus by an animal, their CER result is most easily interpreted as reflecting greater competition from Type I behavior in the "near" condition. Thus, it appears that CER can involve interactions between Type I and Type II processes at both motivational and response levels.

Until recently major emphasis in the analysis of Type I–Type II interactions has been placed on motivational effects. Both Konorski (1967) and Rescorla and Solomon (1967) attempted to understand the effects on instrumental behavior of a CS associated with an *attractive* event, as well as the kind of aversive–attractive interaction represented by the CER, solely in terms of motivational processes. One aim of this chapter is to suggest that this emphasis has been misplaced and that many of the puzzles in this area have arisen because the possibility, and the potential complexity, of behavioral interactions has received insufficient attention. As suggested by the experiment by Karpicke et al., a prerequisite for the analysis of interactions between response systems is some understanding of behavior produced by Type I processes in the absence of any instrumental contingency. This issue is discussed in Section II.

II.   THE DIRECTION AND FORM OF TYPE I BEHAVIOR

The explanation offered for the results of Karpicke et al. depended on the subjects' moving away from a CS associated with shock. Similar withdrawal from a localized CS is also obtained when the stimulus predicts the absence of food (see Hearst, Chapter 2). But what are the direct behavioral consequences of pairing a CS with an attractive event? This question can be divided into two subproblems: what form does the behavior take, and how is it directed?

In most studies using aversive events, the US is a shock that, although usually delivered to the feet, does not have a more specifically located source. In the vast majority of studies using attractive events as USs, the subject is free to move around in its environment, and the US — usually food delivery — occurs at some particular point in that environment. Thus there are two important locations in many such situations, the site of the CS and that of the US. Perhaps for this reason the question of the directiveness of Type I behavior appears to be a more difficult one when attractive rather than aversive USs are involved.

Sign tracking

A great number of experiments using autoshaping procedures have shown that animals approach a localized stimulus predicting the occurrence of an attractive reinforcer (see Hearst, Chapter 2). This kind of directed behavior has been termed "sign tracking" in the review by Hearst and Jenkins (1974), which reported such behavior in a "long box" where the stimulus is some distance from the grain tray. In a chamber of this kind, approaching the CS means less time is available to eat grain than would be the case if the animals stayed close to the tray. More recently Peden, Browne, and Hearst (1977) have shown, by arranging an explicit omission contingency, that sign tracking in such a long box persists even when approach to the CS on a given trial cancels the delivery of grain at the end of that trial. This result makes it quite clear that the behavior is usually the result of Type I conditioning and is not maintained by any implicit response–reinforcer contingency involving Type II conditionng — that is, it is not a result of superstitious conditioning.

Although pigeons have been by far the most popular subjects for this kind of research, demonstrations of sign tracking are by no means confined to this species. If the insertion of a lever into the chamber predicts the delivery of food, then rats will direct behavior toward the lever (e.g., Peterson, Ackil, Frommer, & Hearst, 1972), and this behavior can also persist despite the presence of an omission contingency (e.g., Atnip, 1977). Observations of CS-directed behavior can also be found in reports of studies using the traditional Pavlovian preparation with harnessed dogs. Both Pavlov (1941) and Konorski (e.g., 1967 p. 269) noted that behavior was directed toward the source of noise or light serving as the CS, at least during early conditioning trials. More detailed reports of this kind were provided by Zener (1937) when he used unharnessed dogs in a Pavlovian situation.

Nonetheless, when subjects other than pigeons are studied the strength of sign tracking can be less impressive. If visual stimuli are used as CS for rats, sign tracking is less persistent and is almost completely abolished by the introduction of an omission contingency (e.g., Boakes, 1977). Also observed in such situations is behavior during the CS toward a part of the environment other than the CS location — namely, toward the site of the US. Descriptions of this kind of behavior can also be found in Zener (1937). To contrast this directiveness of behavior toward the US site with sign tracking, the term "goal tracking" has been suggested (Boakes, 1977).

Goal tracking

In a series of experiments in which a visual stimulus signaled the arrival of food to rats, a mixture of sign tracking and goal tracking was obtained (Boakes, 1977). The predominance of one or the other type of behavior was found to depend on a number of factors. For example, an omission contingency applied to goal tracking — recorded in this situation as the operation of a flap in front of the food tray — increased the frequency of sign tracking, while the use of a 50% partial-reinforcement schedule produced very little goal tracking and much more sign tracking than when the CS was followed by the US on every trial.

An obvious question prompted by such results is, Why have effects like these not been observed in autoshaping studies with pigeons? A more productive approach than assuming that pigeons are inherently different in their Type I behavior from most other species is to look at the standard situation for testing pigeons to see whether it contains some potentially critical feature. An earlier failure to obtain very reliable sign tracking in a long box made us stumble onto one such feature. This first use at the University of Sussex of a long box happened to employ a high level of general illumination, which made the light that came on over the grain during reinforcement periods (the traylight used in almost all Skinner box experiments with pigeons) barely discriminable, at least to the human observer. We have recently been systematically investigating the function of the tray light in a series of experiments with pigeons (Boakes & Alam, in preparation).

The long box used in this research is depicted in Fig. 9.2. A special floor section supported in microswitches, the "key pad," was placed by the response key on which the light serving as the CS was projected; and a

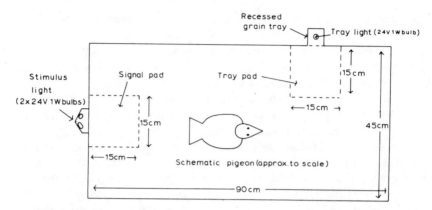

FIG. 9.2.   Plan view of long box used for research described in text showing position of key and tray pads.

similar floor section, the "tray pad," was placed by the aperture giving access to the food tray. One measure of approach to the CS or the US site, was the mean time the appropriate pad was depressed while the CS was present, less the mean time the same pad was depressed during equivalent "dummy" periods in the intertrial interval when the CS was absent. This measure is used in Fig. 9.3, which shows the results from an experiment in this series. The CS was a 6-sec illumination of the key, which was immediately followed by a 5-sec presentation of grain; the mean interval from the end of a reinforcement period to the onset of the CS on the next trial was about 1 min. The 3 groups of 6 pigeons in this experiment differed only in their tray-light conditions: In the "No T/L" group the aperture to the grain tray was never illuminated; in the "Steady T/L" group it was constantly illuminated throughout each session; and for the "Rft T/L" group the standard condition was employed whereby the tray light was on only while grain was available. From Fig. 9.3 it can be seen that sign tracking, or CS approach, developed in this last condition as in similar studies by Jenkins, Hearst, and their colleagues. However, this did not develop reliably in the other 2 groups, although individual subjects in these groups showed sign tracking in early sessions. Instead subjects in the No T/L and Steady T/L groups approached the tray during the CS. Thus, when the standard tray-light condition is not used, the behavior of pigeons resembles that of the aforementioned rats and dogs.

This was also shown by the results of experiments in which, using similar parameters, the effects of 50% and 100% reinforcement were compared. In the first experiment little difference was found between the groups when a standard tray-light condition (Rft T/L in the previous example) was used, but when the tray light was removed (No T/L condition) some subjects in the 100% group began to approach the tray, while none in the 50%

group did so. In a further experiment no tray light was used at any stage: With 100% reinforcement, subjects moved toward the food tray during the CS, but with 50% reinforcement subjects moved away, as shown in Fig. 9.4. This effect of reinforcement probability on goal tracking is exactly the same as that found in rats (Boakes, 1977).

The point of using a long box is that it allows a clear separation between approach to the signal and approach to the reinforcer site. In the conventional pigeon chamber, signal and reinforcer are close together; in approaching one, the pigeon also approaches the other. Consequently one would not expect the effect of tray-light conditions to be as marked when a conventional chamber is used. However, they can be found. In an experiment using a standard autoshaping procedure in a standard chamber, rates of pecking were lower with a Steady T/L condition than with the normal Rft T/L condition, even though the number of trials on which responding occurred was similar in both groups (Alam & Boakes, in preparation). One way of accounting for the reduction in response rate is to assume that, although in front of the key during the CS, a pigeon in the Steady T/L condition spends time looking down toward the tray aperture. Observational evidence is in agreement with this account. Other support comes from the report by Davol, Steinhauer, and Lee (1977) that autoshaping may fail to

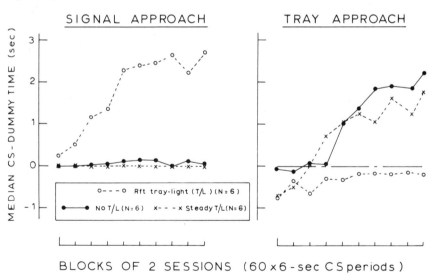

FIG. 9.3.   Effect of tray illumination on signal and tray approach in the long box shown in Fig. 9.2. Times on key or tray pad during CS periods, less time on pads during equivalent Dummy periods during intertrial periods ("CS – Dummy time"), are shown for three groups of pigeons, which differed only in illumination of the food tray: either steady throughout each session (Steady T/L), or during reinforcement periods only (Rft T/L), or not present at any time (No T/L).

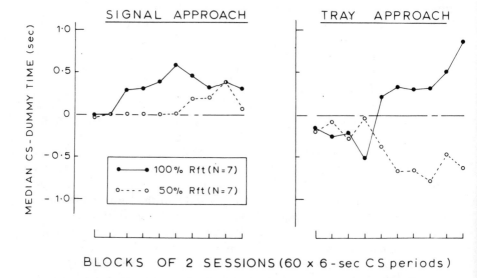

FIG. 9.4. Effect of reinforcement probability on signal and tray approach. CS–Dummy times were calculated as in Fig. 9.3 and are shown for two groups of pigeons, differing only in reinforcement probability, run under No T/L conditions.

occur if no tray light is used, although they offer a different interpretation from the present one.

The varied results from studies of dogs, rats, and pigeons can be summarized as follows. When a relatively brief and localized stimulus event is associated with a reinforcer, two response tendencies may be established. To the extent that the stimulus and reinforcer site are separated in location, sign tracking and goal tracking are incompatible, and goal tracking will interfere with sign tracking unless certain conditions hold. These conditions appear to include a relative increase in the strength of sign tracking from the use of highly salient CS—as in autoshaping studies with rats using lever insertion—and decreases in the strength of goal tracking resulting either from the presence of some salient event close to the US site that is correlated with the presence or absence of reinforcement, or from the use of partial reinforcement.

Diffuse events as conditioned stimuli

So far this section has been concerned with changes in an animal's behavior that arise from preceding each reinforcer by a brief event whose location can be detected by the animal. We consider here the behavioral effects of other kinds of stimuli.

From the immediately preceding discussion, one might expect that where the source of a stimulus paired with an attractive event cannot be located by an animal, only goal tracking will occur. Holland (1977) has recently reported a series of experiments that include detailed recording of the conditioned behavior of rats when diffuse tones or changes in general illumination are followed by food delivery. In his situation most responding was indeed directed toward the food tray. However, some conditioned behavior was not so directed, and systematic differences were found between the behavior patterns elicited by auditory and visual stimuli (see Rescorla, Chapter 4). For example, with the light as CS+, but not with the tone, rats frequently reared—that is, stood up on their hind legs and often pawed the walls— but no consistent orientation of this response was detected. These behavioral differences were seen by Holland as depending on the modality of the stimulus employed, but it is also possible that they reflect differences in the degree to which the source of stimulation could be located.

Further complications are suggested by two experiments with pigeons in which auditory stimuli were used. Schwartz (1973) employed a situation in which tones could be presented from a loudspeaker mounted a few inches from a response key. During an initial phase of training, food was delivered in the presence of a tone, and not in its absence; not unexpectedly, no pecking of the key occurred in this phase. In a subsequent phase the occurrence of food was signaled by a compound stimulus consisting of the tone and a visual stimulus displayed on the response key; this did sustain key pecking. Thereafter the initial condition, whereby only the tone was correlated with food, was reintroduced and now key pecking occurred during the tone, even though illumination of the response key stayed constant. To obtain comparisons across responses, a study by Jenkins (1977) included, as well as key pecking, a head-bobbing response. This was initially shaped using an instrumental procedure whereby reinforcement depended on the response being made during 10-sec periods of noise. Although the response contingency was necessary to establish the response, thereafter its appearance could be controlled by stimulus–reinforcer contingencies alone. When free reinforcement was delivered at the same frequency whether or not the noise was present, no head bobbing occurred; but when such reinforcement was presented only during the noise, head bobbing reappeared.

Clearly this is a topic that deserves much more research, but the foregoing evidence suggests that a diffuse stimulus associated with an attractive event may, when some factor weakens any goal tracking, elicit behavior whose direction is determined by past conditions— as by prior autoshaping involving a localized stimulus, (as in Schwartz's experiment) or by prior instrumental training (as in Jenkin's experiment)— even though these conditions no longer hold.

So far this section has considered the behavioral effects of pairing some occasional and brief event, whether localized or relatively diffuse, with an attractive event. Some of the examples of interactions discussed later in Section IV suggest that specific forms of Type I responding can occur in an unchanging environment where, for example, no information other than being in the experimental chamber can be used to predict the occurrence of reinforcement. The question of whether different mechanisms are involved when long-lasting, as opposed to brief, events serve as CSs was pursued by Konorski late in his career. In his 1967 book Konorski suggested that one needs to distinguish between two kinds of Type I process. He termed these "preparatory" and "consummatory." This distinction is discussed in the final part of this section, where we leave the question of direction and consider the form of Type I behavior.

Preparatory versus consummatory conditioning

A variety of responses, differing widely in direction and form, have been considered here to be the result of Type I processes. Traditionally performance models for classical conditioning have emphasized the role of the reinforcer in determining the form of conditioned behavior, and indeed Konorski's own theory of Type I conditioning was one in which the performance component maintained the principle of stimulus substitution. As both Hearst (Chapter 2) and Mackintosh and Dickinson (Chapter 6) note, this principle contains some ambiguity. In its most restricted form it claims that the CR is identical in form and direction with the UR; in the "surrogate" interpretation that Pavlov occasionally used, the CR closely resembles the UR in form but may be directed toward the CS and not toward the US site; and in the version adopted by Mackintosh and Dickinson, the CR consists of those features of the UR for which there exists concurrent contextual support. Although Konorski does not appear to have discussed these various interpretations, it seems quite clear that he held the most restricted version, based on the virtual identity of CR and UR. He pointed out that in salivary conditioning this identity is such that an observer can fail to detect the point of US onset from observing a subject's behavior. Similarly, his explanation for early orientation toward a CS in Pavlovian preparation in terms of "parasitic" (i.e., superstitious) conditioning—a mistaken explanation, as Hearst suggests—appears to depend on rejection of the possibility that the CR differs in direction from the UR.

In contrast Konorski's distinction (1967) between preparatory and consummatory conditioning can be seen as based, although not explicitly, on the denial that the principle of strict stimulus substitution applies to one kind of Type I process — namely, that typically occurring with stimuli of long duration or with variability in the occurrence of the US — which he termed preparatory conditioning. Thus, he described an association between some particular context and food as producing diffuse general activity, whereas in the Nencki studies using brief stimuli of fixed duration the conditioned response was, as already noted, indistinguishable from the unconditioned response; these last conditions were ones for which it is suggested that consummatory conditioning is most likely to occur.

In his chapter Hearst raises a number of difficulties connected with this preparatory–consummatory distinction; and one could prefer a more fortunate nomenclature. Yet the distinction between two kinds of outcome from classical conditioning does appear to offer a profitable way of resolving some of the problems concerning performance aspects of Type I processes. There are good grounds for holding that consummatory conditioned responses follow the restricted principle of stimulus substitution, as is found in preparations using salivation, leg flexion, and eye blink, whereas the behavior produced by preparatory conditioning may be affected by the specific properties of the US, but is also determined by the context through relationships between features of this context and particular behavior patterns that may have either a learned, innate, or mixed origin. Another way of expressing this view of preparatory conditioning is that, in Garcia's terms, it produces a hedonic shift in the subjective value of the stimuli involved. Garcia and his colleagues have provided in their reports of changing predator–prey relationships the most dramatic examples of discrepancies between conditioned and unconditioned behavior; for example, as a result of consuming one portion of mutton laced with lithium, two wolves who previously seized and killed a sheep with prolonged biting would display submissive behavior toward other sheep (Garcia, Rusiniak, & Brett, 1977).

The distinction between preparatory and consummatory conditioning, together with Konorski's suggestion that a CS can be involved simultaneously in both kinds of process, also provides a useful way of looking at autoshaping studies. Konorski pointed toward a number of differences, in addition to his implicit suggestion that only consummatory behavior obeyed strictly the stimulus-substitution principles. Some of these are listed in Table 9.1. They include differences in speed of acquisition, in sensitivity to partial reinforcement, in the rapidity of extinction, and in the effect of CS duration. The autoshaping research described here and in Boakes (1977) has so far found the same set of differences between sign tracking and goal tracking. When compared with goal tracking, sign tracking develops faster and is more strongly sustained by partial reinforcement. It extinguishes more slowly than goal tracking, so that when reinforcement is removed from an autoshaping procedure with rats (Boakes, 1977) or, in the absence of a tray light, with pigeons (Boakes & Alam, in preparation) the rapid decline of the interfering goal-tracking response produces a temporary increase in the extent to which the animal approaches the CS. Thus, in the light of current

TABLE 9.1
Comparison between contrasting properties of preparatory vs.
consummatory conditioning as discussed by Konorski (1967, Ch. 6) and
those of sign vs. goal tracking (Boakes, 1977; Boakes & Alam, in preparation).

	Preparatory	*Consummatory*	*Sign tracking*	*Goal tracking*
Conditioned behavior	Diffuse "general motor excitement"	Similar to UR	Affected by nature of US and of CS	Similar to UR
Speed of acquisition	Rapid	Slow	Rapid	Slow
Speed of extinction	Slow	Rapid	Slow	Rapid
Partial reinforcement	Increase	Decrease	Increase, or no effect	Decrease
CS duration	Optimal with long duration CSs	Optimal with short CS–onset: US–onset intervals (c. 1 sec)		

evidence, goal tracking may be regarded as a form of consummatory responding and sign tracking, perhaps together with the other forms of behavior that are not directed toward the US-site (as previously discussed), as a form of preparatory behavior. An added complication is that the terminal component of a sign-tracking reponse may follow the principle of stimulus substitution, as in Jenkins and Moore's (1973) demonstration that the form of autoshaped pecking is affected by whether food or water is used as reinforcement, but this may not be a very general result (see Boakes, 1977; Hearst, Chapter 2).

It may be objected that, if there were two forms of Type I conditioning, one would not expect the consistency that is often found across studies of classical conditioning. In the research reported by Wagner (Chapter 3), for example, a similar pattern of results is obtained from experiments using an eye-blink preparation, presumably consummatory, and from those using conditioned suppression, presumably preparatory. However, the dichotomy emphasizes differences in performance, so that in terms of principles of associative learning, preparatory and consummatory conditioning may be indistinguishable. Thus, in studies primarily concerned with associative learning, it may often make little difference whether the experimenter chooses a preparatory or consummatory CR to provide a convenient measure.

This identification of goal tracking as a consummatory CR has been made without justifying the more basic claim that it is based on Type I conditioning. Because a rat or pigeon has to go to the food tray to obtain food and because arriving there just before the pellet or grain reduces the delay before ingestion can begin, instrumental contingencies are clearly present. Perhaps because this leads easily to the assumption that the response is a product of Type II conditioning, this kind of behavior has previously received very little attention. But, as argued earlier, just as it may turn out that changes in behavior caused by a classical procedure may adventitiously involve Type II processes, (i.e., superstitious conditioning), so changes in behavior that appear adaptive to a situation may in fact involve only Type I processes. Deciding this matter in the case of goal tracking is difficult. Because of the fragility of such responses when reinforcement is intermittent, the various techniques such as omission training, that have proved so successful in demonstrating that autoshaping is as product of Type I conditioning (see Hearst & Jenkins, 1974) cannot easily be employed. For the moment the claim that goal tracking is a Type I product has to rest on the kind of parsimony argument that is employed by Mackintosh and Dickinson (Chapter 6) and that was first proposed in almost the exact same context by Morgan (1894). If, after the experimenter or environment has repeatedly paired some previously neutral event with a reinforcer, an animal comes to respond in exactly the same way to the event as it did previously only to the reinforcer, the simplest assumption is that the change is based on the formation of a stimulus–reinforcer association and that no more complex process, such as Type II conditioning need be invoked.

III. SUPERIMPOSITION OF TYPE I CSs ON INSTRUMENTAL BEHAVIOR

Readers familiar with Konorski's 1967 theory will have noted the non-Konorskian tone of the preceding discussion. His theory emphasizes motivational aspects of preparatory conditioning, which is principally viewed as involving an association between a CS and excitation of the drive center appropriate to the reinforcing event. In this context the term "drive" has closer affinity with incentive rather than Hullian approaches to motivational problems by Western psychologists. To some degree Konorski's 1967 account is one in which the motivational components of Type I and Type II processes are identical, and not separable as suggested by Fig. 9.1; and the way in which Type I preparatory conditioning affects Type II behavior is viewed as a relatively direct motivational effect. Because the strength of Type II behavior depends on the level of activation of the common drive center (but see Morgan, Chapter 7), a Type I CS that increases this activity will produce an increase in instrumental behavior, whereas one that inhibits this activity will produce a decrease.

As so far described, Konorski's account differs little from that of Rescorla and Solomon (1967), who suggested that, because a Type I CS associated with an *aversive* US suppresses instrumental responding maintained by an *attractive* reinforcer (CER) and can increase responding maintained by an aversive reinforcer (as with an avoidance baseline), one might expect, on grounds of symmetry, that a Type I CS associated with an attractive US will have the opposite effects. From his prewar research with Miller, Konorski knew that such symmetry did not exist and that in general a Type I CS paired with the delivery of food suppresses Type II responding maintained by the same reinforcer. In more recent Western literature this effect has been termed "positive conditioned suppression." The effect was accommodated within Konorski's 1967 theory by making the motivational system more complex, at least for any behavioral sequence terminating with ingestion. In addition to the excitatory connections between a CS involved in preparatory conditioning and the appropriate drive center, consummatory conditioning is also held to involve a motivational component, but of the opposite kind: Presenting a brief CS immediately before the arrival of food produces a CS–antidrive connection, where the antidrive center has an inherently inhibitory effect on the drive center so that its activation may lead to a decrease in the vigor of instrumental responding (see also Dabrowska, Chapter 12).

Attempts to find motivational explanations for attractive-attractive interactions appear to have been pursued, as suggested earlier, with little consideration of possible behavioral effects. Given the various kinds of Type I behavior described in the previous section, one can examine the results from a variety of studies in which an attractive CS has been superimposed on instrumental responding to see whether they appear to involve interactions at a response level. This section first considers situations in which the animal can move freely, and then the situation in which the animal's movement is severely restricted that provided the basis for Konorski's account.

Studies with freely moving subjects

Effects of superimposing a Type I CS on instrumental behavior that arise from behavioral interactions should vary considerably according to the situation, even to the extent of producing enhancement of the recorded response when the two types of conditioned response are compatible. An effect of this kind was found by LoLordo (1971) using a superimposition procedure. A response-dependent schedule was used to maintain key pecking by pigeons, and occasionally a visual stimulus associated with response-independent reinforcement was projected onto the response key; response rates increased during the stimulus periods. Using a multiple-schedule procedure, Boakes, Halliday, and Poli (1975) similarly found an increase in key pecking during stimulus periods in which additional free reinforcement was delivered, but a decrease in level pressing by rats exposed to the same reinforcement conditions.

Enhancement of instrumental behavior by a superimposed CS appears to occur in situations in which Type I responding directed toward the CS (i.e., some form of sign tracking) supplements behavior maintained by the instrumental contingency (e.g., Schwartz, 1976), thus providing an example of "response additivity," discussed further in Section IV. Where the predominant Type I behavior elicited by the CS, whether sign tracking or goal tracking, is incompatible with the instrumental response, the effect should be one of suppression.

When suppression of instrumental performance is due to interference from sign tracking, the degree of suppression should vary with the distance between the manipulandum and the location of the CS. An effect of this kind was reported by Karpicke et al. (1977) in experiments using rats as subjects. Both lever pressing and chain pulling, maintained by response-dependent presentations of a dipper of sweet milk, were more suppressed by a light signaling additional free dipper operations when the light was some distance from the manipulandum than when it was close. In one condition the light was on the manipulandum, as in LoLordo's (1971) study with pigeons, but neither enhancement nor suppression was detected. The autoshaping results discussed in Section III suggest that this is a situation in which goal tracking as well as sign tracking might be important. Thus, any potential enhancing effect from sign tracking may have been cancelled by interference from goal tracking. Karpicke et al. report an increase in the frequency of dipper approaches during the CS, thus supporting this idea.

When goal tracking provides the main source of competition—as perhaps in situations using nonlocalizable stimuli—the distance between reinforcer site and the manipulandum should prove important. In a further series of experiments, Karpicke (in press) has also found effects of this kind; with the same CS, suppression was greater if additional reinforcement came from a second dipper some distance from the manipulandum than if from the dipper used for response-dependent reinforcement, which was located close to the manipulandum.

The autoshaping results reported in Section III suggest further predictions. For example, the use of partial reinforcement in the classical conditioning part of the procedure should markedly reduce suppression produced by goal tracking but have little effect on suppression produced by sign tracking. Similar differential effects should also be found when CS duration is varied (e.g., Miczek & Grossman, 1971). In contrast Konorski's theory predicts in all situations a similar decrease in suppression when partial reinforcement or long CSs reduce the degree of conditioning of the antidrive center.

The aforementioned studies all employed a free operant baseline to assess the effects of superimposing a Type I CS. Konorski and Miller's early work (see Konorski, 1967, pp. 369-375) also included experiments that used discriminative instrumental training in which leg flexion was reinforced only in the presence of an S+ and in which the effect of a Type I CS was assessed by compounding it with the S+ on test trials. Suppression was also found with this procedure. This was replicated in much later work at the Nencki Institute (e.g., Soltysik, Konorski, Holownia, & Rentoul, 1976; see below), which in addition found no effect of introducing the CSs in the absence of the S+. However, some recent studies employing discriminated instrumental training with freely moving subjects have found facilitation of the instrumental response by Type I CSs presented during intertrial periods — that is, in the absence of the S+ (e.g., Capaldi & Hovancik, 1974; Hovancik, in press). The necessary conditions for obtaining a facilitation effect in situations of this kind are not known. Hence we cannot judge whether the kind of behavioral analysis suggested here is more applicable than the motivational interpretation favored by these authors.

A widely cited study that might appear to raise difficulties for the present approach is the first major report, by Azrin and Hake (1969), of positive conditioned suppression in freely moving animals. In 15 of their 18 subjects, very comparable levels of suppression were obtained even though both a localized visual stimulus and a clicker were used as CSs and though food, water, and intercranial stimulation (ICS) were used as USs. The latter is presumably an event whose source has no perceived location in the external environment. Furthermore they report (1969) that "gross observation of all rats during the CS revealed no pattern of competing responses (p.171)" and go on to describe the variety of behavior seen in different rats during the CS. However the potential complexity of Type I behavior has been appreciated only subsequently; it is now quite plausible that, because the only CS–US combinations used by Azrin and Hake were visual–ICS and auditory–food or auditory–water, sign tracking may have provided the main source of competition in the former case, and goal tracking in the latter. It follows that the use of the auditory–ICS combination, likely to produce little sign tracking and no goal tracking, would have produced far less suppression.

In analyzing interactions in the behavior of freely moving animals, the issue of response compatibility or competition can often be viewed at a physical level alone. If an animal moves toward a stimulus or toward the reinforcer site, one does not have to appeal to presumed incompatibility of responses within the nervous system to explain a reduction in instrumental responding. Arguments from physics are no longer available when considering experiments in which physical constraints ensure that an animals's distance from the manipulandum remains essentially unchanging. Restriction of movement also makes it relatively easy to measure salivation and other more autonomic responses, such as cardiac changes, which direct attention to less peripheral effects. The data providing the basis for Konorski's views on Type I–Type II interactions were obtained almost entirely from this latter kind of situation.

Studies with restrained subjects

For Miller and Konorski the major significance of their discovery that a Type I CS suppressed an instrumentally trained response was the further support it offered for distinguishing between two kinds of conditioning process. In dismissing this distinction, Pavlov (1932) had argued that instrumental procedures produced a connection between the instrumental response and the "alimentary center" by a process of "bidirectional" conditioning (see Mackintosh & Dickinson, Chapter 6). From this it should follow that a stimulus that had acquired the property of exciting the alimentarly center — namely, a Type I food CS — should elicit or augment the instrumental response. Demonstrating the opposite effect was seen by Miller and Konorski as undermining Pavlov's argument.

In his 1948 book Konorski made only a passing reference (p. 220) to these particular prewar experiments. Two years later an effect that appears closely related to positive conditoned suppression was reported and a non-motivational account suggested. Konorski and Wyrwicka (1950) used a transfer of control design to investigate the effect of first establishing a stimulus as a Type I food CS on subsequent acquisition of an instrumental response where the same stimulus now served as an S+. Leg flexion responses to such a stimulus developed slowly, and latencies remained long. The interpretation offered was of a kind familiar by now from this chapter: competition from motor responses established during the initial clasical conditioning phase.

The possibility of motor-response competition in certain contexts was also raised by Konorski in 1967 (see Shettleworth, Chapter 15 for examples), but in considering Type I–Type II interactions he reverted to a motivational interpretation (see Soltysik, 1975). This interpretation split the alimentary mechanism into two parts, drive and antidrive, as previously described. It prompted experimental work on superimposition that was described in full only after Konorski's death (Soltysik et al., 1976). A description of the first, and major, experiment reported in this paper is worth giving, for it provides a good illustration of the experimental approach used at the Nencki Institute. Furthermore, the results are not wholly compatible with the 1967 drive–antidrive analysis of Type I–Type II interactions.

The subjects were 5 naive dogs. Each was placed on a Pavlovian stand within a sound-proofed chamber that contained, immediately in front of the dog, a paddle wheel as manipulandum, an automatic feeder delivering meat portions, and various stimulus sources. The last included a metronome (M+), a loudspeaker delivering 800-Hz (T+) and 1200-Hz (T−) tones, a buzzer (Bz±), and a flashing light (L+). In addition the sound of bubbling water (Bu−) was delivered from behind the dog, and a skin stimulator ("touchie") was mounted on the dog's chest. Salivation and heart rate could be recorded, as well as the instrumental response — a 1.5-cm turn of the circumference of the wheel.

Classical conditioning was employed in the first phase. Initially M+, T+, and L+ were 100% reinforced, Bz± was 50% reinforced, and Bu−

was never followed by food reinforcement. The CS–US interval was gradually extended from 2 sec to 20 sec, and then nonreinforced T− trials were also included. This phase ended when, after 78 to 86 sessions, stable salivation to the positive CSs and little salivation to the negative CSs were obtained.

Instrumental training lasted over 50 sessions, in which the wheel-turn response was shaped, discrimination training with tactile stimulation as the S+ was given, and progression to stable performance on an FR10 schedule was complete. This was followed by a retest phase, in which the manipulandum was removed and the original stimulus reintroduced to check that effects of initial classical training were still intact. A procedure similar to that of Capaldi and Hovacik (1974) was employed in the next phase. With the manipulandum now present again, the various classical stimuli were presented in the intervals between S+ trials. As expected from previous Nencki research, very few instrumental responses were made to the CSs alone. Finally the classical stimuli were tested in the presence of the S+: Each CS was presented for 10 sec, and then the S+ was added for a further 10 sec.

A complete set of results was reported for only 3 of the original 5 subjects. The top half of Fig. 9.5 shows instrumental responding to the various stimuli for these 3 subjects. The lower half gives salivary responses, which were measured concurrently; heart rates are not included because cardiac measurements showed little sensitivity to the experimental manipulations. Figure 9.5 indicates that the main purpose of the study, to obtain a systematic replication under well-controlled conditions of the suppression effect found in earlier Nencki studies, was indeed achieved.

The results count against an orthodox Pavlovian position predicting a positive relationship between activation of the alimentary system, as indexed by salivation, and instrumental responding. The CS+ stimuli, which on their own elicited appreciable salivation (white columns), did not suppress the salivary response to the S+ when compounded with this stimulus, but did suppress the instrumental response. However, as Soltysik et al. point out, the results are not entirely compatible with Konorski's 1967 theory, in which the level of antidrive activation is indexed by salivation. Particularly embarrassing for this theory are the findings that the inhibitory stimuli, CS−, suppressed responding almost as much as the CS+ stimuli and that overall most suppression occurred to the partially reinforced stimuli, CS±, which should have had relatively weak connections to the consummatory center (see also Morgan, Chapter 7).

This complex pattern of results makes it tempting to apply here the kind of behavioral analysis used earlier for superimposition studies with freely moving animals, even though responses directed toward the stimuli or the food dispenser are not physically incompatible with the instrumental

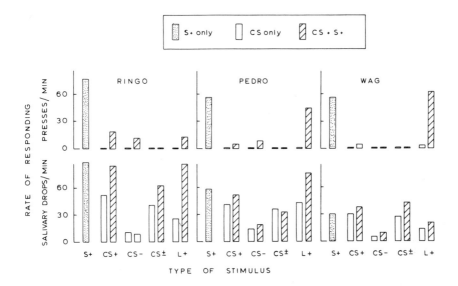

FIG. 9.5. Effects of various Type I food CSs on salivation and instrumental responding. These results are from three dogs in Soltysik et al. (1976), and the present figure is redrawn from Fig. 5 in their paper. Instrumental responding is shown by the histograms above and salivation below. "CS+" indicates the averaged effects of Bz+ and T+; "CS−" the averaged effects of T− and Bu−; "CS±" the effect of the partially reinforced stimulus; and "L+" the effect of the Light CS+.

response in the present situation. "Goal tracking" is clearly present; Soltysik et al. note that "a motivated dog reacts to a CS by focusing its gaze on the feeder" (p. 304). As suggested by results discussed earlier, Type I behavior differing in direction or form (Holland, 1977) may have been elicited by the visual stimulus, thus accounting for the discrepancy between the results for L+ and the other, auditory, stimuli used as positive CSs. Finally, a behavioral analysis does not predict opposite effects from excitatory and inhibitory CSs, for although they may elicit different kinds of Type I behavior, these may equally well compete with the instrumental response.

Soltysik et al. themselves implicitly appeal to response factors to account for the inconclusiveness of these results. They conclude that "in the next studies the attentional aspects of dogs' behavior must be more carefully monitored, and a special analysis should be conducted on the problem of mutual compatability between instrumental response and the *targeting responses to the CSs, manipulandum and feeder*" (p. 305, my italics). However, underlying this comment is a different attitude from that proposed here—namely, that such a behavioral analysis would allow a more direct examination of the more basic suppression effect arising from a motivational interaction. Early in this chapter it was suggested that in the case of aversive–attraction interactions (e.g. CER), the involvement of interactions of a motivational kind is well supported by certain evidence, despite the clear importance of response factors. In the case of attractive–attractive interactions there is no unambiguous evidence indicating that motivational interactions are involved or whether, if all possible response factors were taken into account, the effect of a Type I CS would be to increase to decrease the level of instrumental activity.

It might be argued that on theoretical grounds the complete separation of the motivational components of Type I and Type II processes shown in Fig. 9.1 cannot be correct and that therefore motivational interactions must take place, even if they have not been demonstrated. Without a satisfactory theory of Type II conditioning this argument is hard to judge. It could be that the sole function of the motivational component is the selection of appropriate classes of behavior, which is achieved by inherent inhibitory connections between various systems. Such a view would predict that a CS+ for food interacts with instrumental responding for food only at a response level, but, following Dickinson and Dearing (Chapter 8), suppression by a CS– established by a classical procedure using food reinforcement may arise from both response and motivational interactions.

IV.   SOME BEHAVIORAL PUZZLES

The studies considered in the preceding section involved deliberate manipulation of both stimulus and response contingencies. At the beginning of this chapter it was pointed out that many situations that may appear to provide only one kind of contingency can also contain the other kind. In fact it is difficult to conceive of a situation that does not contain both kinds of relationship: If in a given context reinforcement is made dependent only on the occurrence of some response, then once that response occurs with some frequency the particular context becomes better correlated with that reinforcer than other contexts; whereas if reinforcement is made dependent only on the occurrence of some stimulus, then once some form of behavior consistently occurs to the stimulus a temporal relationship between that behavior and reinforcement is established.

As also noted at the beginning of this chapter, the present approach relies on the assumption that the two kinds of contingency can engage distinguishable types of conditioning process, an assumption that has been maintained here but not justified. It is notoriously difficult to present an unassailable case for this distinction, which is by no means universally held (see Hearst,· Chapter 2). In Chapter 6 Mackintosh and Dickinson suggest that the only direct way of demonstrating its validity is to find an adequate theory for the known phenomena of classical conditioning and then show that this is inadequate to account for the facts of instrumental conditioning. Although direct, this suggestion has the weakness of depending on agreement as to what constitutes an adequate theory of classical conditioning.

An alternative approach to the problem is less direct, but not necessarily less convincing for that reason. As we have seen, one of the arguments for distinguishing Type II from Type I conditioning deployed by Konorski was that, without the distinction, one cannot understand the interacting effects of explicitly arranged stimulus and response contingencies. This section is intended to broaden this indirect argument by considering a number of puzzling effects, where a satisfactory account can be obtained in terms of interactions between Type I and Type II processes stemming from the presence of implicit stimulus–reinforcer contingencies. Otherwise they would seem to remain mysterious.

A considerable merit of the recent debate over "constraints on learning" was the way in which it drew attention to a number of puzzling phenomena that had been largely ignored by more mainstream research in animal psychology. It has clearly had a salutary effect in prompting psychologists to examine the scope of principles based on conventional laboratory paradigms for the study of learning. Claims that there are a variety of distinctive learning processes, perhaps differing widely from species to species, have been based on a variety of findings. Many of these can be viewed as effects resulting from interactions between Type I and Type II processes (cf. Shettleworth, Chapter 15 and LoLordo, Chapter 14). No systematic documentation of this claim is attempted here, but the first two of the five examples discussed here are drawn from studies that have been frequently cited in debates over constraints on learning.

Response-reinforcer specificity

In the volume *Constraints on Learning* (Hinde & Stevenson-Hinde, 1973) Sevenster (1973) describes a study in which male sticklebacks were trained to make either of two responses, which were followed by either of two reinforcers. The responses were swimming through a ring (R) and biting the green tip of a rod (B). The reinforcers were visual exposure to another male, which elicited fighting behavior (F), and exposure to a swollen female, which elicited courtship behavior (C). Both responses occurred at a high rate when fighting was made contingent on their performance. However, although courtship (C) maintained response R at a high rate, it was relatively ineffective in maintaining response B; that is, the combination biting–courtship was found to be a peculiarly ineffectual one, even with extended training.

Various aspects of the data indicated that the basis of this particular response–reinforcer specificity was not associative, but rather there appeared to be some source of interference preventing response learning from being fully reflected in the behavior of the sticklebacks run in this condition. Sevenster investigated the possibility that the source of interference was motivational; that, for example, frequent exposure to a female had a general inhibiting effect on the system underlying biting behavior. Test results suggested that effects operating at this level were of minor importance. His main conclusion was that the major source of interference was conditioned behavior, incompatible with biting, which arose as a consequence of an association between the green tip of the rod and the reinforcer (opportunity to court a female). In other words, the low rate of biting in the BC condition was the result of a competitive behavioral interaction between the effects of Type I and Type II processes.

Misbehavior

The judge presiding over the final proceedings of John Watson's divorce is said to have expressed his belief that Watson, who claimed to be an expert on "behavior," was in fact clearly an expert on "misbehavior." It will be recalled that Watson steadfastly claimed that all learned behavior could be explained in terms of a stimulus-substitution process, most clearly seen in experiments employing classical conditioning procedures. The suggestion here is that the judge was unwittingly expressing some truth and that "misbehavior," of the particular kind described 41 years later by Breland and Breland, is best understood as the result of effects of Type I processes on instrumentally maintained behavior.

Breland and Breland (1961) applied the term "misbehavior" where, in a variety of situations employing different species and different training procedures, instrumental contingencies were found to be relatively ineffective. As an illustration of the present viewpoint, Jenkins and Moore (1973) have argued that most of the Brelands' examples are due to interference from skeletal behavior generated by a Type I process.

Some of the most striking instances come from situations involving token reward. For example, the sequence of response demanded of a raccoon was to fetch a small object (the token) and carry it to a container, into which it had to be inserted. Although a usually reliable shaping method was employed, the animal never came to perform the task rapidly or consistently. Early successes were followed by increasingly frequent occasions on which the token was obtained, but not delivered to the container; instead the raccoon spent a great deal of time kneading the token. Any token reward situation contains an implicit stimulus–reinforcer contingency, whereby the token becomes a stimulus event regularly preceding the occurrence of the reinforcer. Consequently, by means of a Type I process, the token itself may begin to elicit various kinds of behavior. If the latter includes a tendency to retain the token, then this will directly interfere with the behavior "release the token," as demanded by the instrumental contingency.

In the Sussex laboratory a situation has been used in which rats were trained to collect a ball bearing, carry this across the experimental chamber, and drop it into a hole (Boakes, Poli, Lockwood, & Goodall, 1978). Many of the rats in this study displayed the kind of behavior observed by Breland and Breland and were slow to relinquish the balls. One aim was to test a prediction stemming from the analysis offered by Jenkins and Moore: To the extent that behavior generated by a Type I process is determined by the nature of the reinforcer involved, the interaction should vary with different reinforcers. It was found that more misbehavior — measured in terms of slower times for delivering a ball—was obtained with food than with water reward.

The temporal relationship between token and reinforcer was a result of the animal's behavior in this study and thus could not be controlled by the experimenters. This variable can be explicitly manipulated if a classical procedure is employed. In a recent experiment of this kind Boakes and Jeffery (in press) found that approach and seizure of a ball bearing occurred if its arrival in the chamber regularly preceded the delivery of food, but not when the two events were independent. Thus the results of these experiments tend to confirm the analysis of "misbehavior" as a behavioral interaction.

Performance on time-based schedules of reinforcement

The first two examples may suggest that one can find behavioral puzzles amenable to an interaction analysis only in unconventional circumstances. To correct such a possible impression, this third example is based on a situation employed in many laboratories where time-honored reinforcement schedules have been used to study the behavior of an animal in a Skinner box.

One source of confidence in the usefulness of operant methodology was the apparent generality of many of its findings. It appeared not to matter whether, in studying the properties of a given schedule of reinforcement, an experimenter chose a monkey, rat, or pigeon as his subject; food, water, or intercranial stimulation as the reinforcer; bar pressing, wheel turning or key pecking as an operant. The same kind of general pattern of responding is obtained.

Detailed comparisons across such variables within the same study have been relatively rare. One recent paper illustrates that when these *are* carried out, certain systematic differences can be detected. Lowe and Harzem (1977) compared lever pressing by rats and key pecking by pigeons using fixed-interval schedules (FI), together with equivalent response-independent schedules (FT), over a range of intervals from 30 to 120 sec. Detailed comparisons were based on such measures as the time to the first response in each interval (postreinforcement pause) and the subsequent rate of responding (running rate). A strong positive relationship between these two measures was obtained from the rats, such that long postreinforcement pauses tended to be followed by high running rates, but no such relationship was detected in the pigeon data. Also, changes in the programmed intervals produced a greater effect on the distribution of responding by rats than by pigeons. A third major difference resulted from removal of the response contingency by changing from an FI to an FT schedule. Very little change in response rates was obtained from the pigeons, in marked contrast to the rapid decline to a very low rate of responding by the rats.

The explanation for these differences suggested by Lowe and Harzem depends on appealing to the different types of behavior arising in the two situations simply as a result of reinforcement occurring in the particular context. Such Type I behavior would be highly compatible with the response demanded by the FI schedule from the pigeons, but incompatible with the response demanded from the rats. While their hypothesis requires further detailed development and experimental analysis, it offers a way of understanding different performances with the same reinforcement schedule where no plausible alternative appears to exist. Results from studies of DRL schedules (e.g., Hemmes, 1975; Richardson & Clark, 1976), from comparisons between FI and DRL behavior (e.g., Richelle, 1972) and from transitions from a VI to a VT schedule (e.g., Boakes & Halliday, 1975; Rescorla & Skucy, 1969) suggest that it is profitable to regard comparable differences in performance on other time-based schedules of reinforcement in a similar way.

Learning to communicate

The kind of examples cited so far may be familiar to many readers. The next is based on a study that does not appear to be widely known. To test claims that dolphins could communicate with each other at a level of complexity unknown in any other nonhuman species, Bastian (Evans & Bastian, 1969) studied a pair of dolphins separated by an opaque screen running down the middle of their tank. His basic procedure was a familiar left:right successive discrimination.

The two stimuli, a flashing and a steady light, were presented in the "sender" dolphin's half of the tank and could not be seen by the "receiver" dolphin. In the receiver's half were two paddles serving as manipulanda, and only the receiver was required to make a response on each trial. The basic conditions were as follows. On a trial with a flashing light, a response to the left-hand paddle by the receiver produced fish for both animals, but a response to the right was not followed by reward; while on a trial with a steady light, both animals received fish only if a response was made to the right-hand paddle. Thus, performance above chance level (50%) depended on the sender's conveying to the receiver some message in "dolphinese" about her perceptual world.

Early in testing it became clear that dolphinese was inadequate for the situation; performance remained at chance. But after extended exposure to these conditions, the scores began to improve until eventually a correct choice was made on 90% of the trials. Bastian comments that it was as if the dolphins had managed to train each other, but how this happened was unclear.

By viewing the outcome as a result of both Type I and Type II processes, one can gain some idea of what may have been involved. Fig. 9.6 should help to make the following account clear. As soon as the receiver begins to respond on one paddle more frequently than on the other—a preference for the left side, say—one of the stimuli presented to the sender will become a better predictor of reinforcement than the other. This follows from the conditions imposed by the experimenter; with a "flashing-left" condition, a strong left preference will mean that most flashing trials are rewarded, most steady trials unrewarded. If Type I conditioning then occurs in the sender, one stimulus will come to evoke some kind of conditioned behavior more consistently than the other. Provided that some aspect of this conditioned behavior can be sensed by the receiver (e.g., a squeak to the flashing light), there now exists a potential discriminative stimulus for the receiver's instrumental behavior. Once this becomes effective—for example, the receiver goes left only when a squeak is heard, and right otherwise—above-chance performance will begin to appear. The receiver's behavior now imposes response contingencies on the sender's behavior (e.g., flash–squeak: rft; flash–no squeak: no rft; steady–squeak: no rft; steady– no squeak: rft). At the same time, the differential stimulus–reinforcer contingencies that initiated the sender's performance become equated, for now both stimuli equally well predict the occurrence of reinforcement at the end of the trial. Given that these new response contingencies will maintain the sender's performance, so that, for example, the animal does not squeak in the same way to both stimuli, the performance of the pair should become near perfect.

This account may appear elaborate and very hypothetical. Nevertheless its prediction that communication of this simple kind could be learned by any species that is, first, capable of both Type I and Type II conditioning and, second, able to produce as a result of Type I conditioning behavior discriminable by a fellow member of the species has been confirmed for the pigeon (Boakes & Gaertner, 1977) in an experiment designed to simulate Bastian's conditions. The detailed results provided close support for the aforementioned account and failed to suggest any other kind of explanation for the performance of Bastian's dolphins.

SENDER   RECEIVER

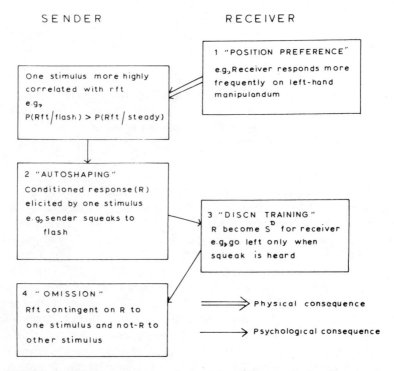

FIG.9.6.   Stages in the development of a simple form of communication. The example shown here is for a condition in which on trials with a flashing light presented to the sender the left response is correct for the receiver, and on trials with a steady light the right response is correct.

Behavioral contrast

The phenomenon whose analysis as a Type I–Type II interaction has been most extensively tested provides the last of these examples. In the context of free operant multiple schedules, the term "positive behavioral contrast" is applied to an increase in response rate occurring in one stimulus component where reinforcement is kept constant, when some intervention produces a decrease in response rate in the other component. Some 5 years ago it was suggested that the effect might be closely related to the phenomenon of autoshaped pecking and result from an additive interaction between the effects of stimulus- and response–reinforcer contingencies. Subsequently a great number of studies have been carried out that bear on this suggestion (see Schwartz & Gamzu, 1977).

The essentials of this response-additivity theory of contrast in pigeons is as follows. During initial baseline conditions, when response-dependent reinforcement is available with equal frequency in the presence of either stimuli, pecking is maintained primarily by a Type II process. The introduction of discrimination conditions, by reducing the frequency of reinforcement in one component, produces a stimulus–reinforcer contingency whereby the positive stimulus becomes more highly correlated with the occurrence of reinforcement than with any other aspect of the situation. If the stimulus is a visual event displayed on the manipulandum for the instrumental response, then this becomes a condition that, by means of a Type I process, will produce pecking directed at the stimulus. Thus, key pecking during positive stimulus periods will be a product of both Type I and Type II processes. Behavioral contrast is seen as essentially identical with the effect obtained when a CS explicitly paired with reinforcement is superimposed in this type of situation — namely, the enhancement effect obtained by LoLordo (1971), described in the previous section.

A considerable amount of evidence has been gathered in support of this theory (see Schwartz & Gamzu, 1977). At the same time various data seemingly embarrassing for the theory, such as that from earlier experiments using multiple schedules that involved errorless, DRL, or DRO procedures, can now be accounted for (e.g., Boakes, Halliday, & Mole, 1976; Halliday & Boakes, 1974; Rilling, 1977). Behavioral contrast therefore provides a particularly apt example for this section. Rather than re-review the evidence and arguments in favor of additivity theory, I wish to draw attention here to an important weakness.

An important element in the theory's success in interpreting results from pigeon experiments is that a great deal has been learned from experiments using autoshaping procedures about how pigeons behave in situations containing explicit stimulus–reinforcer contingencies only. If the results from such studies are used to predict how *rats* should perform on multiple schedules involving a lever-press response, then additivity theory appears to predict that no behavioral contrast should occur. For one thing, in such rat studies their discriminative stimuli are not displayed on the manipulandum and, for another, the instrumental response appears to be less closely related than the pigeons's key peck to the consummatory response to the reinforcer. The difficulty in obtaining behavioral contrast from rats in the same clear and consistent manner as it is obtained in pigeons was at one stage seen as providing further evidence favoring response-additivity theory (e.g., Rachlin, 1973; Schwartz & Gamzu, 1977). Subsequently a number of clear demonstrations of the effect in rats (e.g., Gutman, 1977) appear to have turned the tables, even though it is still not at all clear what situational factors determine whether or not behavioral contrast is obtained; the effect still appears more elusive with rats as subjects.

The autoshaping results discussed in Section II and the interpretation of CS-superimposition effects outlined in Section III suggest that this puzzle over rat-contrast data arises because interactional analyses have only examined the sign-tracking aspect of Type I behavior. Although this appears adequate for pigeon data — for without exception such studies employ a tray-light condition that, as seen in Section II, appears to abolish any goal-tracking tendency — the latter may be an important fact when similar procedures are used to study the discrimination performance of rats. While no studies bearing directly on this suggestion are yet available, it is one that is open to very straightforward tests.

Types of behavioral interaction

These five "puzzles" have been quickly sketched to suggest the range of situations over which interactions may be found. No very detailed account is attempted here, but some comment is required on their relationship to the issues discussed in Section II, where various kinds of Type I behavior were distinguished.

Three of the phenomena may be explained by appealing to the presence of sign-tracking behavior directed toward a discrete stimulus better correlated with reinforcement than with other aspects of the situation. Such behavior can either compete with the instrumental response, as with some instances of misbehavior; supplement the response, as with behavioral contrast in the pigeon; or, as in the communication example, allow the development of response–reinforcer contingencies, which can eventually override the stimulus contingency effects. The possibility that the same kind of stimulus event can also control goal-tracking behavior allows a way of understanding the variability of rat performance on multiple schedules: Conditions likely to produce strong goal tracking incompatible with the instrumental response are likely to produce *induction,* a decreased response rate to the positive stimulus, as opposed to the contrast effect generally found in the pigeon and sometimes in the rat.

The remaining examples come from free-operant situations in which the only environmental event is the occasional delivery of reinforcement. In Section II discussion of experiments by Schwartz (1973) and Jenkins (1977) suggested that even in situations of this kind, Type I behavior may be consistently directed toward a location other than the reinforcer site. Such an assumption is required to explain why, for example, in Sevenster's study sticklebacks directed such behavior toward the green tip of the rod and why the key pecking of pigeons, unlike the lever pressing of rats, is so persistent when a change is made from a response-dependent FI or VI schedule of reinforcement to the equivalent response-independent FT or VT schedule.

V.   SUMMARY AND CONCLUDING COMMENTS

This chapter is based on two assumptions: that two kinds of conditioning process can be distinguished and that an animal's behavior is often the result of interactions between Type I and Type II processes of an associative, motivational, or behavioral kind. Since almost any situation is likely to contain, explicitly or implicitly, both stimulus–reinforcer and response–reinforcer contingencies, the range of conditions under which interactions occur is very wide. One kind of justification for making these two assumptions is that they permit analysis of a variety of poorly understood phenomena. Although some of these may involve associative interactions, where the formation of stimulus–reinforcer associations may block or overshadow response learning, the emphasis here has been on effects that are unlikely to have an associative basis.

Analysis of Type I–Type II interactions depends on understanding the behavioral consequences of Type I conditioning, which, as discussed in Section III, can be complex. Such behavior can be usefully categorized into sign tracking and goal tracking, a distinction that in many respects matches the one made by Konorski on other grounds between preparatory and consummatory conditioning, and both forms may interact with instrumental behavior.

The experimental procedure seen as most directly relevant to the problem consists of establishing a Type I CS and presenting this stimulus during performance of an instrumentally conditioned response. Interpretations of the results from such studies by Konorski and also by Western theorists have usually favored motivational accounts. Such analyses have to contend with the fact that instrumental responding maintained by positive reinforcement is typically suppressed both by stimuli paired with aversive events (CER) and, probably under a more restricted range of conditions, by stimuli paired with attractive events (positive conditioned suppression). Although CER involves motivational interactions, positive conditioned suppression may well involve only response interactions, whether the effect is obtained from freely moving animals or from the situation employed by Konorski and his colleagues.

Several considerations favor a purely behavioral analysis of positive conditioned suppression. It can account for large variations produced by situational variables, such as the nature and location of the stimuli and the relative positions of stimulus sources, reinforcer sites, and manipulandum. From an as-yet-limited amount of data it seems possible to predict the effects of stimuli on instrumental behavior from the behavior displayed to the stimuli when no response contingency is present. In addition it allows a similar account to encompass what happens when stimulus–reinforcer contingencies are explicitly arranged by the experimenter and by phenomena that occur (as discussed on Section IV) as a result of implicit stimulus–reinforcer contingencies arising in situations as diverse as, say, those in which behavioral contrast and misbehavior have been found. Finally, it has the appeal of theoretical symmetry: There is no longer any need for Konorski's special division of the motivational component of conditioning processes involved with food reinforcement into drive and antidrive systems. Instead, by denying that conditioned suppression is a particularly direct way of examining motivational functions, it allows one to treat the motivational components of different systems in a similar manner.

The term "interactions" denotes a two-way process. This chapter has concentrated almost entirely on the effect of Type I processes on instrumental performance, because most relevant research has been in this direction. Very little attention has been paid to the opposite direction, the effects of Type II process on classical performance. At an associative level it is at least feasible that response learning may block or overshadow stimulus learning in the way that Mackintosh and Dickinson describe the converse. Mention was made in Section II of the possibility that behavior initially established by Type II conditioning may come under the control of Type I processes—that is, instrumental responses may join the pool of responses that can occur as a result of preparatory conditioning. Omission training involves imposing response–reinforcer contingencies on Type I behavior, and this can also be viewed as an interaction in the less studied direction; Schwartz and Williams (1972) have suggested a behavioral analysis involving Type I responses and competing responses that are established by superstitious instrumental conditioning, but this is possibly an indequate account.

This final paragraph is directed toward those who may feel that the greater the stress on response factors, the less interesting the topic becomes. Such reader may judge that attempts to analyze interactions of a behavioral kind miss the heart of the matter of animal psychology, the mind of an organism. Two comments are called for. First, as Halliday (Chapter 1) and Shettleworth (Chapter 15) point out, two traditions may be distinguished, that of understanding the adaptiveness of behavior and that of understanding associative learning. A full analysis of Type I–Type II interactions is very crucial for the first aim. As for the second, theories of associative learning depend on correct inferences from behavior. Karpicke (in press) has pointed out that the results of a CER study by Testa (1975), which have been cited in this volume and elsewhere as demonstrating that associations are more easily acquired between events that are spatially contiguous, can equally well be explained in terms of differences in Type I behavior produced by varying locations of CS and US. A thorough understanding of behavior produced by stimulus–reinforcer contingencies and of Type I–Type II interactions is required to avoid drawing incorrect inferences about the nature of associative learning. In steering away from the former enticement of response-centered theories of learning there can loom the danger of viewing a living organism as a static logic machine. A general context for the ideas discussed in this chapter is a belief, which Konorski plainly held, that one should take great care to avoid running aground on either shore and, while recognizing that a brain is capable of highly complex and flexible operations, to remember that it is part of a living organism that moves through and acts on its environment.

ACKNOWLEDGMENTS

The comments of Tony Dickinson and Alan Silberberg on an earlier version of this chapter were very much appreciated. The experiments described in Section II were from collaborative research undertaken with Seemeen Alam and were made possible by the technical support provided by R. F. Shrimpton and his colleagues.

REFERENCES

Alam, S., & Boakes, R. A. *Effects of tray and chamber illumination in autoshaping and omission training of the pigeon.* Manuscript in preparation.

Atnip, G. W. Stimulus- and response-reinforcr contingencies in autoshaping, operant, classical and omission training procedures in rats. *Journal of the Experimental Analysis of Behavior,* 1977, *28,* 59-70.

Azrin, N. H., & Hake, D. F. Positive conditioned suppression: Conditioned suppression using positive reinforcement as the unconditioned stimuli. *Journal of the Experimental Analysis of Behavior,* 1969, *12,* 167-173.

Boakes, R. A. Performance on learning to associate a stimulus with positive reinforcement. In H. Davis & H. M. B. Hurwitz (Eds.), *Operant-Pavlovian interactions.* Hillsdale, N. J.: Lawrence Erlbaum Associates, 1977.

Boakes, R. A., & Alam, S. *The tray-light effect: CS and US site approach by pigeons.* Manuscript in preparation.

Boakes, R. A., & Gaertner, I. Development of a simple form of communication. *Quarterly Journal of Experimental Psychology,* 1977, *29,* 561-575.

Boakes, R. A., & Halliday, M. S. Disinhibition and spontaneous recovery of response decrements produced by free reinforcement in rats. *Journal of Comparative and Physiological Psychology,* 1975, *88,* 436-446.

Boakes, R. A., Halliday, M. S., & Mole, J. Successive discrimination training with equated reinforcements frequencies: Failure to obtain behavioral contrast. *Journal of the Experimental Analysis of Behavior,* 1976, *26,* 65-78.

Boakes, R. A., Halliday, M. S., & Poli, M. Response additivity: Effects of superimposed free reinforcement on a variable-interval baseline. *Journal of the Experimental Analysis of Behavior,* 1975, *23,* 177-191.

Boakes, R. A. & Jeffery, G. Autoshaping and misbehavior. *Ricerche di Psicologia,* in press.

Boakes, R. A., Poli, M., Lockwood, M. J., & Goodall, G. A study of misbehavior: Token reinforcement in the rat. *Journal of the Experimental Analysis of Behavior.* 1978, *29,* 115-134.

Breland, K., & Breland, M. The misbehavior of organisms. *American Psychologist,* 1961, *16,* 202-204.

Brown, P. L., & Jenkins, H. M. Auto-shaping of the pigeon's key peck. *Journal of the Experimental Analysis of Behavior,* 1968, *11,* 1-8.

Capaldi, E. D., & Hovancik, J. R. Effects of Partial *vs* consistent reward in non-contingent pairings of stimuli with reward. *Animal Learning and Behavior,* 1974, *2,* 39-42.

Davol, G. H., Steinhauer, G. D., & Lee, A. The role of preliminary magazine training in acquisition of the autoshaped key peck. *Journal of the Experimental Analysis of Behavior,* 1977, *28,* 99–106.

Dickinson, A., & Pearce, J. M. Inhibitory interactions between appetitive and aversive stimuli. *Psychological Bulletin,* 1977, *84,* 690–711.

Evans, W. E., & Bastian, J. Marine mammal communication: Social and ecological factors. In Y. T. Anderson (Ed.), *The biology of marine mammals.* London: Academic Press, 1969.

Estes, W. K., & Skinner, B. F. Some quantitative properties of anxiety. *Journal of Experimental Psychology,* 1941, *29,* 390–400.

Garcia, J., Rusiniak, K. W., & Brett, L. P. Conditioning food–illness aversions in wild animals: *Caveant canonici.* In H. Davis & H. M. B. Hurwitz (Eds.), *Operant–Pavlovian Interactions.* Hillsdale, N. J.: Lawrence Erlbaum Associates, 1977.

Gutman, A. Positive contrast, negative induction and inhibitory stimulus control in the rat. *Journal of the Experimental Analysis of Behavior,* 1977, *27,* 219–233.

Halliday, M. S., & Boakes, R. A. Behavioral contrast without response rate reduction. *Journal of the Experimental Analysis of Behavior,* 1974, *22,* 453–462.

Hearst, E., & Jenkins, H. M. *Sign-tracking: The stimulus reinforcer relation and directed action.* Austin, Texas: The Psychonomic Society, 1974.

Hemmes, N. S. Pigeons' performance under differential reinforcement of low rate schedules depends on the operant. *Learning and Motivation* 1975, *6,* 344–357.

Herrnstein, R. J., & Sidman, M. Avoidance conditioning as a factor in the effects of unavoidable shocks on food-reinforced behavior. *Journal of Comparative and Physiological Psychology,* 1958, *51,* 380–385.

Hinde, R. A., & Stevenson-Hinde, J. (Eds.), *Constraints on Learning.* New York: Academic Press, 1973.

Holland, P. C. Conditioned stimulus as a determinant of the form of the Pavlovian conditioned response. *Journal of Experimental Psychology: Animal Behavior Processes, 1977, 3,* 77–104.

Hovancik, J. R. The effect of deprivation level during non-contingent pairings and instrumental learning on subsequent instrumental performance. *Learning and Motivation,* in press.

Jenkins, H. M. Sensitivity of different response systems to stimulus–reinforcer and response–reinforcer relations. In H. Davis & H. M. B. Hurwitz (Eds.), *Operant–Pavlovian Interactions.* Hillsdale, N. J.: Lawrence Erlbaum Associates, 1977.

Jenkins, H. M., & Moore, B. R. The form of the autoshaped response with food or water reinforcers. *Journal of the Experimental Analysis of Behavior.* 1973, *20,* 163–181.

Karpicke, J. Directed approach responses and positive conditioned suppression in the rat. *Animal Learning and Behavior,* in press.

Karpicke, J., Christoph, G., Peterson, G., & Hearst, E. Signal location and positive *vs* negative conditioned suppression in the rat. *Journal of Experimental Psychology: Animal Behavior Processes,* 1977, *3,* 105–118.

Konorski, J. *Conditioned reflexes and neuron organisation.* Cambridge: Cambridge University Press, 1948.

Konorski, J. *Intergrative activity of the brain.* Chicago: University of Chicago Press, 1967.

Konorski, J., & Miller, S. The effect of unconditioned and conditioned excitors on motor conditioned reflexes. *Comptes Rendus des Séances de la Société de Biologie,* 1930, *104,* 911–913.

Konorski, J., & Wyrwicka, W. Research into conditioned reflexes on the second type. I. Transformation of conditioned reflexes of the first type into conditioned reflexes of the second type. *Acta Biologiae Experimentalis,* 1950, *15,* 193–204.

Leslie, J. C. Effect of food deprivation and reinforcement magnitude on conditioned suppression. *Journal of the Experimental Analysis of Behavior,* 1977, *28,* 107–116.

LoLordo, V. M. Facilitation of food-reinforced responding by a signal for response independent food. *Journal of the Experimental Analysis of Behavior,* 1971, *15,* 49–55.

Lowe, C. F., & Harzem, P. Species differences in temporal control of behavior. *Journal of the Experimental Analysis of Behavior,* 1977, *28,* 189–201.

Miczek, K. A., & Grossman, S. P. Positive conditioned suppression: Effects of CS duration. *Journal of the Experimental Analysis of Behavior,* 1971, *15,* 243–247.

Miller, S., & Konorski, J. Sur une forme particuliere des reflexes conditionnnels. *Comptes Rendus des Séances de la Société de Biologie,* 1928, *99,* 1155–1157.

Morgan, C. L. *An introduction to comparative psychology.* London: Scott, 1894.

Mowrer, O. H. *Learning theory and personality dynamics.* New York: Ronald Press, 1950.

Pavlov, I. P. The reply of a physiologist to psychologists. *Psychological Review,* 1932, *39,* 91–127.

Pavlov, I. P. *Conditioned reflexes and psychiatry (Vol. 2).* New York: International Publishers, 1941.

Peden, B. F., Browne, M. P., & Hearst, E. Persistent approaches to a signal for food despite food omission for approaching. *Journal of Experimental Psychology: Animal Behavior Processes,* 1977, *3,* 377–399.

Peterson, G. B., Ackil, J., Frommer, G. P., & Hearst, E. Conditioned approach and contact behavior towards signals for food or brain-stimulation reinforcement. *Science,* 1972, *77,* 1009–1011.

Rachlin, H. Contrast and matching. *Psychological Review,* 1973, *80,* 217–234.

Rescorla, R. A., & Skucy, J. C. Effect of response-independent reinforcers during extinction. *Journal of Comparative and Physiological Psychology,* 1969, *67,* 381–389.

Rescorla, R. A., & Solomon, R. L. Two-process learning theory: Relationships between Pavlovian conditioning and instrumental learning. *Psychological Review,* 1967, *74,* 151–182.

Richardson, W. K., & Clark, D. B. A comparison of the treadle-press and key-peck operants in the pigeon: Differential reinforcement of low rate schedule of reinforcement. *Journal of the Experimental Analysis of Behavior,* 1976, *26,* 237–256.

Richelle, M. Temporal regulation of behavior and inhibition. In R. A. Boakes & M. S. Halliday (Eds.), *Inhibition and learning.* London: Academic Press, 1972.

Rilling, M. Stimulus control and inhibitory processes. In W. K. Honig & J. E. R. Staddon (Eds.), *Handbook of operant behavior.* Englewood Cliffs, N. J.: Prentice-Hall, 1977.

Schwartz, B. Maintenance of key pecking by response-independent food presentation: The role of the modality of the signal for food. *Journal of the Experimental Analysis of Behavior,* 1973, *20,* 17–22.

Schwartz, B. Positive and negative conditioned suppression in the pigeon: Effects of the locus and modality of the CS. *Learning and Motivation,* 1976, *7,* 86–100.

Schwartz, B., & Gamzu, E. Pavlovian control of operant behavior. In W. K. Honig & J. E. R. Staddon (Eds.), *Handbook of operant behavior.* Englewood Cliffs, N. J.: Prentice-Hall, 1977.

Schwartz, B., & Williams, D. R. The role of the response–reinforcer contingency in negative automaintenance. *Journal of the Experimental Analysis of Behavior,* 1972, *17,* 351–362.

Scobie, S. R. Interaction of an aversive Pavlovian conditional stimulus with aversively and appetitively motivated operants in rats. *Journal of Comparative and Physiological Psychology,* 1972, *79,* 171–188.

Sevenster, P. Incompatibility of response and reward. In R. A. Hinde & J. Stevenson-Hinde (Eds.), *Constraints on learning.* New York: Academic Press, 1973.

Skinner, B. F. "Superstition" in the pigeon. *Journal of Experimental Psychology,* 1948, *38,* 168–172.

Soltysik, S. Post-consummatory arousal of drive as a mechanism of incentive motivation. *Acta Neurobiologiae Experimentalis,* 1975, *35,* 447–474.

Soltysik, S., Konorski, J., Holownia, A., & Rentoul, T. The effect of conditioned stimuli signalling food upon the autochthonous instrumental response in dogs. *Acta Neurobiologiae Experimentalis,* 1976, *36,* 277–310.

Testa, T. J. Effects of similarity of location and temporal intensity pattern of conditioned and unconditioned stimuli on the acquisition of conditioned suppression in rats. *Journal of Experimental Psychology: Animal Behavior Processes,* 1975, *104,* 114–121.

Zener, K. The significance of behavior accompanying conditioned salivary secretion for theories of the conditioned response. *American Journal of Psychology,* 1937, *50,* 384–403.

10 Extinction, Inhibition, and Differentiation Learning

Kazimierz Zielinski
Nencki Institute of Experimental Biology

I.   INTRODUCTION

The problem of conditioned inhibition—or, to use Pavlovian terminology, the problem of internal inhibition—occupied a central position in the theory of instrumental conditioning developed by Jerzy Konorski. Over the years his understanding of the role of inhibition in classical and instrumental conditioning changed considerably. Three periods in his development of the concept may be distinguished. The first covers the 11 years of experimental work on Type II conditioned reflexes, starting from the early papers published in 1928 (Miller & Konorski, 1928) and ending in September 1939, the date of the destruction of the Nencki Institute and of Konorski's escape from Warsaw at the time of the German occupation. The most important date for the second period was 1948, the year in which his book *Conditioned Reflexes and Neuron Organization* was published. Until around 1960, the experimental work in the Department of Neurophysiology of the Nencki Institute was conducted according to the theoretical lines presented in this book, and we may tentatively accept 1960 as a division between the second and third periods. During the next period (1960-1973) the scientific interests of Konorski and the research undertaken by his group expanded considerably, a fact that was reflected in his next book, *Integrative Activity of the Brain: An Interdisciplinary Approach* (1967). To the very end of his life Konorski presented papers in which he sought to extend his concept of inhibition to encompass a growing body of facts.

Western experimental psychologists have long neglected the problem of conditioned inhibition. Significant research in this area only began to appear in Western journals in the 1960s. Some theoretical papers published over the last decade seem to be important for further conceptualization of the role of inhibition in learning and conditioning. Rapidly increasing attention to a given scientific problem may result in many misunderstandings or even mistakes. However, a knowledge of other traditions in the field may help overcome such a danger.

The present chapter is limited to a consideration of Konorski's main thoughts on the role of inhibition in conditioning, developed during the first and the second periods of his scientific activity. There are many reasons why such a limitation should be made. One of the most important is related to the fact that Konorski's early papers are relatively unknown to Western psychologists, and even his 1948 volume passed almost unnoticed in the USA (See Halliday, Chapter 1; Konorski, 1974; Mowrer, 1976).

II.   EXTINCTION AND CONDITIONED INHIBITION — SAME OR DIFFERENT PROCESSES?

A. Definitions of Learning

In his popular textbook Kimble defined learning as a relatively permanent change in a behavioral potentiality that occurs as a result of *reinforced* practice (Kimble, 1961, p. 2). Thus, according to this definition, changes in observed behavior due to the presentation of a conditioned stimulus (CS) that is systematically unpaired with an unconditioned stimulus (US) are not the result of some learning process, but rather affect only performance of a previously acquired conditioned response (CR). In another paper Kimble explicitly stated (1964) that "this definition does a fairly satisfactory job of distinguishing learning from such processes as adaptation, fatigue, *experimental extinction,* changes in motivation and maturation which also lead to changes in behavior (p. 33, italics mine)." Such a view of learning differs from that of Pavlov (1927), who assumed that some learning is involved in the process of extinction of a previously acquired CR. Similarly, Spence (1937) and Hull (1943, 1950, 1952), in order to explain differentiation learning, assumed that reinforcement given in the presence of one discriminative stimulus results in an increase of its excitatory potential, whereas not reinforcing the response in the presence of the other discriminative stimulus builds up an inhibitory potential. According to such conditioning–extinction theories, the tendency to respond and the tendency not to respond were both a result of conditioning

Pavlov's, Spence's, and Hull's concept of the mechanisms responsible for changes in behavior occurring when a CS is systematically presented without the US are not commonly shared by modern students of conditioned inhibition. In a recent paper Hearst concluded (1972):

An inhibitory stimulus was defined by Hearst et al. as a multidimensional environmental event that as a result of conditioning (in this case based on some *negative* correlation between presentation of the stimulus and subsequent occurrences of another event or outcome, such as "reinforcement") develops the capacity to decrease performance below the level occurring when that stimulus is absent [pp. 6–7].

However, later in the text he stated (1972):

Decremental effects due to changes in the initiating or maintaining variables themselves (e.g. *complete removal of reinforcement,* decreases in the frequency or amount of reinforcement, lowered drive, withdrawal or variation of S+) seem more parsimoniously described in terms of manipulations of only one type of factor (excitatory) [p. 9; italics added].

Thus, Hearst does not necessarily consider experimental extinction as a conditioning procedure leading to the transformation of an excitatory into an inhibitory stimulus.

B. Pavlov's terminology of inhibition

Pavlov and his pupils distinguished between two basic situations in which performance of a learned conditioned response may be temporarily or permanently suppressed due to some experimental manipulation independent of changes in motivation, fatigue, or adaptation (cf., Konorski & Miller, 1933, p. 27-44). If an extraneous agent is presented concurrently with a CS, the CR is either abolished or reduced in magnitude. Such an extraneous factor evokes its own unconditioned response (UR) (e.g., an orienting response), and because of competition between two reflex arcs in the central nervous system (CNS), the measured response is partially or totally inhibited. Pavlov labeled such inhibition as "external inhibition" (and the acting extraneous agent as an "external inhibitor") to stress that the arc of the established conditioned reflex was intact, but that performance of the response was inhibited by excitation of other nervous centers. The second method for suppressing a conditioned response involves a change in the relationship between the important events in a conditioned procedure: the contingency between CS and US. This change may be accomplished by (1) a complete cessation of US presentation when the CS is given, (2) increasing the time period between CS and US onsets, and (3) not presenting a US when a new stimulus — different from the original CS, but receiving generalized excitatory properties — is given. In the first case, experimental extinction occurs; in the second, one may observe a phenomenon called "inhibition of delay"; and in the third, "differentiation" or "differential inhibition" develops. The term "conditioned inhibition" was reserved by Pavlov for a special variety of the last procedure, consisting of nonreinforced presentations of the original CS preceded and usually accompanied by a new stimulus. This new stimulus in the compound was labeled by Pavlov as a "conditioned inhibitor" (Pavlov, 1927).

C. Some procedures for developing inhibitory reflexes

According to Konorski the same basic principle is responsible for changes in behavior occurring during experimental extinction, inhibition of delay, and differentiation learning. This principle was purely connectionistic (Konorski, 1968, p. xii). At the start of this discussion, Konorski analyzed the extinction of an unconditioned reflex, the orienting reflex, and only then did he go on to draw analogies with both the process of CR acquisition and the phenomenon of experimental extinction. In the same way that a new stimulus, when repeatedly presented alone, becomes a signal that no other biologically important change in the environment will follow, a CS presented repeatedly without a previously accompanying reinforcing event or outcome comes to signal a lack of the previous reinforcement (see Wagner, Chapter 3). In the process of experimental extinction, the previous positive correlation between the CS and the reinforcing event is transformed into a negative correlation and not a lack of correlation between these two events. This process has to be distinguished from the gradual weakening of a previously acquired CR due to a long-lasting break in the evocation of the reflex (Konorski, 1948, pp. 83, 101, 113, and elsewhere).

A series of experiments followed that aimed at confirming Konorski's view of experimental extinction as a process of transformation from an excitatory conditioned reflex (or a "positive conditioned reflex," to use Konorski's terminology) to an inhibitory conditioned reflex. However, it was expected that this transformation would not eliminate previously established intercentral connections, because the application of an extraneous stimulus together with the extinguished CS leads to "disinhibition" of the inhibitory reflex and the CR is again performed (Pavlov, 1927). An increase in motivation and, in general, of the arousal level of the CNS results in similar disinhibition of extinguished conditioned reflexes. From this Konorski concluded that new plastic changes are superimposed on the old one (Konorski, 1948, pp. 134–135).

1. Signaling the offset of the US

Experiments with dogs showed that, depending on the time relationships between onsets and offsets of the CS and US, either an excitatory or an inhibitory conditioned reflex may be established (Konorski, Lubinska & Miller, 1936). In a typical conditioning procedure, where CS onset precedes the time when food reinforcement is given, a salivary conditioned reflex to the CS is established. An excitatory CR may also be obtained when onset of the CS follows the presentation of food, provided that the CS offset does not coincide with the termination of the act of eating but signals an additional portion of food. The CS presented regularly in such circumstances evokes both salivation and motor activity typical of excitatory classical conditioned alimentary reflexes when tested in sporadic trials without any food presentation. However, the CR evoked by such a CS may be easily extinguished if the additional portion of food is no longer presented, so that now the CS offset coincides with the end of the act of eating. It is possible to differentiate two CSs with onsets that always follow the US onset. With such a procedure, offset of one CS occurs before the end of the act of eating (such a CS acquires excitatory properties), and offset of the second coincides with the end of eating (this CS acquires inhibitory properties). It was shown later that a CS applied just before the end of eating acquires signaling properties and evokes the instrumental responses of ceasing to eat (Zbrozyna, 1957, 1958).

These experiments provided strong arguments against the Pavlovian notion of "inductory relations" between different centers of the CNS and formed the basis for the hypothesis that the mechanism responsible for inhibitory conditioned reflexes is (1948) "formation of inhibitory connections between two centers as the result of coincidence in excitation of the conditioned (i.e., stimulus) center with fall of excitation in the unconditioned (i.e., stimulus) center (p. 252)."

2. "Acute" and "chronic" extinction

The next step in understanding the phenomena of experimental extinction came from experiments in which the course of various kinds of transformation of conditioned reflexes—initially extinction and restoration of salivary reflexes — were investigated. Two procedures of extinction, "acute" and "chronic," were used (Konorski & Szwejkowska, 1950). In each dog several stimuli of different modalities were used within the same daily session, all being reinforced in a similar manner as far as duration of the CS–US interval, and the quality and quantity of the food US were concerned. After salivary responses to all CSs were well-established, one CS was subjected to experimental extinction. The "acute" extinction procedure consisted of successive nonreinforced presentations of a given CS, unaccompanied by presentations of other CSs and/or the food US within the run of extinction trials. Konorski was aware that a response decrement obtained with this procedure may reflect a decrease in the alimentary drive or in the general expectancy of food, rather than true extinction of the specific CR to a given CS. Thus, the "chronic" extinction procedure was used as a rule. In this procedure one CS was given without food reinforcement only once a day, in a context in which other CSs were presented, and reinforced during the same daily session. The "chronic" extinction procedure differs from the regular Pavlovian differentiation procedure in one aspect only—namely, the chronically extinguished CS does not constitute a "pair" with some other reinforced stimulus, because all CSs differ on many dimensions. It was found that the more salient the stimulus, the slower the extinction. However, restoration of the CR did not depend on the quality of the stimulus. CSs with a long history of reinforcement were more resistant to extinction than newly introduced CSs. Generally, chronic extinction is a slow process, whereas restoration of the salivary CR is rapid, and after a few reinforcements the CR is performed with the same magnitude as before extinction (Konorski & Szwejkowska, 1950; cf. Konorski 1967, p. 317). Similar asymmetry has also been observed when classically conditioned defensive reflexes are subjected to chronic extinction and restoration (Konorski & Szwejkowska, 1952; cf. Konorski, 1967, p. 118-119). It may be argued that this asymmetry in the speed of extinction and of restoration processes was related to the procedure used, for the CS has been presented among other reinforced CSs, against an "excitatory background." Thus, these findings were confronted with data obtained in other experiments in which the CS subjected to extinction was applied in separate extinction sessions (each consisting of two presentations of a CS without the food US), and these sessions were interspersed among normal sessions in which all the other excitatory CSs were presented. With this procedure, named "chronic extinction against an inhibitory background," extinction was also far slower and more irregular than the process of CR restoration (Szwejkowska, 1950). Detailed analysis showed that the amount of salivation during the second half of the CS subjected to extinction was similar in both procedures, whereas during the first half of the CS greater salivation was observed when the CS that was subjected to extinction was presented among excitatory CSs, than when the same CS was given without reinforcement in separate experimental sessions. Moreover, salivation evoked by any excitatory CS presented on the trial following the extinguished CS was consistently reduced.

All of these data indicate that the magnitude of the CR, either excitatory or inhibitory, depends on two factors: the relationship between the CS and the US established during training and the background against which a given CS is presented. Interference between excitatory and inhibitory conditioned reflexes is subject to considerable change in the course of consolidation of the reflexes, and more attention ought to be paid to such transient processes.

3. The "Primary inhibitor"

Transformation of excitatory into inhibitory reflexes and vice-versa is strongly dependent on the previous history of the stimulus. If after the shaping of the salivary CR to one CS, another stimulus of a different modality (or the same modality, but differing from the excitatory CS on many dimensions) has been introduced into the experiment and over a long series of trials is never paired with food, such a stimulus later on may be transformed into an excitatory CS only with great difficulty. A stimulus with such a history was labeled by Konorski a "primary inhibitor." A primary inhibitor transformed into an excitatory CS evokes a salivary CR of lower magnitude (even after a long series of reinforcements) when compared with a newly introduced stimulus paired with a similar number of food reinforcements. On the other hand, an excitatory CS that was previously a "primary inhibitor" extinguishes very easily, and the magnitude of the salivary CR decreases rapidly after several applications of such a CS without a food US (Konorski & Szwejkowska, 1952; cf. Konorski, 1967, p. 320).

Thus, with respect to asymmetry during experimental extinction and during restoration of excitatory properties, all stimuli may be divided into two classes. However, no sharp border between these classes exists. Acquisition of excitatory properties is easiest when a newly introduced stimulus is presented among other excitatory CSs; is more prolonged when inhibitory as well as excitatory CSs are presented in the same experimental session; and is most difficult when a given stimulus has previously been repeatedly presented without food reinforcement. The less similar a "primary inhibitor" is to the excitatory CS applied in the same session, the more difficult is its transformation into an excitatory CS (Konorski & Szwejkowska, 1952; Szwejkowska, 1959; cf. Konorski, 1967, p. 320).

These findings were compared with data obtained in Pavlov's laboratories, where it was shown that the "principle of the primacy of first training" was valid, not only for the analysis of various procedures of experimental extinction, but also for evaluating Pavlovian differentiation and for understanding the effects of drastic changes in the CS–US interval (Konorski & Szwejkowska, 1952). As shown by Bignami (1968), these findings have similar importance for contemporary discussions concerning stimulus control after differentiation learning with and without errors, because the "primary inhibitor" procedure is similar to methods introduced by Terrace (1963) for errorless differentiation learning.

Experiments on primary inhibition forced Konorski to change his previous view on the conditions necessary for the development of conditioned inhibitory reflexes. In his revision Konorski concluded that (1967) "the inhibitory CR is not established by the *decrease* of excitation in the US center following the operation of the CS, but rather by the *lack* of excitation in this center during its operation (p. 323)." However, to this statement should have been added the following condition: providing that in the same situation another CS (or CSs) is followed by an increase of excitation in the US center. This addition was not discussed in Konorski's book; nevertheless, the experiments described earlier clearly indicate that a primary inhibitor acquires its unique properties when repeatedly presented without the food US on an "excitatory background" (cf. Wagner & Rescorla, 1972, p. 319).

4. The "Conditioned inhibitor"

According to Konorski, inhibitory conditioned reflexes established as a result of experimental extinction, of training with a long CS–US interval, or of differential learning are based on both excitatory and inhibitory influence on the unconditioned stimulus center. In contrast, with a conditioned inhibitor, the unconditioned center is under purely inhibitory influences (Konorski, 1948, pp. 154–157, 252).

This distinction again bears a direct relationship to the problem of learning of conditioned inhibitory reflexes with and without errors. If no special precautions are taken, training of a conditioned inhibitor (a stimulus that, on a given trial, signals that the excitatory CS will not be followed by a reinforcing event) involves numerous errors. For instance, in our experiments with Soltysik on shaping a conditioned inhibitor for avoidance bar pressing in cats, many errors of commission to the inhibitory compound were observed (Soltysik & Zielinski, 1962). Depending on the time relationships between the two elements of the compound and the stage of training, these errors were restricted either to the action of the conditioned inhibitor (CI) or to the excitatory (CS) component. Extinction of responses to the CS presented in compound with the CI markedly influenced response latencies to the excitatory CS presented alone, and resulted in a dramatic drop in avoidance performance after prefrontal lobectomy (Zielinski, 1972). Thus, in this form of conditioned inhibitory reflex, interactions between excitatory and inhibitory tendencies conditioned to different environmental events also exist. During presentation of the stimulus serving as the CI, the unconditioned stimulus center is under not only inhibitory, but also excitatory influences, whose strength depends on the amount of generalization from the excitatory CS to the CI. The view of a conditioned inhibitor as a unique sort of conditioned inhibitory reflex is incorrect.

D. Stimulus-intensity effects on excitatory and inhibitory CRs

All the aforementioned experimental data and considerations support Konorski's view on extinction of the CR as an active process controlled by number of parameters independent of changes in motivation. The speed of extinction depends on certain conditions that were in action long before the reinforcing event or outcome has been withdrawn.

It seems that one of the sources of the skepticism shown by Western psychologists toward such a point of view was negative or inconclusive attempts to discover a dependency between CR strength and CS intensity in experiments employing extinction procedures on test trials (e.g., Gray, 1965). To separate stimulus-intensity effects on the learning of a conditioned reflex from those on performance of the acquired response, factorial designs had been introduced in which separate groups of subjects were conditioned with CSs of different intensities and were then divided into subgroups extinguished with each CS intensity. Response strength was measured during the extinction phase of such experiments. Difference in extinction between subgroups trained with different CS intensities, counterbalanced over CS intensities used in extinction, were taken as a measure of CS-intensity effects on the learning process. On the other hand, differences between subgroups extinguished with different intensities and counterbalanced over CS intensities used in acquisition were considered to provide a measure of CS-intensity effects on performance (e.g., Grant & Schneider, 1948, 1949). If we accept the view of extinction as an active conditioning process, possible differences between subgroups given extinction with different CS intensities cannot simply be considered as a measure of CS-intensity effects on performance of the previously established CR. If CS intensity has an effect on the acquisition of an excitatory CR, this parameter ought to have similar effects on the acquisition of an inhibitory CR. Thus, the measure of CR strength in a series of extinction trials is confounded by two opposite effects of CS intensity: The more intense the CS (within certain limits, of course), the greater the strength of the excitatory CR resulting in their resistance to extinction; and the more intense the CS, the better the conditions for acquisition of the new inhibitory properties of the CS.

This hypothesis was verified in an experiment employing the Estes–Skinner "conditioned emotional response" (CER) technique (Estes & Skinner, 1941). The CER was subjected to acute extinction by drastically prolonging the CS, given once a day without a shock US. On the first extinction trial, rats trained and extinguished with a more intense CS suppressed their ongoing food-motivated behavior longer than did rats trained and extinguished with a less intense CS. However, on the second extinction trial rats trained and extinguished with the more intense CS showed little suppression and resumed barpressing shortly after onset of the CS (see Fig. 10.1). Rats trained and extinguished with a less intense CS maintained suppression much longer after the CS onset on the second extinction trial (Jakubowska & Zielinski, 1976a, 1978). Thus, we may conclude that an important parameter, stimulus intensity, which influences learning and performance of excitatory CRs, has similar effects on the acquisition by a CS of inhibitory properties.

A similar effect of CS intensity on the development of inhibition of delay has also been demonstrated. After response suppression with a CER procedure reached asymptote, the slow process of attenuation begins. This process is due to weakening of the suppressing properties of the onset of the CS (Zielinski, 1966). Further, the more intense the CS, the more rapid and more pronounced are the differences in suppression between the early and final periods of the CS (Zielinski & Walasek, 1977). The effects of CS inten-

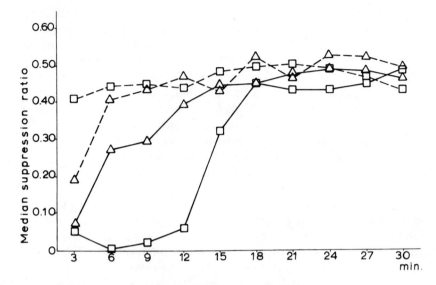

FIG. 10.1. The course of CER extinction on the first 2 extinction days. Suppression ratios were computed for 10 consecutive 3-min periods of the prolonged CS action. Squares represent the 80 dB group; triangles indicate 50 dB group. Solid lines designate the first extinction day, and dashed lines the second extinction day. From Jakubowska & Zielinski (1978).

sity on the course of differentiation learning are more complex and are discussed in the following sections.

E. Kimble's definition of learning — some extensions

On the basis of the foregoing considerations and the experimental findings of Konorski and his colleagues, we postulate that the usual understanding of Kimble's definition of learning ought to be changed. We fully agree that learning is a result of *reinforced* practice. However, in the case of conditioned inhibitory stimuli, such reinforced practice may be restricted to the excitatory CSs applied in the same experimental situation. All conditioning procedures involve the interaction of excitatory tendencies conditioned to one environmental event and inhibitory tendencies conditioned to other environmental events. As far as the procedure of successive differentiation of either classically conditioned or instrumental reflexes is concerned, this problem has been recognized by researchers interested in measures of the stimulus control of behavior. Special techniques were developed to permit control of the effects of antecedent reinforcement and nonreinforcement on responding to both excitatory and inhibitory stimuli (Jenkins, 1965). Differentiation learning without errors reduces generalization of the CR from excitatory to inhibitory CSs in a manner similar to procedures used to produce "primary inhibitor" properties. However, reciprocal interactions of excitatory and inhibitory tendencies conditioned to different environmental events, including background cues, are not totally abolished.

III.   DIFFERENTIATION LEARNING AS PARALLEL SHAPING OF DIFFERENT CONDITIONED REFLEXES[1]

A. Konorski's definition of differentiation

The meaning of the term "differentiation" has been explained by Konorski (1967):

> We should emphasize that a sharp boundary must be drawn between two quite different, although often confused phenomena which we shall refer to respectively as *discrimination* of stimulus-objects, and their *differentiation*. We shall call discrimination a *purely perceptive* process, even not necessarily involving different associations or different responses established to each discriminated stimulus-object. On the other hand, by differentiation we shall denote, in agreement with the Pavlovian usage of the word, the *utilization* of discriminated stimuli for different responses of the organism. Whereas differentiation obviously presumes the discriminability of the two stimuli used, the converse is not necessarily true [p. 93].

The distinction between these two phenomena has been also recognized by some students of the stimulus control of behavior. Bloomfield expressed a view as follows (1969):

> One aspect of discrimination learning concerns the animal's *classification* of the stimuli; another concerns what the animal *does* in the presence of each stimulus. In the presence of the stimulus which signals nonreward (S−), for instance, the animal could refrain from doing what was rewarded during S+; or, alternatively, it could do something quite specific that happened to conflict wth making the S+ response [p. 224].

We fully agree with these statements, and thus the use of the term "differentiation" as a synonym for the method of "Pavlovian differentiation" is incorrect. We note subsequently the importance of the last sentence of this quotation from Bloomfield for understanding differentiation involving instrumental conditioning.

B. CS, US, and CR relationships in classical and instrumental reflexes

In light of the discussion in previous sections of this chapter, we must distinguish between excitatory and inhibitory classically conditioned re-

---

[1]Some aspects of this section have been presented at a Memorial Session for Jerzy Konorski of the Conference on "Brain and Behavior," held in Liblice near Prague (in 1975) and published previously (Zielinski, 1976).

flexes. With excitatory classically conditioned reflexes, the change in behavior occurring to the CS is a result of previous CS and US pairings. By analogy, one might say that with inhibitory classically conditioned reflexes, the change in behavior occurring to the CS is a result of the previously negative correlation between the CS and US. A stimulus that has acquired inhibitory properties in a classical conditioning situation signals that the US will not be presented during and/or shortly after its action. Moreover, in some situations the same environmental event may possess two signaling properties, which, in the case of a delayed conditioning procedure, are dependent on the lapse of time from its onset. When delayed salivary reflexes are well-established, the CS onset has inhibitory properties and signals that *now* food will be not given, whereas the latter part of the same CS has excitatory properties and signals that food *is about to be* presented. In classically conditioned reflexes the CR changes neither the time nor the probability of the US occurrence.

In the case of instrumental conditioned reflexes, relationships between conditioned stimulus, conditioned response, and unconditioned stimulus are more complex. Summarizing their data obtained in experiments on dogs, Konorski and Miller (1933, p. 96) distinguished four possible relationships between CS, CR, and US that may be programmed in instrumental conditioning. These four possibilities constitute the conditions necessary for the development of four varieties of Type II conditioned reflexes and are presented in Table 10.1. The terms "attractive" and "aversive" US will serve better than the terms "positive" and "negative" US, which were used in Konorski's table (1948) to denote the character of the biologically important events used in the shaping of instrumental reflexes.

By means of classically conditioned reflexes, the subject may adapt to an inevitable chain of events, whereas by means of instrumental reflexes the subject may change the probability of occurrence of a US. This is illustrated in Fig. 10.2, in which relationships between the CS and US are presented for excitatory and inhibitory classically conditioned reflexes and for the four varieties of instrumentally conditioned reflexes. For instrumental conditioned reflexes the time relationships between CS, CR, and US are presented in the form used by Konorski and Miller in their early period of experimental work (Konorski & Miller, 1933, 1936).

In all his periods of experimental and theoretical activity, Konorski was aware that any learning situation involves the interaction, not only between excitatory and inhibitory tendencies, but also between different types and varieties of conditioned reflexes. Describing the simplest procedure for reward training, consisting of presenting food every time a leg flexion occurred, Konorski and Miller explicitly stated that the experimental situation acquires the properties of evoking an excitatory classically conditioned alimentary reflex (Konorski & Miller, 1933, p. 73 and elsewhere). Active leg

TABLE 10.1
Four varieties of conditioned reflexes of the second type,
after Konorski (1948, p. 232).

Variety	Compound s + r	Stimulus s applied separately	Unconditioned reinforcing stimulus	Form of conditioned reflex second type
First	Reinforced	Nonreinforced	Attractive	s ⟶ r
Second	Nonreinforced	Reinforced	Attractive	s ⟶ ~r
Third	Nonreinforced	Reinforced	Aversive	s ⟶ r
Fourth	Reinforced	Nonreinforced	Aversive	s ⟶ ~r

s, external stimulus
r, movement applied, or corresponding proprioceptive compound stimulus
~r, movement antagonistic to r.

flexions occur at that time when, because of the instrumental contingency, the excitatory properties of the experimental situation become partially inhibited. In fact, the main concern of the two extensive publications by Konorski and Miller was the analysis of interactions between classically conditioned and instrumental reflexes in the course of shaping and stabilization of the four varieties of instrumental reflexes (Konorski & Miller, 1933, 1936).

In more contemporary psychological literature a detailed analysis of interactions between excitatory classically conditioned and instrumental defensive reflexes has been published (Black, 1971). However, the role of inhibitory classically conditioned reflexes in the training of instrumental reflexes remains unexplored in Western psychological laboratories. Despite the fact that the experimental data collected by Konorski and Miller, and later by a team headed by Konorski, gave a lot of information on the role of conditioned inhibitory reflexes under various learning procedures, analysis of this topic is not presented in this chapter.[2] Rather, we restrict our analysis of the interactions between different types and varieties of conditioned reflexes to situations in which two different sporadic stimuli are presented on separate trials and in which two explicitly different modes of behavior are required. To establish these different modes of behavior, *only one* kind of biologically important event, the US, is used. Thus, only the analysis of differentiation tasks involving "homogeneous conditioned reflexes" (cf. Konorski, 1948, p. 109) is undertaken here.

---

[2]Indeed, an extensive anaysis is contained in the 1936 paper by Konorski and Miller published in Russian, which is now being translated and annotated for publication in *Acta Neurobiologiae Experimentalis.*

282   ZIELINSKI

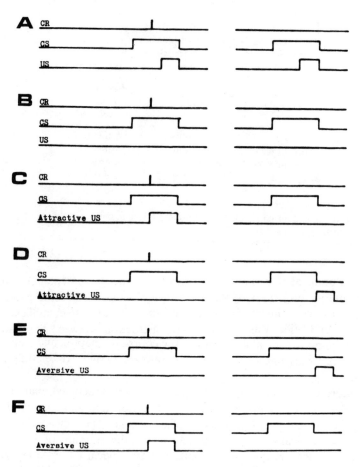

FIG. 10.2   Relationships between the CS and presentation of the US for
both the occurrence (left side) and nonoccurrence (right) of the CR in
classically conditioned reflexes and in four varieties of instrumental reflexes. The time course in this and in the following figures is from left to right.
A. Excitatory classically conditioned reflex. B. Inhibitory classically conditioned reflex. C. Reward training. D. Omission training. E. Avoidance
training. F. Punishment training.

C. Pavlovian differentiation

The procedure of Pavlovian differentiation consists of the presentation of
sporadic stimuli on separate trials, Typically two stimuli are used, but if the
number of stimuli employed is greater, two of them constitute a "pair."
One conditioned stimulus, termed "positive" (CS⁺), is always paired with
the US, and the second, termed "negative" (CS⁻), is never paired with the
US (e.g., Konorski, 1967, p. 327–328). Consequently, trials on which CS⁻ is
presented are termed "negative" and the others "positive" trials.

From the preceding discussion and inspection of Fig. 10.3, one may

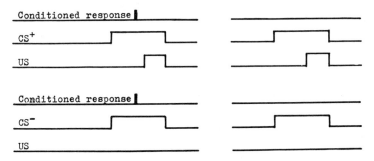

FIG. 10.3.   Relationships among CSs, CR occurrence, and US presentation on positive (upper) and negative (lower) trials presented in the course of the differentiation of the classically conditioned reflexes. From Zielinski (1976).

easily infer that during Pavlovian differentiation two reflexes are trained: the excitatory classically conditioned reflex to the CS^+ and the inhibitory classically conditioned reflex to the CS^-. The nature of the task does not change when the CS^+ signals a probability of US presentation (p_1) less than 1.0 and the CS^- signals a probability of US presentation (p_2) greater than zero, provided that $p_1 > p_2$ and that this difference would be perceived by a subject.

In the case of classically conditioned reflexes, the procedure for differentiation learning is straightforward. The excitatory and inhibitory conditioned stimuli must be presented on separate trials, because each of them signals a definite probability of US presentation and the subject's behavior cannot change these probabilities.

D. Differentiation learning involving instrumental conditioning

When instrumental conditioning is employed in differentiation learning, the CSs may be presented either on the same trial (simultaneous differentiation) or on separate trials (successive differentiation). The same or different overt responses to the CSs may be required. Let us further limit our analysis to situations in which only one overt response is required to one stimulus, whereas lack of this response to the second stimulus is accepted as correct performance (go–no go differentiation).

1. Procedures of "symmetrical" and "asymmetrical" go-no go differentiation

In the differentiation of alimentary instrumental reflexes, performance of a specified movement on positive trials is a necessary condition for food presentation. On negative trials two different procedures are possible. Either the CS^- may never be paired with food or the correct response to the CS^-—namely, nonoccurrence of a specified movement—is reinforced by food at about the time the CS^- is terminated. The first procedure is more common.

The second procedure has been employed in some experiments on monkeys (e.g., Gross, 1963; Gross & Weiskrantz, 1962; Weiskrantz & Mishkin, 1958). Because food reinforcement was available on both "positive" and "negative" trials, this procedure is termed "go–no go differentiation with symmetrical reinforcement." However, for a long time investigators were not aware of the possible consequences for learning and retention of employing one, rather than the other, of these two procedures. The differences between the two procedures were first recognized and investigated experimentally by Dabrowska (Dabrowska, 1971, 1972; Dabrowska & Szafranska-Kosmal, 1972). She introduced the term go–no go differentiation with asymmetrical reinforcement for the procedure in which food on negative trials is never presented.

The relationships between the occurrence of the CSs, US, and performance of the instrumental response in the two procedures are shown in Fig. 10.4.

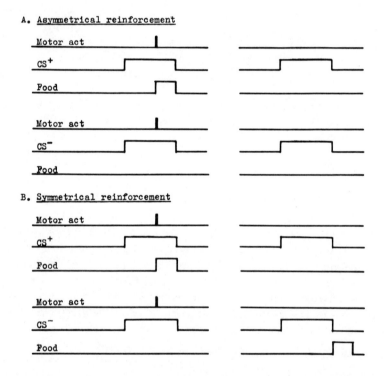

FIG. 10.4. Relationships among the CSs, occurrence of the required CR, and presentation of the food US on two kinds of trials presented in the course of the go–no go differentiation of alimentary instrumental reflexes with asymmetrical (A) and symmetrical (B) procedures of reinforcement. From Zielinski (1976).

Let us analyze the contingencies presented in Fig. 10.4, comparing them with the different types and varieties of conditioned reflexes shown in Fig. 10.2. We may easily discover that on positive trials, with both asymmetrical and symmetrical procedures the first variety of instrumental reflexes is employed. On negative trials in the asymmetrical procedure, the classical conditioning paradigm is in operation, and the relationships between CS, US, and CR are the same as on the negative trials of Pavlovian differentiation. On the other hand, in symmetrical go–no go differentiation, the second variety of instrumental reflexes is employed on negative trials. Thus, asymmetrical and symmetrical procedures differ with respect not only to the opportunity of receiving food reinforcement on negative trials, but also to the hetero- or homogeneity of the types of conditioning procedures that are in action on positive and negative trials.

A complete analogy exists in go–no go differentiation for defensive instrumental reflexes. With asymmetrical reinforcement the relationships between the CS, CR, and US are the same on negative trials as in differentiation of classical conditioned reflexes. In the symmetrical procedure the fourth variety of instrumental conditioning is in operation on negative trials (cf. Figs. 10.2 and 10.5). In our experiments with asymmetrical and symmetrical differentiation of instrumental defensive reflexes, modification of the third variety of instrumental reflexes was employed on positive trials, and a comparison of Fig. 10.2 with Fig. 10.5 shows this deviation from Konorski's scheme. This modification was used because termination of the CS immediately after performance of the avoidance response makes the learning process easier, as does requirement of the escape response if the subject fails to avoid shock. It should be mentioned, however, that because of this modification the procedure shown in Fig. 10.5, Part A is asymmetrical only with respect to the relationships between instrumental response performance and application of the noxious US. Because, in the course of learning, the CS$^+$ acquired fear-evoking properties (according to the laws of classical conditioning paradigms), termination of the CS$^+$ on positive trials implies additional, secondary reinforcement of the instrumental response. These fear-evoking properties generalize from the CS$^+$ to the CS$^-$; thus performance of the instrumental response on negative trials is "punished" by continuation of the CS$^-$ after the error is committed. Experimental data indicate that if the avoidance reflexes are firmly established, either punishment of the instrumental response by shock (i.e., change from the third to the fourth variety of instrumental procedure) or prolongation of the CS beyond the moment of performance of the instrumental response are necessary conditions for extinction, differentiation, and development of conditioned inhibition of the avoidance reflex (Soltysik & Zielinski, 1962).

From the preceding discussion we may conclude that differentiation learning (or, to be exact, go–no go differentiation, since only these pro-

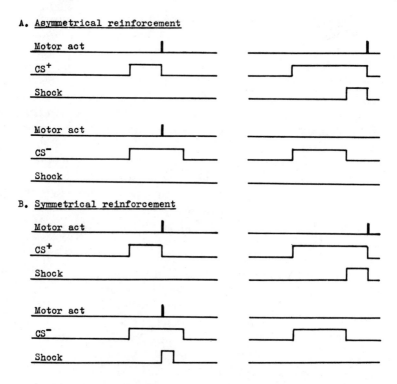

FIG. 10.5.   Relationships among the CSs, occurrence of the required CR, and application of the aversive US on two kinds of trials presented in the course of the go–no go differentiation of defensive instrumental reflexes with asymmetrical (A) and symmetrical (B) procedures of reinforcement. From Zielinski (1976).

cedures are analyzed in this chapter) consists of the parallel shaping of two different conditioned reflexes, each of them being a result of the particular contingencies used in training. Of course, other procedures of go–no go differentiation than those presented in Figs. 10.3, 10.4, and 10.5 are possible. For instance, Konorski and Miller used a differentiation procedure in which the second variety of instrumental conditioning, the method of omission training, was established on "positive trials" and in which contingencies of inhibitory classical conditioned reflexes were employed on "negative trials" (Konorski & Miller, 1936, pp. 210–213). However, in other experiments they established differentiation that would now be labeled as go–no go differentiation with symmetrical reinforcement, in which omission training was used on "negative trials." Thus, it follows from this example that terms such as: "positive trial" "negative trial" and "positive CS," "negative" or "inhibitory CS" give us no information about the contingencies used in a differentiation task.

3. Inhibitory conditioning as conditioning of an opposite behavioral tendency

This entire interpretation is in agreement with Konorski's understanding of the term "differentiation." The same pair of discriminatory stimuli may be differently used by a subject; and even if the "correct response" to the CS^- is nonperformance of the response required to the CS^+, the task may be solved in different ways depending on the contingencies imposed by the experimenter.

Using a different line of analysis from that presented in this chapter, Konorski examined all of his experimental data again and came to the conclusion that (1967) "both the terms 'internal inhibition' and 'inhibitory CR' are not adequate and should be abandoned. The CR (i.e., conditioned reflex) established by food non-reinforcement of a given stimulus is based on two *excitatory* CR arcs (p. 32)." We do not present here the model proposed by Konorski (see Dabrowska, Chapter 12) that lies behind this radical restatement, but the essential assumption is that any conditioning procedure uses inhibitory interconnections between antagonistic centers of the brain inborn or developed in ontogeny and that only excitatory connections are established in the course of conditioning. In accordance with this model (1967), "the extinction of the CR (i.e., conditioned reflex) does not lead to its *transformation* in the precise meaning of the word, but rather to the establishment of a new, opposite CR side by side with the old one. As a consequence, two antagonistic CRs are actually formed to the CS, either one taking the upper hand (p. 333)."

Further research will clarify whether the model proposed by Konorski is correct. However, we agree with the foregoing statement on the nature of extinction process. The course of extinction and the final outcome have to be different, depending on what kind of a new conditioned reflex is established. This was confirmed by experiments in which procedures of go–no go differentiation with asymmetrical and symmetrical reinforcement were compared.

3. Parameters influencing symmetrical and asymmetrical differentiation

In experiments on alimentary conditioned reflexes in dogs, Dabrowska showed that after medial prefrontal lesions (removal of gyrus pregenualis), asymmetrically reinforced differentiation was transiently impaired, whereas retention of symmetrically reinforced differentiation was perfect. In contrast, after lateral prefrontal lesions (removal of gyrus orbitalis), symmetrically reinforced differentiation was almost permanently impaired, whereas asymmetrically reinforced differentiation was only transiently impaired when short (15-sec instead of 60-sec) intertrial intervals were used. The character of this disturbance was also different with these two procedures. With the asymmetrical procedure, prefrontal lesions increased the number of errors on negative trials only, whereas with the symmetrical procedure lesioned dogs produced errors both to CS⁺ and to CS⁻. On the basis of these data it was concluded that go–no go differentiations with asymmetrical and symmetrical reinforcement are quite different tasks and depend on different structures (Dabrowska & Szafranska-Kosmal, 1972; Dabrowska, Chapter 12).

In differentiation of an avoidance reflex, the two procedures shown in Fig. 10.5 produce different effects on learning. Both with differentiation of classical defensive conditioned reflexes (Zielinski, 1965) and with differentiation of defensive instrumental reflexes with asymmetrical reinforcement (Kowalska, Dabrowska, & Zielinski, 1975a; Zielinski & Czarkowska, 1973), the rate of learning depends on the quality and/or intensity relationships between the two conditioned stimuli. The differentiation is easier if the CS⁺ is a more salient, and the CS⁻ a less salient stimulus, than vice-versa. When the symmetrical reinforcement procedure is used in go–no go differentiation of defensive reflexes, effects of quality and/or intensity relationships between the CSs on the rate of learning and number of errors are markedly reduced (Kowalska, Dabrowska, & Zielinski, 1975b; Kowalska & Zielinski, 1976).

Direct comparison of the course of go–no go differentiation with asymmetrical and symmetrical reinforcement of instrumental defensive reflexes was made by Kowalska (1977). Easily discriminable (click vs. tone) or similar (70 dB white noise vs. 50 dB white noise) CSs were used, and either the more or the less salient stimulus was the CS⁺ for different groups of dogs. The interaction between saliency of the CS⁺ and the reinforcing procedure was highly significant. The length of training to criterion was affected by neither the quality nor the intensity of the CS⁺ when symmetrical go–no go differentiation was employed (see Fig. 10.6).

These differences in learning may be easily understood if one takes into account that with the symmetrical procedure two active, but antagonistic movements have to be performed on positive and negative trials. Thus with the symmetrical procedure, response generalization is markedly reduced, and each stimulus influences the strength of its "own" instrumental response. Conversely, with the asymmetrical procedures a specific instrumental response is required only on "positive trials," and parameters having effects on the strength of this CR influence responding on both positive and negative trials.

Konorski suggested that stimulus generalization influencing the course of differentiation is presumably restricted to these tasks, in which CS⁻ is introduced only after the CR to the CS⁺ has been established. When both stimuli are introduced into training at the same time, stimulus generalization is markedly reduced (Konorski, 1967, pp. 345–348). Konorski's hypothesis was not confirmed in experiments performed on rats trained in differentia-

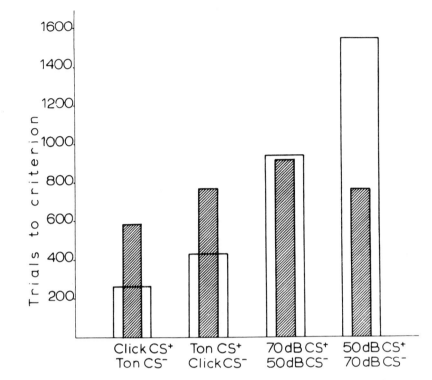

FIG. 10.6. Mean number of trials to 90% criterion of correct performance for eight groups of dogs trained in go–no go differentiation with asymmetrical (wide open bars) and symmetrical (narrow, crossed bars) reinforcement of the instrumental defensive reflexes.

tion of classically conditioned defensive reflexes (Jakubowska & Zielinski, 1976b; Zielinski & Jakubowska, 1977a). The amount of stimulus generalization was found to depend on the contingencies employed on negative trials rather than on the time of CS⁻ introduction into training.

Because of the identity of the contingencies on negative trials, differentiation of classically conditioned reflexes and go–no go differentiation with asymmetrical reinforcement have many features in common. On the other hand, Dabrowska (1972) suggested that with the symmetrical procedure, the mechanisms first proposed by Wyrwicka (1960) for differentiation of the two different overt responses (i.e., right-leg—left-leg differentiation) are in operation.

IV.   CONCLUDING COMMENTS

All the data and considerations presented in this chapter indicate that the analysis of procedures used to develop inhibitory conditioned reflexes must be based on identification of the contingencies employed. The difficult task is to identify the contingencies involved in differentiation learning, especially if we take into account the following circumstances.

First, any instrumental reflex is based on, and accompanied by, the corresponding classically conditioned drive reflex (Konorski, 1967; Konorski & Miller, 1933; Mowrer, 1947; Solomon & Wynne, 1953). Similarly, differentiation learning of any instrumental reflex, independent of the reinforcing procedure, is accompanied by differentiation of the corresponding classically conditioned drive reflex. Second, differentiation between the CS$^+$ and the background cues is typically accomplished on the basis of asymmetrical reinforcement, even in the case of symmetrical reinforcement of instrumental responses performed to sporadic conditioned stimuli. Depending on the stage of extinction of intertrial responses and on the discriminability of the CSs from the background, these two differentiation learning processes may interact in different ways. Third, it was shown that different parameters influence acquisition of the instrumental responses and its further consolidation (Zielinski & Jakubowska, 1977b). In the course of learning, the significance of CSs changes considerably; similarly, at consecutive stages of differentiation various mechanisms may be used for suppressing performance of the previously acquired responses to the negative stimulus (Soltysik & Zielinski, 1962; Terrace, 1972).

In summary, it must be stressed that the problem of conditioned inhibition intrigued Konorski throughout his 45 years of research activity. Some important implications of the early versions of Konorski's concept remain unexplored. Recent experimental data indicate that these ideas still possess a great power of prediction and may be of considerable interest for contemporary students of conditioning and learning.

ACKNOWLEDGMENT

I am greatly indebted to Dr. J. Brennan for interesting discussions during the preparation of this paper and for corrections of its English style.

REFERENCES

Bignami, G. Stimulus control: contrast, peak shift, and Konorskian model. Psychonomic Science, 1968, *11*, 249–250.

Black, A. H. Autonomic aversive conditioning in infrahuman subjects. In F. R. Brush (Ed.), *Aversive conditioning and learning.* New York: Academic Press, 1971.

Bloomfield, T. M. Behavioral contrast and the peak shift. In R. M. Gilbert & N. S. Sutherland (Eds.), *Animal discrimination learning.* London: Academic Press, 1969.

Dabrowska, J. Dissociation of impairment after lateral and medial prefrontal lesions in dogs. *Science,* 1971, *7,* 1037–1038.

Dabrowska, J. On the mechanism of go–no go symmetrically reinforced tasks in dogs. *Acta Neurobiologiae Experimentalis,* 1972, *32,* 345–359.

Dabrowska, J., & Szafranska-Kosmal, A. Partial prefrontal lesions and go–no go symmetrical reinforced differentiation test in dogs. *Acta Neurobiologiae Experimentalis,* 1972, *32,* 817–834.

Estes, W. K., & Skinner, B. F. Some quantitative properties of anxiety. *Journal of Experimental Psychology,* 1941, *29,* 390–400.

Grant, O. A., & Schneider, D. E. Intensity of the conditioned stimulus and strength of conditioning. I. The conditioned eyelid response to light. *Journal of Experimental Psychology,* 1948, *38,* 690–696.

Grant, O. A., & Schneider, D. E. Intensity of the conditioned stimulus and strength of conditioning. II. The conditoned galvanic skin response to an auditory stimulus. *Journal of Experimental Psychology,* 1949, *39,* 35–40.

Gray, J. A. Stimulus intensity dynamism. *Psychological Bulletin,* 1965, *63,* 180–196.

Gross, C. G. Effect of deprivation on delayed response and delayed alternation performance by normal and brain operated monkeys. *Journal of Comparative and Physiological Psychology,* 1963, *56,* 48–51.

Gross, C. G., & Weiskrantz, L. Evidence for dissociation of impairment on auditory discrimination and delayed response following lateral frontal lesions in monkeys. *Experimental Neurology,* 1962, *5,* 453–476.

Hearst, E. Some persistent problems in the analysis of conditioned inhibition. In R. A. Boakes & M. S. Halliday (Eds.), *Inhibition and learning.* London: Academic Press, 1972.

Hull, C. L. *Principles of behavior.* New York: Appleton-Century-Crofts, 1943.

Hull, C. L. Simple qualitative discrimination learning. *Psychological Review,* 1950, *57,* 303–313.

Hull, C. L. *A behavior system.* New Haven: Yale University Press, 1952.

Jakubowska, E., & Zielinski, K. Conditioned stimulus intensity effects on acute and chronic extinction of CER. *Activitas Nervosa Superior,* 1976, *18,* 15–16. (a)

Jakubowska, E., & Zielinski, K. Differentiation learning as a function of stimulus intensity and previous experience with the CS⁺. *Acta Neurobiologiae Experimentalis,* 1976, *36,* 427–446 (b).

Jakubowska, E., & Zielinski, K. Stimulus intensity effects on acute extinction of the CER in rats. *Acta Neurobiologiae Experimentalis,* 1978, *38,* 1–10.

Jenkins, H. O. Measurement of stimulus control during discriminative operant conditioning. *Psychological Bulletin,* 1965, *64,* 365–376.

Kimble, G. A. *Hilgard and Marquis' conditioning and learning.* New York: Appleton-Century-Crofts, 1961.

Kimble, G. A. Categories of learning and the problem of definition, In A. W. Melton (Ed.), *Categories of human learning.* New York: Academic Press, 1964.

Konorski, J. *Conditioned reflexes and neuron organization.* Cambridge: Cambridge University Press, 1948.

Konorski, J. *Integrative activity of the brain: An interdisciplinary approach.* Chicago: University of Chicago Press, 1967.

Konorski, J. *Conditioned reflexes and neuron organization.* New York: Hafner, 1968. (Facsimile reprint of the 1948 edition with a new foreword and supplementary chapter).

Konorski, J. Jerzy Konorski. In G. Lindzey (Ed.), A history of psychology in autobiography (vol. 6). Englewood Cliffs, N.J.: Prentice-Hall, 1974.

Konorski, J., Lubinska, L., & Miller, S. Elaboration of conditioned reflexes in cerebral cortex inhibited due to "negative induction," *Acta Biologiae Experimentalis*, 1936, *10*, 297-330. (Polish)

Konorski, J., & Miller, S. The foundation of physiological theory of acquired movements, motor conditioned reflexes. Warsaw: Ksiaznica Atlas TNSW, 1933. (Polish)

Konorski, J, & Miller, S. Conditioned reflexes of the motor analyser. Trudy Fiziologicheskikh Laboratorii Akademika I. P. Pavlova, 1936, *6*, 119-288. (Russian)

Konorski, J., & Szwejkowska, G. Chronic extinction and restoration of conditoned reflexes. I. Extinction against the excitatory background. *Acta Biologiae Experimentalis*, 1950, *15*, 155-170.

Konorski, J., & Szwejkowska, G. Chronic extinction and restoration of conditioned reflexes. III. Defensive motor reflexes. *Acta Biologiae Experimentalis*, 1952, *16*, 91-94.

Kowalska, D. Effects of partial prefrontal lesion in dogs on differentiation and reversal learning of the avoidance reflex. Unpublished doctoral dissertation, Nencki Institute, Warsaw. (Polish)

Kowalska, D., Dabrowska, J., & Zielinski, K. Effects of partial prefrontal lesions in dogs on go–no go avoidance reflex differentiation and reversal learning. *Acta Neurobiologiae Experimentalis*, 1975, *35*, 549-580. (a)

Kowalska, D., Dabrowska, J., & Zielinski, K. Retention of symmetrical go–no go avoidance differentiation after prefrontal lesions in dogs. *Bulletin de L'Académie Polonaise des Sciences, Série des Sciences Biologiques*, 1975, *23*, 487-494. (b)

Kowalska, D., & Zielinski, K. Asymmetrical versus symmetrical go–no go avoidance differentiation in dogs. *Activitas Nervosa Superior*, 1976, *18*, 16-18.

Miller, S., & Konorski, J. Sur une forme particuliere des réflexes conditionnels. *Comptes Rendus des Séances de la Société de Biologie*, 1928, *99*, 1155-1157.

Mowrer, O. H. On the dual nature of learning. A re-interpretation of "conditioning" and "problem solving." *Harvard Educational Review*, 1947, *17*, 102-148.

Mowrer, O. H. Jerzy Konorski memorial address. *Acta Neurobiologiae Experimentalis*, 1976, *36*, 249-276.

Pavlov, I. P. [Conditioned reflexes. An investigation of the physiological activity of the higher nervous activity of the cerebral cortex] (G. V. Anrep, Trans.). London: Oxford University Press, 1927.

Solomon, R. L., & Wynne, L. C. Traumatic avoidance learning: Acquisition in normal dogs. *Psychological Monographs*, 1953, *67*, 1-19.

Soltysik, S., & Zielinski, K. Conditioned inhibition of the avoidance reflex, *Acta Biologiae Experimentalis*, 1962, *22*, 157-167.

Spence, K. W. The differential response in animals to stimuli varying within a single dimension. *Psychological Review*, 1937, *44*, 430-444.

Szwejkowska, G. The chronic extinction and restoration of conditioned reflexes. II. The extinction against an inhibitory background. *Acta Biologiae Experimentalis*, 1950, *15*, 171-184.

Szwejkowska, G. The transformation of differentiated inhibitory stimuli into positive conditioned stimuli. *Acta Biologiae Experimentalis*, 1959, *19*, 151-159.

Terrace, H. S. Discrimination learning with and without "errors." *Journal of the Experimental Analysis of Behavior*, 1963, *6*, 1-37.

Terrace, H. S. Conditioned inhibition in successive discrimination learning. In R. A. Boakes & M. S. Halliday (Eds.), *Inhibition and learning*. London: Academic Press, 1972.

Wagner, A. R., & Rescorla, R. A. Inhibition in Pavlovian conditioning: Application of a theory. In R. A. Boakes & M. S. Halliday (Eds.), *Inhibition and learning*. London: Academic Press, 1972.

Weiskrantz, L., & Mishkin, M. Effects of temporal and frontal cortical lesions on auditory discrimination in monkeys. *Brain*, 1958, *81*, 406-414.

Wyrwicka, W. An experimental approach to the problem of mechanism of alimentary conditioned reflexes, Type II. *Acta Biologiae Experimentalis*, 1960, *20*, 137–146.

Zbrozyna, A. W. The conditioned cessation of eating. *Bulletin de L'Academie Polonaise des Sciences, Série des Sciences Biologiques*, 1957, *5*, 261–265.

Zbrozyna, A. W. On the conditioned reflex of the cessation of the act of eating. I. Establishment of the conditioned cessation reflex. *Acta Biologiae Experimentalis*, 1958, *18*, 137–162.

Zieliński, K. The influence of stimulus intensity on the efficacy of reinforcement in differentiation training. *Acta Biologiae Experimentalis*, 1965, *25*, 317–335.

Zieliński, K. "Inhibition of delay" as a mechanism of the gradual weakening of the conditioned emotional response. *Acta Biologiae Experimentalis*, 1966, *26*, 407–418.

Zieliński, K. Effects of prefrontal lesions on avoidance and escape reflexes. *Acta Neurobiologiae Experimentalis*, 1972, *32*, 393–415.

Zieliński, K. Konorski's classification of conditioned reflexes: implications for differentiation learning. *Activitas Nervosa Superior*, 1976, *18*, 6–14.

Zieliński, K., & Czarkowska, J. Go–No go avoidance reflex differentiation and its retention after prefrontal lesion in cats. *Acta Neurobiologiae Experimentalis*, 1973, *33*, 467–490.

Zieliński, K., & Jakubowska, E. Auditory intensity generalization after CER differentiation training. *Acta Neurobiologiae Experimentalis*, 1977, *37*, 191–205.

Zieliński, K., & Jakubowska, E. Temporal and stimulus intensity factors in avoidance reflex acquisition in cats. *Activitas Nervosa Superior*, 1977, *19*, 14–16.

Zieliński, K., & Walasek, G. Stimulus intensity and CER magnitude: dependence upon the type of comparison and stage of training. *Acta Neurobiologiae Experimentalis*, 1977, *37*, 299–309.

11 Brain Mechanisms in the Inhibition of Behavior

Jeffrey A. Gray
J. N. P. Rawlins
J. Feldon
University of Oxford

It is a dishonorable tradition to conduct physiological psychology in virtual isolation from psychological theory, and indeed from experimental results gathered in other branches of psychological research. This myopia affects not only physiological psychology itself, where all too often clever men must struggle desperately to comprehend a series of random behavioral observations that have in common only that they follow the same insult to the brain, but also general psychological theory, which clings too long to concepts already demonstrated to be false in physiological experiments. To this tradition Jerzy Konorski was an honorable and almost unique exception. In his work, thinking about brain and about behavior is a single, indissoluble activity. If a theory about brain function does not fit the facts of purely behavioral experiments, that theory has to be changed; and, conversely, if a theory based on behavioral data cannot make sense out of the results of experiments in physiological psychology, it too must be discarded. Thus we cannot doubt that he would have regarded it as a legitimate enterprise to see how one of his own, largely behavioral, theories fares when it is brought into contact with the results of experiments in physiological psychology.

I. KONORSKI'S THEORY OF INHIBITION

In Konorski's 1948 theory of inhibition, a radical distinction was drawn between the inhibition of classical and instrumental responses, paralleling the distinction between the acquisition of these two kinds of behavior. According to this theory (see Rescorla, Chapter 4) inhibition in classical conditioning is due to the development of connections from the "center"for the inhibitory stimulus to the US center (today's more cognitive vocabulary would substitute "representation" for "center," but mean the same thing). When activated these connections then serve to raise the threshold for excitation of the US center by impulses emanating from the CS center. In contrast, the suppression of instrumental responses was attributed by Konorski (1948, pp. 239–241) to a mechanism that, by the time we reach his 1967 book, is termed "motor act inhibition." As described in 1967 this mechanism works "by establishing connections linking the CS with the antagonistic movement. This inhibition takes place when the performance of a given movement in response to a given stimulus does not lead to the satisfaction of drive, whereas its non-performance does (p. 458)." Both in the 1948 and the 1967 book, motor-act inhibition is said to underlie the elimination of punished responses and of responses subjected to appetitive omission training (in which a reward is made contingent on not performing a specified response).

Konorski's 1948 position is one that we find congenial. It bears a number of similarities to a theory that has guided our own work — notably, its distinction between classical and instrumental conditioning and its identification of punishment and reward omission as the two major conditions for engaging the mechanism of instrumental inhibition (Gray, 1975, 1976). However, the 1948 theory of inhibition was considerably modified in Konorski's 1967 book. This modification occurred at two major points.

First, the particular mechanism underlying the inhibition of classically conditioned responses was now said to be that of "drive inhibition." This mechanism was described as follows (1967):

> When in a situation in which a given US is repeatedly presented—that is, in which a given drive is dominant—a "neutral" stimulus is repeatedly presented without being followed by this US, then the gnostic units representing this stimulus form connections with the gnostic units representing the *lack of the US* and with the *antidrive* units. In this way the negative CR is established to that stimulus, being antagonistic to the positive CR established to the stimuli followed by the given US [p, 344, Konorski's italics].

Second, instead of one mechanism of inhibition of instrumental responses, there were now said to be three: motor act inhibition, as in 1948; drive inhibition (which appears to be identical with the drive inhibition operative in classical conditioning, although Konorski is not completely explicit about this); and "retroactive inhibition." Appropriate behavioral situations for the production of the third kind of inhibition are ones in which (1967)

> the incorrect instrumental response performed in the presence of the type II CS is not followed by satisfaction of drive and in consequence the animal performs some other movements. If one of these movements signals the termination of drive, then after a

number of such trials the animal will learn to perform that movement at once in the presence of the given CS... the first movement is then retroactively inhibited by the second movement [p. 449].

The change from a raised threshold of excitation in the US center to drive inhibition as the mechanism underlying the inhibition of classical CRs leaves intact the essential similarities between Konorski's 1948 views and our own, as previously noted. But the second change does not. Instead of a neat division between classical and instrumental inhibition, we now have an asymmetrical situation in which there are three mechanisms of instrumental inhibition, of which one (drive inhibition) is identical with the unique form of classical inhibition. There is an odd imbalance in the kinds of reasons given by Konorski for the two changes. Whereas the first — the change to drive inhibition as the mechanism underlying the disappearance of classical CRs — is backed up by cogent experimental arguments (Konorski, 1967, pp. 316–323), the second — the addition of "reciprocal" and "drive" to "motor-act" inhibition in the instrumental case — is justified in the most cursory manner by the consideration that all three kinds of inhibition are *logically* possible (p. 448).

It is possible to minimize the extent to which Konorski's 1967 theory of inhibition differs from his 1948 theory. For example the chief role played by the concept of retroactive inhibition in the 1967 theory is to ensure that only the response immediately preceding instrumental reinforcement is learned; and this sort of process is not normally thought of as having anything to do with the *inhibition* of responses. As regards drive inhibition, this plays a role in Konorski's theory of instrumental inhibition only to the extent that instrumental learning is based on classical conditioning and thus would be sensitive to inhibition of the classical CR. If one were to adopt this line of argument (which, however, is not definitely implied by Konorski's 1967 text), those features of the 1948 theory that are similar to our own would still be essentially present in the 1967 theory: a major difference between classical and instrumental inhibition (based on drive and motor-act inhibition, respectively); and the engagement of instrumental inhibition by punishment and the omission of reward.

II. THE EVIDENCE FROM PHYSIOLOGICAL PSYCHOLOGY

If either interpretation of Konorski's theory of inhibition is a correct description of the way in which organisms go about eliminating various kinds of behavioral responses, it is a reasonable inference that there ought to exist in the brain separate kinds of process, one corresponding to each of the kinds of inhibition he distinguishes and therefore involved in the tasks that require the appropriate kind of inhibition. It follows that if one were able to interfere selectively with one of these processes but not with the others, one ought to observe impaired performance only in the tasks said by Konorski to depend on that particular kind of inhibition. The trouble with this argument, of course, is that the enormity of the "if" makes it very asymmetrical: It is inherently unlikely that random interventions in the brain (which, by and large, are all we have) will light neatly on one of Konorski's behavioral processes, so that the failure to find a predicted pattern of behavioral change will necessarily be inconclusive; a positive outcome of such a search, on the other hand, would be a striking confirmation of Konorski's theory.

One way to strengthen the significance of a negative outcome in such a case is to repeat the search with a different intervention in the functioning of the brain. But this by itself would add very little, given the low a priori probability that any such search will land on the right set of brain processes. But suppose that the two interventions are very different from each other, that they both produce clear-cut impairments in behavioral inhibition, that the patterns of change each produces are very similar, and that this pattern of change is unlike the pattern predicted by Konorski both in what it puts together and in what it splits apart. Then one would surely begin to doubt the value of the particular groups of tasks proposed by Konorski, and therefore of the principles on which these are based.

This, then, is the strategy we adopt. We test Konorski's principles of classifying "inhibitory" tasks against the available literature on two very different physiological treatments. One of these treatments is pharmacological—the administration of any of the class of minor tranquilizing drugs (i.e., the barbiturates, the benzodiazepines, and alcohol); the other is neurological — the making of large lesions in the septal area or in the hippocampus.

This is not the place to review in detail the very large number of experiments that have been concerned with the behavioral effects of the minor tranquilizers and of septal and hippocampal lesions. We have done this job elsewhere for the drugs (Gray, 1977); and Gray and NcNaughton (in preparation) are in the process of doing it for both kinds of lesion (thus adding to the many existing reviews of this literature — e.g., Altman, Brunner, & Bayer, 1973; Caplan, 1973; Dickinson, 1974; Lubar & Numan, 1973; Nadel, O'Keefe, & Black, 1975); so we hope the reader will accept the conclusions summarized in Table 11.1. Where no references are given in this table, a large number of studies deal with the point at issue, and they are in substantial agreement.

The first point to notice about Table 11.1 is the concordance between the behavioral effects of the minor tranquilizers, septal lesions, and hippocampal lesions. This is not just an artifact of the particular behavioral situations included in the table. On the contrary, these have been chosen in the light of

TABLE 11.1

Type of Konorski Inhibition	Task	Effects of		
		MT*	Septal Lesion	Hippocampal Lesion
Retroactive	Simultaneous discrimination with intratrial correction	0	0	0
	Reversal learning	-	-	-
	Successive discrimination	-	-	-
Drive	Extinction of rewarded response	-	-	-
	Conditioned suppression	?0	?0	0
	Classical conditioned inhibition	?	?0 (Moore, Ch. 5)	0 (Moore, Ch. 5)
Motor Act	Simultaneous discrimination without intratrial correction	0	0	0
	Passive avoidance	-	-	-
	Omission learning	?	-(Atnip & Hothersall, 1975)	?
	DRL	-	-	-

0: No systematic effect   -: Generally an impairment   ?: Insufficient data

*minor tranquilizers

their bearing on Konorski's theory, together with the knowledge that there are drug and lesion data relevant to them. Indeed, it is possible to draw up a much longer table of the similarities between the three treatments we are considering, and the number of discrepant results is surprisingly few (see Gray, in press). Furthermore, in all cases in which the effects of septal and hippocampal lesions are in agreement with each other, the effects of the minor tranquilizers are the same again. This pattern of results is good support for the hypothesis (Gray, 1970) that the minor tranquilizers alter behavior by some kind of action on the septohippocampal system.

The second point to note about Table 11.1 is that it offers little support for the rather complex classificaiton of inhibitory tasks offered by Konorski in 1967. The tasks set out in this table have been allocated to "retroactive," "drive," or "motor-act" inhibition in accordance with specific remarks made about each of them in the 1967 text (pp. 325, 448–458). However, it must be admitted that some of these allocations are more or less arbitrary. Thus simple extinction of an appetitve instrumental response is said by Konorski (1967, p. 450) to depend on all three of the instrumental inhibitory processes: "retroactive" early on in extinction, then "drive," and finally "motor act." It is clear, however, that Konorski regarded drive inhibition as the most important mechanism operating during simple extinction, and for this reason we have allocated extinction to this category. The same is true of successive discrimination. The capacity to inhibit responding on a schedule of differential reinforcement of low rate (DRL) is not explicitly discussed by Konorski, but it seems reasonable to regard it as sharing important features with appetitive omission training, which is paradigmatic (along with punishment) for motor-act inhibition.

Even with these reservations about the correctness of the allocation of different tasks to the different categories of Konorski's 1967 theory of inhibition, there is little hint in Table 11.1 of any selective action of the minor tranquilizers or of septal or hippocampal lesions on any one of these categories. Each of Konorski's types of inhibition provides at least one instance of impairment and one of no change after the three physiological treatments shown. The preponderance of cases of impairment in the table taken overall suggests that one would do better to join his three types into a more unitary theory of inhibition. A second possibility is that one should return to his simpler 1948 theory. To see this, consider the exceptional cases in Table 11.1 in which no impairment is seen after minor tranquilizers or septal or hippocampal lesions.

The first group of exceptions consists of simultaneous discriminations, learned with or without intratrial corrections. This exception is not surprising, for there is little reason to suppose that inhibitory processes of any kind play an important role in this kind of discrimination learning (see Mackintosh, 1974, p. 568). In this connection it is worth noting that, when impairments in simultaneous discrimination have been reported after any of the treatments listed in Table 11.1 (most often after hippocampal lesions), this has usually been because the need for behavioral inhibition has been stepped up by training the animal against an initial response bias, against some prior learnt strategy of responding, or against response tendencies elicited by interfering stimuli present in the experimental situation (Gray & McNaughton, in preparation). The second group of exceptions is more interesting: classical conditioned inhibition (see Moore, Chapter 5) and conditioned suppression. These are the only classical conditioning tasks that figure in Table 11.1 Thus Konorski's 1948 theory is perhaps correct: There may be a fundamental difference between classical and instrumental inhibition. And it is possible that minor tranquilizers and septal or hipocampal lesions affect only the latter.

III.   THE PHYSIOLOGY OF PUNISHMENT AND CONDITIONED SUPPRESSION

The oddest feature of Table 11.1 is the discrepancy that it reveals between experiments using response-dependent shock as a punishment (producing passive avoidance) and experiments using signaled, response-independent shock (producing conditioned suppression). The behavioral effects of these two procedures are so similar that it has long been a moot question whether the response suppression produced by punishment is a species of conditioned suppression, whether the response suppression occurring in a conditioned-suppression experiment is a species of punishment, or whether there are two different things going on in the two types of experiment (e.g., Gray, 1975, p. 242; Mackintosh, 1974, p. 295). These questions are specific instances of the more general question of whether classical and instrumental conditioning are two fundamentally different learning processes. (Konorski, 1948; Gray, 1975) or whether they are each special cases of a more general learning process, as held for example by Mackintosh (1974) and Bindra (1976), though in very different ways. Thus, if we could find a physiological scalpel able to separate conditioned suppression and punishment, this might have a quite general psychologcal importance. It is therefore worth examining more closely the effects of the minor tranquilizers and of septal and hippocampal lesions on these two forms of response suppression.

There is no reason to dwell on the effect of the minor tranquilizers on punishment-produced response suppression. It is quite clear that these drugs reduce such suppression across a wide variety of conditions (Gray, 1977). These conditions include both tasks that have a large spatial element (e.g., in the alley) and those that do not (e.g., in the Skinner box), a distinction whose relevance will become clear in a moment. With regard to the effects of hippocampal lesions on this form of response suppression, we must be a little less hasty, for the belief that this insult to the brain reduces passive avoidance, previously widely accepted (e.g., Altman et al., 1973; Douglas,

1967) has recently been questioned (Nadel et al, 1975). A similar question has been raised by Caplan (1973) in relation to the same belief about septal lesions (e.g., Lubar and Numan, 1973; McCleary, 1966). It is partly because of these doubts that we (Gray & McNaugton, in preparation) have undertaken yet another review of this vast literature. To do so we have examined every report we could find dealing with septal or hippocampal lesions from 1960 to the end (so far) of 1975. Our conclusions are summarized below.

With regard to septal lesions there can be little doubt that they grossly impair the ability of an animal to withhold a punished response. This finding has been reported when the punished response is locomotion toward food or water, step through to a more attractive environment, step down from a platform to the ground, bar pressing for food or water, a previously acquired shock-escape or shock-avoidance response, a consummatory response of feeding or drinking, or simple movement. This variety of tasks —some invoving positive reinforcers such as food or water, but many not— makes Caplan's (1973) claim that the septal effect on passive avoidance is an artifact of heightened incentive motivation (which may be a further consequence of the lesion) untenable. It is also not possible to apply to these data the spatial hypothesis, used by O'Keefe and Nadel (in press) to account for the hippocampal syndrome (see later), because step-down, step-through, bar-pressing, and punished consummatory behavior or punished movement are minimally concerned with space. The fact that the spatial hypothesis does not apply to the septal syndrome must make us suspicious about applying it to the hippocampal syndrome. According to O'Keefe and Nadel (in press) the spatial functions attributed to the hippocampus involve the operation of the hippocampal theta rhythm and, as is well known (Gray, 1971; Stumpf, 1965), this is abolished after septal lesions.

The case for a general hippocampal deficit in the withholding of punished responses is only slightly less strong than the case for a septal one. According to O'Keefe and his collaborators (Black, Nadel, & O'Keefe, 1977; Nadel et al., 1975; O'Keefe & Nadel, in press), however, this deficit is a secondary consequence of the true impairment produced by hippocampal lesions. This is said to be an inability to form cognitive maps of the spatial environment or to use them to effectively guide behavior in a spatially complex task. To support this view, these workers have argued that the hippocampal dificit in withholding punished responses is confined to those cases in which the punished response is locomotor and takes the animal to a dangerous "place" (a term that is given a rather technical definition within the O'Keefe and Nadel spatial theory). Our own review of the literature (Gray & McNaughton, in preparation) does not confirm this generalization.

It is not in dispute that passive avoidance is impaired by hippocampal lesions when responses performed in a spatially extended situation (such as a runway or two adjacent boxes) are first learned for positive reinforcement and then punished with shock. The results of step-down experiments support the spatial hypotheses: Only one out of four experiments of this kind found a deficit in response inhibition in hippocampal animals. But, contrary to the claim made by Black et al. (1977), this negative picture does not hold when one looks at step-through experiments: Five out of nine experiments of this kind have shown a hippocampal deficit. Experiments on punished drinking are not clear-cut. If only experiments that have produced major (total or dorsal) hippocampal damage are considered, there are three reports of a deficit in response inhibition, as against the same number of reports of no change. It would therefore seem that, although the hippocampal deficit in passive avoidance is not as robust or as ubiquitous as the septal deficit, it is certainly not confined to situations with a pronounced spatial element. Thus the view that the hippocampal deficit in passive avoidance is a secondary consequence of a more fundamental impairment in spatial ability is untenable. Even if the spatial theory could account for the passive avoidance deficit in hippocampal animals, there are many other data that this theory cannot handle—for example, increased resistance to extinction of bar pressing and impaired reversal learning in several nonspatial modalities.

We have labored our defense of the old view that hippocampal lesions produce a deficit in inhibition against the assault launched on it by the O'Keefe camp because, if they were right, the whole strategy of this paper would be wrong: One cannot evaluate a theory of inhibition by appealing to a lesion that disrupts spatial analysis. But in our view it is reasonable to conclude that hippocampal lesions, like septal lesions and the minor tranquilizers, impair an animal's ability to withhold a response if a punishment is made contingent on it. What are the effects of these treatments on conditioned suppression?

There is no doubt that the minor tranquilizers are more effective in alleviating punishment-produced response suppression than conditioned suppression (Gray 1977). Indeed, there is no good evidence that either the barbiturates or alcohol alter the latter at all. When it comes to the benzodiazepines the position is more complicated. There is some evidence that these drugs reduce on-the-baseline conditioned suppression—that is, when the CS and the shock US are both presented to the animal while it is performing the rewarded operant. But these findings are not substantiated when an off-the-baseline procedure is used—that is, when the CS–shock pairings are carried out separately and only the CS is presented to the animal while it is performing the operant response (Gray, 1977). So the possibility remains open that, even with the benzodiazepines, the apparent disinhibiton of conditioned suppression is due to an alleviation of suppression produced by adventitious punishment (which can occur in the on- but not the off-the-baseline procedure).

Given the general similarities between the effects of the minor tranquilizers, on the one hand, and those of septal and hippocampal lesions, on
the other, we have attempted to confirm the apparent differences between
punishment and conditioned suppression in our review of the latter
literature (Gray & McNaughton, in preparation). Five experiments have
used large lesions of the septal area (more discrete lesions are considered
later) and have examined conditioned suppression, in every case on the
baseline. Two of them found no effect (Duncan, 1972; Molino, 1975).
Brady and Nauta (1955) did find a deficit, but only at 3 days postoperatively; at 60 days postoperatively, there was no effect. Harvey Lints,
Jacobson, and Hunt (1965) also found a deficit, but only with a 1-mA
shock, not with a 0.5 mA. However, these 2 positive outcomes are suspect,
for they used water reward for the baseline operant, whereas the other 2
studies used food; and it is known that large septal lesions increase water intake and that increased incentive motivation decreases conditioned suppression (Haworth, 1971). A final experiment was performed by Dickinson
(1975), who made an explicit comparison between the effects of septal lesions on the response suppression produced by response-dependent and
response-independent shock, and found an equal alleviation of suppression
in the 2 cases. Dickinson's (1975) data are very convincing; nonetheless, it is
clear that the case for a reduction in response-independent response suppression in septal animals is less strong than for a reduction in response-
dependent suppression. The hippocampal literature is even less susceptible
to such an interpretation. Only off-the-baseline conditioned suppression has
been used with hippocampal animals. In 3 experiments this has yielded all 3
possible results: no change from controls, decreased suppression, and increased suppression (Antelman & Brown, 1972; Freeman, Mikulka, &
D'Auteuil, 1974).

There seems, then, to be at least a prima facie case for supposing that
response suppression produced by punishment involves a process that is not
involved in conditioned suppression and that is susceptible to injections of
minor tranquilizers and septal or hippocampal lesions. Conceivably this process is the response contingency itself, the heart of instrumental as opposed
to classical conditioning.

IV. WITHIN-ANIMAL COMPARISON OF PUNISHMENT
AND CONDITIONED SUPPRESSION

To examine this problem more closely we have used a within-animal design
(modeled after the technique introduced by Church, Wooten, and
Matthews, 1970) to make a direct comparison between the vulnerability of
punishment and conditioned suppression, respectively, to minor tranquilizers and to septal lesions. Church et al. showed that the rat is capable

of forming a discrimination between a stimulus in the presence of which an operant response is punished with shock, and one in the presence of which noncontingent shocks are presented against the operant baseline. We modified their procedure by using only one manipulandum, instead of two, thus making the comparison between punishment and conditioned suppression even closer.

The rats were first trained on a random-interval (RI) 64-sec schedule of sucrose reinforcement for bar pressing. They were then given 4 "intrusion" periods, each 3 min long, at random intervals during a 1-hr daily session in the Skinnner box. In two of these periods a punishment contingency was in operation, and an associated stimulus (S_p) was present throughout the 3 min; in the other 2, a noncontingent shock of the same intensity was presented in association with a second stimulus (S_c). In both kinds of intrusion period, the same RI schedule, independent of the one controlling sucrose delivery but also with a mean value of 64 sec, was used: in S_p, to set up a shock contingent on the next response; in S_c, to actually deliver the shock. Rats find this discrimination very difficult, but eventually about two-thirds of our subjects learn it, in the sense that suppression is consistently greater in one of the 2 stimuli than the other; among the animals that do form a discrimination, suppression is usually greater in S_p than in S_c.

Our first experiment with this design used 3 rats which consistently suppressed more in S_p than in S_c and looked at the effects on this discrimination of 5 mg/kg chlordiazepoxide versus saline, each given in an ABBA design to each rat over a period of 20 days. Fig. 11.1 presents our results. They show that chlordiazepoxide reduced punishment-induced response suppression more than conditioned suppression and abolished the animal's discrimination between these 2 conditions. This abolition was not merely a ceiling effect, because the rats still showed significant response suppression under the drug.

If one wished to substantiate the hypothesis that chlordiazepoxide alleviates *only* punishment-induced suppression, the ideal result would be for this drug to reduce suppression in the punishment component to a level equal to that seen in the conditioned-suppression component of the multiple schedule, while not changing the latter at all. One could then argue that punishment-induced suppression contains an element of classically conditioned suppression (unaffected by the drug) and an element of response-contingent suppression (attacked by the drug) (Church et al., 1970; Gray, 1975, p. 244). Our result is less ideal than this, but is nonetheless consistent with the hypothesis just stated.

The within-animal design inevitably has the problem that there is likely to be generalization between S_p and S_c, especially when as, in these experiments, one does one's best to ensure that all other factors (apart from the rules governing delivery of shock) are constant between these two stimuli. And even if this were not so, there is the problem (using on-the-baseline condi-

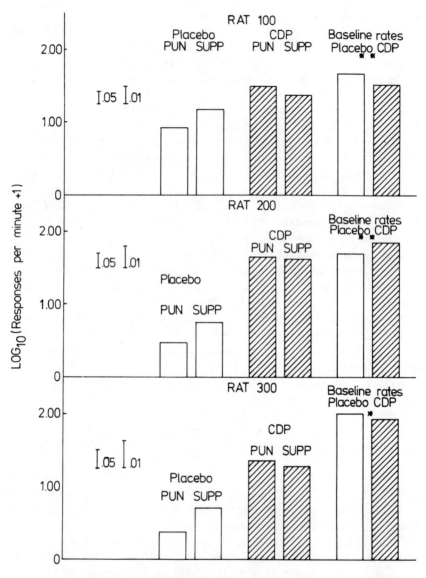

FIG. 11.1. Effects of chlordiazepoxide (CDP) HCl, 5 mg/kg i. p., on a within-animal discrimination between signaled punishment (PUN) and conditioned suppression (SUPP) in three rats. Bar markers to the left indicate significance levels of differences between response rates during any pair of the four intrusion periods making up the Drug x Condition (i.e., PUN vs. SUPP) interaction, shown as the four columns to the left. The two columns on the right show baseline rates in drug and placebo: ** 0.01; * 0.05.

tioned suppression) of possible adventitious punishment. If either of these factors affected the performance of our animals, we would expect the result obtained. Alleviation of suppression in both components to the extent that either is affected by punishment contingencies (whether real, imagined, or generalized) should result in greater disinhibition in S_p than in S_c, but some disinhibition in the latter stimulus; in some continued suppression in both stimuli; and in the abolition of the discrimination. Whether this is in fact what took place will require further experiments comparing the effects of chlordiazepoxide on punishment and conditioned suppression when these are the only conditions to which the animal is exposed; and also comparing its effects in the on- and off-the-baseline procedures for producing conditioned suppression. Experiments along these lines are currently in progress in our laboratory.

Encouraged by these results we proceeded to examine the effects of small, discrete lesions with the septal area on behavior in this same task. To understand the kind of lesion we aimed to produce, some preliminaries are necessary concerning the organization of the septohippocampal system.

Although the septum contains a number of separate nuclei, there appears to be a major functional division between the medial and the lateral septal areas. The medial septal area (i.e., the medial septal nucleus and the nucleus of the diagonal band) contains the pacemaker cells that control the hippocampal theta rhythm (Stumpf, 1965). The major input from the medial septal area to the hippocampus travels in the fimbria and fornix. The major subcortical projection from the hippocampus, in its turn, goes to the lateral septal area, traveling by way of the fimbria (Andersen, Bland & Dudar, 1973; Raisman, Cowan, & Powell, 1966). Thus the medial septal area is a major source of input to the hippocampus and the lateral septal area a major recipient of its output.

Our lesions were intended selectively to destroy either the medial septal area or the dorsal part of the lateral septal area. For a medial septal lesion, we used the smallest current, passed through an electrode lowered exactly on the midline, which in preliminary experiments had abolished the hippocampal theta rhythm almost completely; for a lateral septal lesion, we used the largest current, passed through each of two electrodes lowered 0.7 mm lateral to the midline, which produced only slight reductions in the amplitude of theta. In the experiment on punishment and conditioned suppression, the success of our intended lesions was initially verified electrophysiologically, using the aforementioned criteria of change in theta, and subsequently histologially.

Our experiment was directed toward several issues. First, we were interested in following up McGowan, Hankins and Garcia's (1972) report that lateral, but not medial, septal lesions impair conditioned suppression. Second, we wished to see whether the impairment in the discrimination between punishment and conditioned suppression, produced by chlordiazepoxide, could also be produced by septal lesions. Third, we wished to test the hypothesis (Gray, 1970) that the minor tranquilizers alter behavior through an action on the septohippocampal system, by seeing whether the effects of one of these drugs (chlordiazepoxide) would be altered in animals that had undergone major disruption of this system.

There were 3 main phases in the experiment. In the first, all animals received the same shock intensity; in the second phase shock intensity was varied for each animal to achieve similar levels of suppression in all of them; in the third phase, while maintaining the same condition as in Phase 2, each animal was tested under chlordiazepoxide (5 mg/kg) and a placebo in an ABBA design. In the analysis of our results, we have avoided the usual suppression ratios, for two reasons: They obscure the actual changes in response rate seen in the baseline and suppression components, and they are not suited for parametric statistical analysis. Instead, we analyzed response rates separately for the baseline and for the punishment and conditioned-suppression intrusion periods. Rates during the two intrusion periods were also submitted to analysis of covariance, with baseline response rate as the covariate. These covariance analyses achieve the main purpose of the conventional suppression ratio—that of allowing for differences in baseline response rate in assessing the degree of response suppression.

Several differences between the effects of medial and lateral septal lesions emerged in our results.

First, the medial lesion produced a significant increase in baseline response rates, whereas the lateral lesion was without effect on these.

Second, the lateral lesion clearly reduced suppression in both the punishment and the conditioned-suppression intrusion periods. This result emerged both in Phase 1, in which lateral animals had significantly higher response rates than controls in both intrusion periods; and in Phase 2, in which they required significantly higher levels of shock to suppress to the same extent as controls. The medial lesion, in contrast, had no effect on suppression in Phase 1: Rates in the two intrusion periods were significantly higher in medials than in controls, but this difference was eliminated in the analysis of covariance. In Phase 2, however, the medial animals did require a slightly higher level of shock to suppress to the same degree as controls. This part of our results, then, essentially replicates the finding of McGowan et al. (1972) that conditioned suppression is reduced by lateral but not (or barely) by medial septal lesions, and extends this finding to the case of punishment-produced suppression. Our results also replicate Dickinson's (1975) finding that septal lesions impair both conditioned suppression and punishment-produced suppression.

Third, the medial lesion clearly *enhanced* the rat's ability to discriminate between the punishment and conditioned-suppression intrusion periods. This effect emerged both in the learning of the discrimination, which occurred significantly sooner in the medials than in the controls, and in the asymptotic size of the discrimination (as measured in each animal by the excess of responding in S_c relative to reponding in S_p). The lateral lesions, in contrast, had no significant effect on the ability to discriminate between the two kinds of intrusion periods — although there was a trend, visible in several aspects of the data, for lateral animals to be poorer than controls.

In the final phase of the experiment, we replicated our earlier findings with chlordiazepoxide. The drug alleviated both punishment-produced and conditioned suppression, but the former more than the latter. In consequence, the discrimination between the two intrusion periods was abolished or substantially reduced in all groups. There was no interaction between the drug's effect and the animal's lesion condition.

These results pose a number of problems for the hypotheses that guided our experiments and, we suspect, for most alternative views of septal function.

First, the fact that lateral, but not medial, septal lesions severely disrupt punishment-produced response suppression is difficult to intergrate with the view that punishment and nonreward act upon a common physiological system (Gray, 1967; Konorski, 1948; Wagner, 1966), for there is evidence that response suppression caused by the withdrawal of reward is impaired by *medial* septal lesions. We deal with this point further in the next section.

Second, our results weaken the hypothesis (Gray, 1970) that the site of action of the minor tranquilizers is in the septohippocampal system. The effect of chlordiazepoxide on response suppression was unlike that of either lesion in detail. The medial septal lesion, indeed, had the opposite effect to that of the drug, in that it increased the animal's ability to discriminate response-dependent from response-independent shock. The lateral septal lesion fares better from this point of view, in that, like the drug, it alleviated response suppression and showed a (nonsignificant) tendency to weaken the discrimination between punishment and conditioned suppression. However, the drug interacted with neither lesion in the final phase of the experiments; and it would surely have been expected to do so, had we destroyed an important site for its action.

Third, our results pose a new puzzle concerning the behavioral effects of the medial septal lesion and of the loss of theta by which it is accompanied. What underlying change is likely to have produced an enhancement in the animal's ability to discriminate response-dependent from response-independent shock? We know of no existing theory that could have predicted this result.

Finally, we return to one of the main themes of this paper: the relationship between inhibition in classical conditioning and instrumental inhibition. Our results with chlordiazepoxide are encouraging for the notion that the two kinds of conditioning require separate processes of inhibition, and that only (or mainly) the instrumental kind is affected by this drug; but our results, and Dickinson's (1975) earlier ones, using septal lesions do not suggest that the septal area is differentially involved in the two kinds of inhibition.

V.   THE PHYSIOLOGY OF THE PARTIAL REINFORCEMENT EXTINCTION EFFECT

So far we have been primarily concerned with the aspect of Konorski's (1948) theory of inhibition that postulated separate processes for response suppression in the classical and instrumental cases. We turn now to consider in more detail a second feature of that theory: the functional equivalence of punishment and nonreward in engaging motor-act inhibition.

This part of Konorski's approach to the problem of inhibition dates from his earliest work, and thus precedes similar postulates made by Western workers (Gray, 1967; Miller, 1964; Wagner, 1966) by over 30 years. From the notion of a functional equivalence between punishment and nonreward, it is possible to go a step further and postulate a physiological equivalence (Gray, 1967) — namely, that a common physiological system mediates at least some of the behavioral effects of these two operations. The best candidate for this role is certainly the septohippocampal system (Gray, 1970); as shown in Table 11.1, lesions to both the septal area and the hippocampus reduce the efficiency with which an animal suppresses behavior followed by either punishment or nonreward.

However, as we have just seen, there is some reason to suppose that this physiological equivalence between punishment and nonreward breaks down when, instead of large lesions to the septal area, one makes discrete lesions of the lateral or medial septal nuclei. In our experiments on punishment and conditioned suppression, the disinhibition of suppressed responding was much greater after lateral than after medial septal lesions, confirming earlier findings (McGowan et al., 1972). In contrast, the existing evidence strongly suggests that the effective lesion for altering responses to nonreward lies in the medial, not the lateral septal area (Butters & Rosvold, 1968; Donovick, 1968; Gray, Quintao, & Araujo-Silva, 1972).

However, we have recently completed some experiments on the role of the septal area in the partial reinforcement extinction effect (PREE) in the straight alley that reveal a much more complex picture than this. Earlier work (Gray et al., 1972; Henke, 1974) had shown that septal lesions attenuate or may even completely abolish the PREE, raising resistance to extinction after continuous reinforcement (CRF) and reducing it after partial reinforcement (PRF). Gray et al. (1972) concluded that this effect was due to damage to the medial septal area, but our new results show that this conclusion was erroneous.

Our more recent experiments have used the lesions described earlier. Most of them have been run at one trial per day, a design that we have shown to maximize the attenuation of the PREE produced by both sodium amobarbital and chlordiazepoxide (Feldon, Guillamon, Gray, DeWit & McNaughton, in press). Using these parameters we have found that medial septal lesions do not reduce the PREE. On the contrary they increased resistance to extinction after both CRF and PRF, and even tended to increase the difference between the two reinforcement conditions. This facilitation of the PREE emerged more clearly when we ran an experiment using short intertrial intervals (3–5 min) and only 6 acquisition trials. These parameters failed to produce a PREE in the controls, but there was a substantial PREE in animals that had sustained a medial septal lesion (Fig. 11.2). In contrast to these results with the medial lesion, lateral lesions clearly abolished the PREE in the 1-trial-a-day paradigm, doing so by reducing resistance to extinction after PRF. When differences in acquisition asymptotic running speed were taken into account, the lateral lesion did not alter resistance to extinction after CRF in the 1-trial-a-day experiment (Fig. 11.3).

These experiments make it clear that both the medial and the lateral septal area are concerned in some way with behavioral responses to nonreward. The medial septal area appears to be concerned with the elimination of a response that is consistently nonrewarded, as during extinction, but not with the learning about inconsistent nonreward on which the PREE depends. The lateral septal area appears to have the converse properties: It is involved in the PREE but not in straightforward resistance to extinction. It is evident, therefore, that any comparison between the roles played by these two structures in responses to nonreward, on the one hand, and to punishment, on the other, must have due regard to the exact parameters of the two kinds of experiment.

It is certainly not easy to make any direct comparison between our own experiments on punishment and conditioned suppression and those on the PREE. Specially designed parallel experiments are needed, in which, as far as possible, the only difference is the use of punishment in one case and nonreward in the other. We are at present attempting to set up such experiments, both in the Skinner box and in the alley. For the moment, in our view, the possibility that the septohippocampal system mediates behavioral inhibition produced by either punishment or nonreward remains open.

VI. CONCLUSION

This chapter has been concerned mainly with two features of Konorski's theory of inhibition, both of which were more prominent in his 1948 book than in the *Integrative Activity of the Brain* (1967).

The first is the possibility that the process responsible for the inhibition of classical CRs is different from the one responsible for the inhibition of instrumental response. This possibility receives some support from the existing literature on the effects of minor tranquilizers on punishment and conditioned suppression, and from our own experiments (previously described) using chlordiazepoxide. Our experiments using septal lesions, however, like those of Dickinson (1975), do not suggest any difference between the two kinds of response—except perhaps at one point: Medial septal lesions *improved* the animal's ability to learn a discrimination between response-dependent and response-independent shock.

Our second main theme has been the possibility that a common physiological system mediates the behavioral effects of both punishment and nonreward. We have not reviewed the considerable body of evidence in favor of this proposition from experiments using minor tranquilizers (Gray,

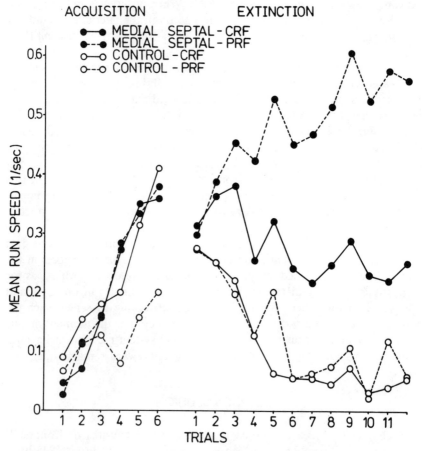

FIG. 11.2.   Course of acquisition and extinction of a running repsonse at a 4-min intertrial interval as a function of continuous (CRF) or partial (PRF) reinforcement in animals with medial septal lesions or sham operations. All acquisition trials were run on the first experimental day, all extinction trials on the second.

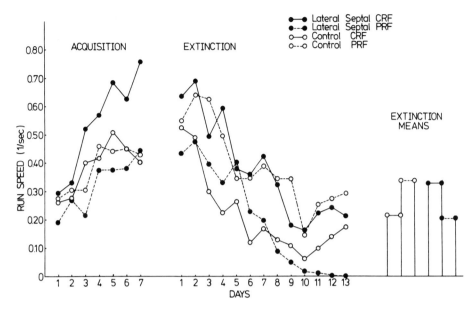

FIG. 11.3.   The course of acquisition and extinction of a 1-trial-a-day running response as a function of continuous (CRF) or partial (PRF) reinforcement in animals with lateral septal lesions or sham operations. The histogram to the right shows speeds averaged across the 13 days of extinction.

1977). Experiments using large septal lesions have also supported this view. But experiments using lesions localized in the medial or lateral septal areas — including our own, described in this chapter — offer difficulties for it. Whereas lateral septal lesions increase resistance to punishment, the medial septal lesions increase resistance to extinction. However, our experiments also showed that the effects of lateral and medial septal lesions on responses to nonreward in the alley depend on the particular schedule of reward and nonreward used. In particular, lateral septal lesions reduce or abolish the PREE, while medial septal lesions do not reduce and sometimes even increase it. Thus it would be premature to abandon the hypothesis that the septal area treats punishment and nonreward as essentially equivalent events.

   The role of the medial septal area (and perhaps of the hippocampal theta rhythm that it controls) emerges as particularly enigmatic, both in our experiments on punishment and in those on nonreward. Animals that have been deprived of this structure are improved in their ability to discriminate between response-dependent and response-independent shock, and in their sensitivity to those features of a PRF schedule that produce the PREE. These two effects may not be unconnected. Whereas the PREE is almost invariably found in instrumental tasks, the comparable effect is rarely if ever seen in classical conditioning. It is not clear whether this is a trivial or a fundamental difference between the two learning paradigms (Gray, 1975; Mackintosh, 1974). If one assumes for a moment that it is fundamental, one might deduce from our experiments that the chief role of the medial septal area is to blur the distinction between classical and instrumental conditioning: The animal can better tell which mode he is operating in when this part of the brain has been removed. There must be a better interpretation of our data, but we do not as yet see it.

ACKNOWLEDGMENTS

Our research was supported by the United Kingdom Medical Research Council. J. F. held the Kenneth Lindsay Scholarship awarded by the Anglo-Israel Association. We thank Neil McNaughton for help with the statistical analyses and for allowing us to quote from an unpublished review (Gray & NcNaughton, in preparation). We also thank the editors for their helpful comments on the first draft of the paper.

REFERENCES

Altman, J., Brunner, R. L., & Bayer, S. A. The hippocampus and behavioral maturation. *Behavioral Biology*, 1973 *8*, 557–596.
Andersen, P., Bland, B. H., & Dudar, J. D. Organization of the hippocampal output. *Experimental Brain Research*, 1973, *17*, 152–168.
Antelman, S. M., & Brown, T. S. Hippocampal lesions and shuttlebox avoidance behavior: A fear hypothesis. *Physiology and Behavior*, 1972, *9*, 15–20.
Atnip, G., & Hothersall, D. Response suppression in normal and septal rats. *Physiology and Behavior*, 1975, *15*, 417–421.
Bindra, D. *A theory of intelligent behavior.* New York: Wiley, 1976.
Black, A. H., Nadel, L. & O'Keefe, J. Hippocampal function in avoidance learning and punishment. *Psychological Bulletin*, 1977, *84* 1107–1129.
Brady, J. V., & Nauta, W. J. H. Subcortical mechanisms in emotional behavior: The duration of affective changes following septal and habenular lesions in the albino rat. *Journal of Comparative and Physiological Psychology*, 1955, *48*, 412–420.
Butters, N., & Rosvold, H. E. Effects of caudate and septal nuclei lesions on resistance to extinction and delayed alternation. *Journal of Comparative and Physiological Psychology*, 1968, *65*, 397–403.
Caplan, M. An analysis of the effects of septal lesions on negatively reinforced behavior. *Behavioral Biology*, 1973, *9*, 129–167.
Church, R. M., Wooten, C. L., & Matthews, T. J. Discriminative punishment and the conditioned emotional response. *Learning and Motivation*, 1970, *1*, 1–17.
Dickinson, A. Response suppression and facilitation by aversive stimuli following septal lesions in rats: A review and model. *Physiological Psychology*, 1974, *2*, 444–456.
Dickinson, A. Suppressive and enhancing effect of footshock on food-reinforced operant responding following septal lesions in rats. *Journal of Comparative and Physiological Psychology*, 1975, *88*, 851–861.

Donovick, P. J. Effects of localized septal lesions on hippocampal EEG activity and behavior in rats. *Journal of Comparative and Physiological Psychology,* 1968, *66,* 569–578.

Douglas, R. J. The hippocampus and behavior. *Psychological Bulletin,* 1967, *67,* 416–442.

Duncan, P. M. Effect of septal area damage and baseline activity levels on conditioned heart-rate response in rats. *Journal of Comparative and Physiological Psychology,* 1972, *81,* 131–142.

Feldon, J., Guillamon, A., Gray, J. A., Dewitt, H., & McNaughton, N. Sodium amylobarbitone and responses to nonreward. *Quarterly Journal of Experimental Psychology,* in press.

Freeman, F. G., Mikulka, P. J., & D'Auteuil, P. Conditioned suppression of a licking response in rats with hippocampal lesions. *Behavioral Biology,* 1974, *12,* 257–263.

Gray, J. A. Disappointment and drugs in the rat. *Advancement of Science,* 1967, *23,* 595–605.

Gray, J. A. Sodium amobarbital, the hippocampal theta rhythm and the partial reinforcement extinction effect. *Psychological Review,* 1970, *77,* 465–480.

Gray, J. A. Medial septal lesions, hippocampal theta rhythm and the control of vibrissal movement in the freely moving rat. *Electroencephalography and Clinical Neurophysiology,* 1971, *30,* 189–197.

Gray, J. A. *Elements of a two-process theory of learning.* London: Academic Press, 1975.

Gray, J. A. The behavioural inhibition system: A possible substrate for anxiety. In M. P. Feldman & A. Broadhurst (Eds.), *Theoretical and Experimental Bases of the Behaviour Therapies.* London: Wiley, 1976, 3–41.

Gray, J. A. Drug effects on fear and frustration: Possible limbic site of action of minor tranquilizers. In L. Iversen, S. Iversen, & S. Snyder (Eds.), *Handbook of Psychopharmacology* (Vol. 8). New York: Plenum Press, 1977, pp. 433–529.

Gray, J. A. A neuropsychological theory of anxiety. In C. E. Izard (Ed.), *Emotions and Psychopathology.* New York: Plenum Press, in press.

Gray, J. A., & McNaughton, N. A. Comparison between the behavioral effects of septal and hippocampal lesions. Unpublished Manuscript, Oxford University, 1978.

Gray, J. A., Quintao, L., & Araujo-Silva, M. T. The partial reinforcement extinction effect in rats with medial septal lesions. *Physiology and Behavior,* 1972, *8,* 491–496.

Harvey, J. A., Lints, C. E., Jacobson, L. W., & Hunt, H. F. Effects of Lesions in the septal area on conditioned fear and discriminated instrumental punishment in the albino rat. *Journal of Comparative and Physiological Psychology,* 1965, *59,* 37–48.

Haworth, J. T. Conditioned emotional response phenomena and brain stimulation. Unpublished doctoral dissertation, Manchester University, 1971.

Henke, P. G. Persistence of runway performance after septal lesions in rats. *Journal of Comparative and Physiological Psychology,* 1974, *86,* 760#767.

Konorski, J. *Conditioned reflexes and neuron organization.* Cambridge: Cambridge University Press, 1948.

Konorski, J. *Integrative activity of the brain.* Chicago: Chicago University Press, 1967.

Lubar, J. F., & Numan, R. Behavioral and physiological studies of septal functions and related medial cortical structures. *Behavioral Biology,* 1973, *8,* 1–25.

McCleary, R. A. Response-modulating functions of the limbic system: Initiation and suppression. In E. Stellar & J. M. Sprague (Eds.), *Progress in Physiological Psychology* (Vol. 1). New York: Academic Press, 1966.

McGowan, B. K., Hankins, W. G., & Garcia, J. Limbic lesions and control of the internal and external environment. *Behavioral Biology,* 1972, *7,* 841–852.

Mackintosh, N. J. *The psychology of animal learning.* London: Academic Press, 1974.

Miller, N. E. The analysis of motivational effects illustrated by experiments on amylobarbitone. In H. Steinberg (Ed.), *Animal behaviour and drug action.* London: Churchill, 1964.

Molino, A. Sparing of function after infant lesions of selected limbic structures in the rat. *Journal of Comparative and Physiological Psychology,* 1975, *89,* 868–881.

Nadel, L., O'Keefe, J. & Black, A. A critique of Altman, Brunner and Bayer's response-inhibition model of hippocampal function. *Behavioral Biology,* 1975, *14,* 151–162.

O'Keefe, J. & Nadel, L. *The hippocampus as a cognitive map.* Oxford, England: Oxford University Press, in press.

Raisman, G., Cowan, W. M., & Powell, T. P. S. An experimental analysis of the efferent projections of the hippocampus. *Brain,* 1966, *89,* 83–108.

Stumpf, Ch. Drug action on the electrical activity of the hippocampus. *International Review of Neurobiology,* 1965, *8,* 77–138.

Wagner, A. R. Frustration and punishment. In R. M Haber (Ed.), *Current research on motivation.* New York: Holt, Rinehart and Winston, 1966.

12

Cortical Mechanisms and The Inhibition of Instrumental Responses

Jadwiga Dabrowska
Nencki Institute of Experimental Biology

I.   INTRODUCTION

This chapter is concerned with a particular part of the cortex, the prefrontal area, and the manner in which lesions within this area affect performance in situations involving the conditioning of instrumental responses. Research on the role of the prefrontal cortex in conditioning has shown that it is possible to distinguish between several smaller areas within this region, because their removal impairs the performance of different tasks. It has been suggested that separate areas within this region are involved in different functions, for analysis of the tasks has shown that they involve different mechanisms. Two apparently similar tasks are affected in opposite ways by lesions in two areas in dogs: the pregenual areas, situated on the medial aspect of the prefrontal cortex in front of the anterior cingulate gyrus, and the orbital areas, situated on the lateral aspect of the prefrontal cortex in front of the presylvian fissures; these are shown in Fig. 12.1.

Two explanations for these effects are compared in this chapter. According to one of these, the disappearance of an instrumental response to a stimulus in whose presence the response is not reinforced is due to a *direct* inhibitory connection between the stimulus and the relevant drive center. This explanation is based on Konorski's original scheme (1964). According to the other explanation, the inhibitory effect is *indirect* in that the stimulus becomes connected to the related antidrive center; this follows from Konorski's subsequent scheme (1972b). Different predictions may be derived from these two hypotheses, and these were tested in an experiment described in the latter part of the chapter.

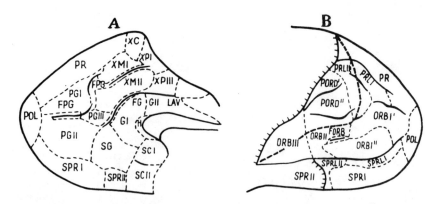

FIG.12.1.   Medial (A) and Lateral (B) aspects of the prefrontal cortical areas in dog, according to Kreiner (1966). The heavy interrupted line in B shows areas lying underneath the presylvian fissure.

The two tasks of interest here are go–no go differentiation with asymmetrical and with symmetrical reinforcement. These tasks and the impairments produced by prefrontal ablations are now described.

A. Asymmetrical differentiation

The procedure used in this task consists of reinforcing a response when it occurs in the presence of a positive stimulus and withholding reinforcement in the presence of the negative stimulus, whether or not the response occurs. Once the differentiation has been acquired, the positive stimulus can be viewed as excitatory and the negative stimulus as inhibitory for the drive center as well as for the instrumental response. Thus Konorski (1972a) called it "drive–no drive differentiation." Ablation of the lateral prefrontal cortex did not disturb this differentiation, whereas it was moderately impaired after removal of the medial prefrontal cortex. The impairment was of a disinhibitory kind: The dog performed the instrumental response to both stimuli; so only the inhibitory response to the negative stimulus was disturbed (Brutkowski & Dabrowska, 1966).

B. Symmetrical differentiation

With a "symmetrical differentiation" procedure, reward follows performance of an instrumental response in the presence of one stimulus (S1), but in the presence of a second stimulus (S2) is delivered only if the response does not occur. Thus, no reward occurs on S1 trials in which no response is made, or on S2 trials in which the response is made. Performance on this tasks was unaffected by removal of the medial prefrontal cortex, but severe

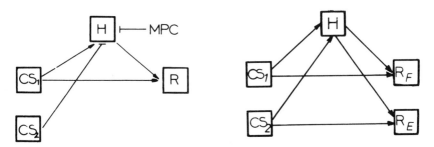

FIG.12.2. Block models of the mechanisms of go, no-go asymmetrical (left-hand side) and symmetrical (right-hand side) differentiation. CS_1 and CS_2: centers of the discriminative stimuli; R: instrumental response centers; R_F: flexor response; R_E: extensor response centers; H: hunger system; MPC: medial prefrontal cortex; arrows: excitatory connections; stopped lines: inhibitory connections.

impairment followed ablation of the lateral prefrontal cortex (Dabrowska, 1971, 1972). It is interesting that the latter impairment was not a disinhibitory kind: The dogs made errors to both stimuli, and their behavior changed during the course of retention training. Two phases, one of responses to both stimuli and another of failure to respond to either stimulus, were separated by a period of chaotic performance. The results obtained from dogs with lateral prefrontal lesions suggested that in symmetrical differentiation there is probably not a single instrumental response, that is elicited by one stimulus and inhibited by the other, but two responses: a flexor response (R_F) and an extensor response (R_E), as shown in Fig. 12.2. Both stimuli activate the hunger-drive center, and thus its level of activation cannot determine which response is to be performed. This can be determined only by direct connections between the cortical centers of the stimuli and those of the instrumental response. Konorski (1972a) called this task "motor act differentation."

In summary, it appeared that the lateral prefrontal cortex of dogs is involved in the correct operation of direct connections — that is, the defining properties of stimuli — while the medial prefrontal cortex indirectly affects the appetitive aspect of the stimuli through modulation of the level of excitation in the hunger system.

II. MECHANISM OF THE DISINHIBITION SYNDROME FOLLOWING MEDIAL PREFRONTAL LESIONS

The physiological mechanism of drive–no drive differentiation and the functional significance for this task of the medial prefrontal cortex seemed to be understood, as previously outlined. However, Konorski's later concept of antidrive center led us to reexamine this issue.

Brutkowski (1964), Brutkowski and Dabrowska (1966), and Konorski (1967, 1972a, 1972b) explained the impairment of asymmetrical differentiation by medial prefrontal lesions as a release of drive functions from cortical inhibitory control. However, each author suggested different physiological mechanisms to account for this release. Before comparing these suggestions, it is important to consider Konorski's views on the inhibition of instrumental responding.

During asymmetrical differentiation of two *similar* stimuli, an instrumental response is initially obtained to both stimuli. However, because responding to one of the stimuli is never reinforced, this gradually disappears. Disappearance of the salivary response to this stimulus occurs at the same time as that of the instrumental response. In 1964 Konorski suggested two possible mechanisms for explaining inhibition of an instrumental response, based on consideration of the centers most likely to be involved in inhibitory training. As shown in Fig. 12.3, there are two possible pathways for an inhibitory effect: first, via a direct CS–R pathway, and second, via a CS–D pathway. From the first, it follows that a stimulus should have dual properties: excitatory along CS–D and inhibitory along CS–R. This would give the stimulus a conflicting relationship to the instrumental response, for excitation of the drive center by a stimulus tends to elicit the response, whereas the same stimulus acting directly through the CS–R pathway should inhibit the response. Consequently this first possibility would not produce stable inhibition of the instrumental response. On the other hand, the second possibility, formation of an inhibitory connection between the stimulus and the drive center, would be sufficient to inhibit the response, for inactivation of the drive center blocks performance of the response.

In the following discussion of alternative accounts of the effects of medial prefrontal lesions on performance of an asymmetrical differentia-

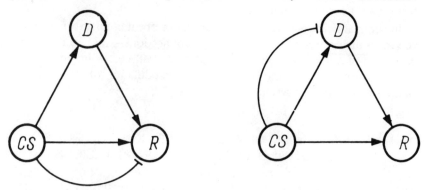

FIG.12.3. Two possible schemes illustrating inhibition of an instrumental response, according to Konorski's (1964) original hypothesis. D: drive center. Other symbols are as for Fig. 12.2.

tion, the term "excitatory CS" is used for a stimulus that has excitatory connections with both the drive center and the response, while "inhibitory CS" is used for a stimulus that inhibits activity in the drive center.

A. Direct drive inhibition hypothesis

The disinhibition syndrome refers to performance of the instrumental response to both inhibitory and excitatory CSs. Brutkowski (1964) argued that the medial prefrontal cortex MPC (pregenual areas) in dogs sends an inhibitory connection to the primary alimentary areas (hypothalamus) and that damage to this region releases the drive functions from cortical inhibitory control. However, the influence of an inhibitory CS may also be addressed directly to the alimentary hypothalamic center. If this is so, the hypothalamic center is affected by two inhibitory influences: the unconditioned one, mediated by the medial prefrontal cortex, and the conditioned one, mediated by the inhibitory CS. When the first influence is removed by ablation of the medial prefrontal cortex, the inhibitory connection between the inhibitory CS center (indicated in the fig. as CS_2) and the drive center (CS_2-H) is too weak to appreciably inhibit the excited hunger center. As a result, the instrumental response will occur until the inhibitory connections between the CS_2 and H become stronger (Dabrowska, 1972); see Fig. 12.4.

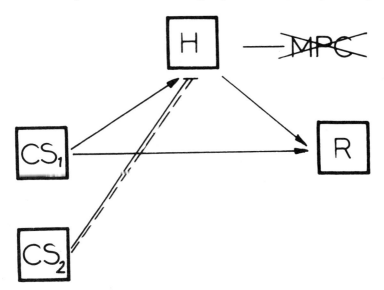

FIG.12.4. Block model of the mechanism underlying compensation for the postoperative disinhibition syndrome, according to the "direct drive inhibition" hypothesis. The stopped line indicates additional inhibitory connections established during postoperative retention training.

Thus, with further training, this impairment is compensated for by additional inhibitory connections between the same structures that had been engaged in establishing the conditioned reflex before the operation.

B. Indirect drive inhibition hypothesis

The second hypothesis followed from Konorski's (1967) conclusion that the "affective aspect of animal's activity involves two diametrically opposite states: drives and anti-drives (p. 49)." He suggested that there are three hunger states based on different neural mechanisms. The first, satiation, is caused by excitation of the hypothalamic center of satiation by humoral factors. The second, the antihunger state, arises by tasting food in the mouth and is elicited by excitation of antihunger units scattered among hunger units located in the lateral hypothalamus. These two states are unconditioned. The third, the antidrive state, arises when a stimulus signals that food is never delivered during its presentation. This state is conditioned, and Konorski suggested that it was more similar to satiation than to the state elicited by food in the mouth. The center representing this antihunger state is the highest one and is located (1972b)" in the prefrontal extension of the limbic system, i.e. in the medial prefrontal cortex in dogs (p. 331)." This area becomes connected with centers of all external stimuli that signal the unavailability of food, and its activation exerts an inhibitory influence on humoral centers (Konorski, 1972b, p. 331).

Figure 12.5 shows a block model of possible connections formed between brain centers during asymmetrical differentiation training. An excitatory stimulus, CS_1, has a direct connection to H, and excitation in H elicits an instrumental response, R, because at the beginning of training this response

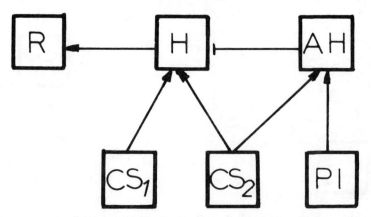

FIG.12.5.   Block model of the mechanism for asymmetrical differentiation, according to the "indirect drive inhibition" hypothesis. PI: primary inhibitor; AH: antihunger center, situated in the medial prefrontal cortex.

to the CS_1 was rewarded by food. There is no direct connection between CS_1 and R, because during go–no go training in this situation enhancement of the excitation level of the hunger center is sufficient to evoke the instrumental response. An inhibitory stumulus, CS_2, similar to CS_1, has double connections to the hunger center because of generalization from CS_1, and to the antihunger center, because CS_2 was never rewarded. A dissimilar stimulus, for which there is no generalization from CS_1 and that was never rewarded by food, has a connection only to the antihunger center. Such a stimulus is called a "primary inhibitor" (PI). The difference between Konorski's two concepts is as follows. According to the first, brain centers representing conditioned stimuli may have excitatory or inhibitory connections directly addressed to the hunger system. Both these connections are formed during conditioning. According to the second, brain centers representing conditioned stimuli have only excitatory connections, but they may be addressed to the two different hunger subsystems: excitatory CSs to the primary hunger system and inhibitory CSs to the antihunger center, located in the medial prefrontal areas. These excitatory connections are conditioned. Thus, inhibition of the instrumental response is evoked indirectly, through the unconditioned connection existing between antihunger and hunger centers.

Turning to the effects of prefrontal lesions on inhibitory conditioned reflexes, Konorski (1972b) suggested that hunger-drive functions are released from cortical inhibitory control after ablation of the medial prefrontal cortex, but he assumed that the physiological mechanism of this release

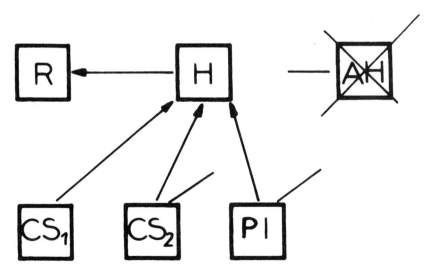

FIG.12.6.   Block model of the "drive disinhibition" syndrome, according to the "indirect drive inhibition" hypothesis.

is different from that postulated by Brutkowski and Dabrowska (1966). According to Konorski (1972a, 1972b) the disinhibitory syndrome occurs after removal of the medial prefrontal region only in response to stimuli that had double connections established to both hunger and antihunger centers before the operation; see Fig. 12.6. When the antihunger center is removed, existing connections between the CS_2 center and the hunger center are revealed, and the instrumental response to the CS_2 may be observed. On the other hand, if a primary inhibitor is applied, the disinhibitory syndrome is not observed because such a stimulus has never had any connection to the hunger center.

III.   EMPIRICAL IMPLICATIONS OF THE TWO HYPOTHESES

After discussing this problem with Konorski at the end of 1972, we decided to perform the following experiment. Four groups of dogs were trained on asymmetrical differentiation. Similar stimuli were used for Groups I and II, while Groups III and IV were given dissimilar stimuli (stimuli that evoked weak or no generalization). When they reached criterion, the dogs in Groups I and III were operated on, while the dogs in Groups II and IV served as unoperated controls. In all operated animals, the medial prefrontal cortex was ablated bilaterally. After the operation the dogs were tested for task retention. At the end of retention training, reversal learning involving both stimuli was introduced in all groups.

Because such a procedure of conditioned-reflex transformation had not been investigated before in dogs, predictions were based on the results of Konorski and Szwejkowska (1950, 1952) and Szwejkowska (1959), who used a different method of transformation. These studies did not include a two-way transformation of two stimuli—inhibitory into excitatory CS and vice-versa—but only a one-way transformation—excitatory into inhibitory CS, or inhibitory CS into excitatory CS—in any given experiment. In the first example of a one-way transformation, a conditioned response was formed to several stimuli, and subsequently only one of them was transformed into an inhibitory CS. This transformation of an excitatory CS into an inhibitory CS took a long time, whereas resistance to restoration of the original function of the stimulus was minimal. In the second example, when the conditioned response to an excitatory CS was established, two new stimuli were introduced that were never reinforced. One of them was more similar than the other to the reinforced excitatory CS. The resistance to extinction of the more similar CS was greater than that of the less similar CS. On the other hand, when both these stimuli were reinforced after completion of the extinction procedure, resistance to conditioning of the more similar stimulus was weaker than the resistance to conditioning of the less similar stimulus. When a completely dissimilar stimulus was employed, which did not evoke generalization from the excitatory CS and which was never reinforced by food, this CS was extremely resistant to transformation into an excitatory stimulus. The animal even refused to take food in the presence of such a stimulus. This last type of CS was called a "primary inhibitor."

Taking into account the foregoing findings on single transformations, we made the following predictions about two-way reversal learning in normal animals.

1. Reflexes based on similar stimuli evoking strong generalization may be regarded as primary excitatory reflexes. During differentiation training, one of these stimuli remains excitatory, while the second stimulus is transformed into an inhibitory CS. Such a transformation should be difficult and should take a long time. During reversal learning the primary excitatory CS should be transformed slowly into an inhibitory CS, while the inhibitory CS should very easily be transformed into an excitatory CS. The course of such reversal learning should be similar to the course of differentiation learning.

2. Differentiation learning involving dissimilar stimuli should be rapid, because of weak generalization between the stimuli. No transformation of an excitatory into an inhibitory CS is involved at this stage. But during reversal learning, transformation of the excitatory CS into an inhibitory CS, as well as the reverse process, should take a long time. The transformed reflexes to such stimuli should be weak and reversal more difficult, taking longer to develop than reversal learning of similar stimuli.

Let us consider predictions concerning reversal learning in operated dogs that arise from the different accounts of inhibitory conditioning described in the previous section.

A. Direct drive inhibition hypothesis

1. Postoperative retention

If the cortical inhibitory area, MPC, is removed, the excitation level of the drive center is higher than before the operation, because the hunger functions have been released from cortical inhibitory control. Inhibitory connections between CS_2 and H established before the operation are too weak to inhibit this enhanced excitation of the drive center. Hence the disinhibitory syndrome should appear. Disinhibition should be observed with both similar and dissimilar pairs of stimuli. Compensation of the impairment should take place during postoperative training, because of the development of additional inhibitory connections between the CS_2 center and H (see Fig. 12.4). When this has occurred, the animal should stop performing the instrumental response to the nonreinforced stimulus.

2. Postoperative reversal learning of similar stimuli

When the medial prefrontal cortex is removed, transformation of an inhibitory stimulus into an excitatory one should be immediate, whereas extinction of an excitatory stimulus should be difficult and laborious. Although in general reversal learning in normal and medial dogs should be similar, the duration of reversal learning should be slightly longer in medial animals than that of initial training and reversal in normal animals, because medial animals lack the cortical inhibitory control that modulates the primary alimentary center, as illustrated in Fig. 12.7.

3. Postoperative reversal learning of dissimilar stimuli

The direct drive inhibition hypothesis predicts symmetry in reversal learning after ablation of the medial prefrontal cortex when dissimilar stimuli are employed. This follows because the process of transforming a primary excitatory CS into an inhibitory CS takes a similar amount of time to that taken by a transformation in the opposite direction.

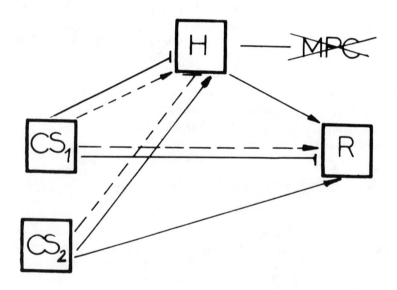

FIG.12.7. Block model of the postoperative reversal learning, according to the "direct drive inhibition" hypothesis. Solid lines: actual strong connections; interrupted lines: previous connections established during differentiation training.

B. Indirect drive inhibition hypothesis

1. Postoperative retention

The extent of disinhibition should be directly related to the degree of stimulus generalization occurring during the acquisition stage of differentiation training, because the greater the generalization between CS_1 and CS_2, the stronger the connection between CS_2 and H. If the antihunger center is removed, the disinhibitory syndrome should be observed only when similar stimuli are used, because of double connections existing between the inhibitory stimulus center and the hunger and antihunger centers. After ablation of the latter center, only connections to the hunger center remain to evoke an instrumental response. However, compensation of this impairment is very difficult to explain on the basis of the scheme presented in Fig. 12.6, because of the absence of an antidrive region through which the inhibitory stimulus, CS_2, could affect the hunger center. Konorski suggested that certain other structures could compensate for this defect, although such compensation should take longer than the compensation suggested by the direct drive inhibition hypothesis.

If dissimilar stimuli are used, the inhibitory stimulus center (primary inhibitor) possesses only excitatory connections to the antihunger center, which is unable to elicit either motor excitement in general or the particular instrumental response, whether or not the level of hunger drive excitation is high. According to this idea, removal of the antihunger center should not affect postoperative retention of an asymmetrical differentiation when dissimilar stimuli are used. The effect of removing this brain area, with which a primary inhibitor is connected, should be to abolish the inhibitory properties of such a stimulus and make it equivalent to an indifferent stimulus to which the animal had been habituated before the operation.

2. Postoperative reversal learning of similar stimuli

When similar stimuli are used, transformation of the inhibitory stimulus into an excitatory CS should be immediate because of double connections existing between the inhibitory stimulus center and the hunger and antihunger centers. When the antihunger center is removed, primary excitatory connections between the inhibitory CS and the hunger center are abolished, and instrumental responding to this stimulus should be observed. Transformation of the excitatory stimulus into an inhibitory CS should not be possible, because of the absence of the antihunger center. This is illustrated by Fig. 12.8.

3. Postoperative reversal learning of dissimilar stimuli

After ablation of the antihunger center, reversal learning of two dissimilar stimuli should be extremely difficult for two reasons. First, a

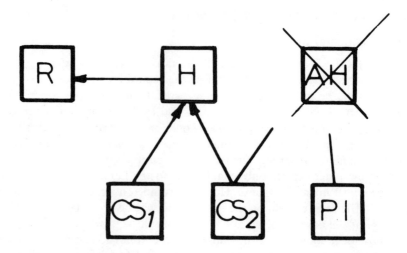

FIG.12.8.   Block model of postoperative reversal learning, according to the "indirect drive inhibition" hypothesis.

primary inhibitor can be transformed into an excitatory stimulus only very slowly. Second, a primary excitatory stimulus cannot be transformed into an inhibitory CS, for the antihunger center to which this CS should be addressed has been removed.

IV.   DETAILS OF THE EXPERIMENT

A. Procedure

We trained 31 mongrel dogs in an asymmetrical differentiation task employing auditory stimuli. For 14 of the dogs, similar stimuli (1000-Hz and 700-Hz tones) and for the remaining dogs dissimilar stimuli (1000-Hz tones and a clicker) were used. Each session contained 20 stimulus presentations, 10 of the reinforced stimulus (S+) and 10 of the nonreinforced stimulus (S−), in balanced order. When the instrumental response—placing the right forepaw on the foodtray — was made to S+ within 5 sec, food was presented and the stimulus terminated. During the 5-sec presentation of S−, neither performance of this response nor withholding it were reinforced.

Animals were trained to a criterion of 95 correct responses to both stimuli in 100 consecutive trials. When all animals had reached this criterion, the medial prefrontal cortex was ablated in 18 dogs, while the remainder served as unoperated controls. Group I (N = 8), consisting of operated dogs, and Group II (N = 6), consisting of control dogs, received the similar stimuli; Group III (N = 10) consisting of operated dogs, and Group IV (N = 7), consisting of control dogs, received the dissimilar stimuli. A recovery period of 8 days was allowed for the operated dogs. An identical rest period was given to the control dogs.

All animals were trained to criterion once more (retention test), and then the values of the stimuli were reversed. During this reversal training, reward was now dependent on the occurrence of the response to the previously negative stimulus and was not delivered in the presence of the previously positive stimulus. Reversal training lasted until the animals reached the same criterion, 95 out of 100.

During the course of reversal training, some animals ceased to respond, and if an animal had made no response in 2 consecutive sessions, a special procedure designed to restore responding was introduced. During these sessions only the rewarded stimulus occurred, and its presentation was prolonged beyond the usual 5 sec until a response was made, whereupon reward was given. If no response was made within 20 sec, a free reward was given. This procedure was continued until an animal had responded with a latency of less than 5 sec to 20 successive presentations of the stimulus. Regular reversal training was then resumed on the next session.

On completion of reversal training, histological verification of the lesions was performed on the experimental dogs.

B. Results

Figure 12.9 shows examples of the extent of the smallest and largest lesions. The smallest lesions (dog 115) included total bilateral ablation of pregenual areas I and III. Pregenual gyrus II was removed partially. Areas XM I and II and the medial part of the proreal gyrus were also only partially bilaterally removed. The largest lesion (dog 56) included total bilateral ablation of the medial prefrontal cortex. In addition to this, the cortex of the subgenual area was totally removed, and very small but bilateral damage was also made to the most anterior part of the cingulate gyri.

Although the extent of the cortical damage varied among subjects, the extent of lesions invading the other prefrontal regions, such as the proreal, subproreal, and XM areas, did not affect the magnitude of impairment. These data support previous observations (e.g., Brutkowski & Dabrowska, 1966) that the crucial areas for inducing the disinhibitory syndrome are the pregenual gyri, which were almost completely removed in the present experiment.

Figure 12.10 presents the number of trials, and Fig. 12.11 the number of errors made by individual subjects in each of the groups during differentiation, retention, and reversal learning. In general, during acquisition of the initial differentiation task, animals given the condition using similar stimuli (Groups I and II) produced more errors and took more trials to criterion than those in the condition involving dissimilar stimuli (Groups III and IV).

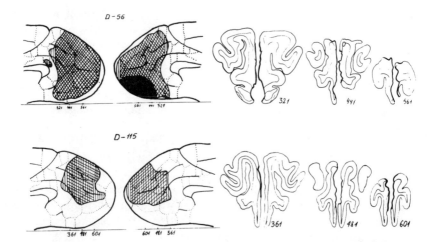

FIG. 12.9.   Extent of the smallest and largest lesions in the experimental dogs.

However, large individual differences were observed among subjects, especially during the differentiation of dissimilar stimuli. This suggests that the inhibitory stimulus was really a primary inhibitor for only a few of the animals in Groups III and IV, whereas in the other dogs in these groups the stimuli were more or less generalized. In this case the degree of stimulus generalization during initial differentiation training (lower white columns) should be related to the degree of postoperative impairment (hatched columns) or to the duration of reversal learning (black columns).

The experimental results may be summarized as follows:

1. There is slight or moderate impairment of postoperative retention of similar, as well as dissimilar, stimulus differentiation.

2. Degree of impairment is not related to the quality of stimuli, to the difficulty of preoperative training, or to the degree of stimulus generalization. This result of in agreement with the "direct drive inhibition" hypothesis, while providing difficulties for the "indirect drive inhibition" hypothesis.

3. No differences were observed in reversal learning (black columns) between similar and dissimilar stimuli in control dogs in terms of number of trials and errors (Group II vs. Group IV). Individual differences in control animals were not related to the duration of preoperative conditioning nor to the degree of stimulus generalization.

This result differs from that obtained by Konorski and Szwejkowska (1952) and Szwejkowska (1959). The difference is probably caused by differences in the procedures and stimuli employed, which were discussed in Section III.

4. Reversal learning of dissimilar stimuli in operated dogs did not differ from reversal learning in control dogs (Groups III vs. IV). The individual differences were not related to the difficulty of preoperative acquisition nor to the magnitude of postoperative impairment in retention. This result confirms the "direct drive inhibition" hypothesis.

5. Reversal learning of similar stimuli was somewhat more difficult in operated dogs than in the other groups. However, this difficulty was not related to the length of preoperative acquisition nor to the magnitude of postoperative impairment of retention. This result is also in agreement with the "direct drive inhibition" hypothesis.

These results clearly support the "direct drive inhibition" hypothesis to the extent that every prediction is experimentally confirmed. However, a very interesting phenomenon was observed during reversal learning that was not predicted by either of the 2 hypotheses. It appeared that, regardless of the quality of stimuli employed, 2 different patterns of responding were observed during reversal training. The first pattern, which will be termed "response strategy," consisted of responding to both stimuli. The second

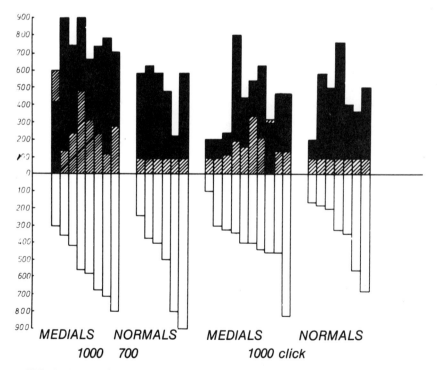

FIG.12.10.   The number of trials to criterion for operated (medial) and control dogs during differentiation (white columns below the abscissa), retention (hatched columns above abscissa), and reversal training (black columns).

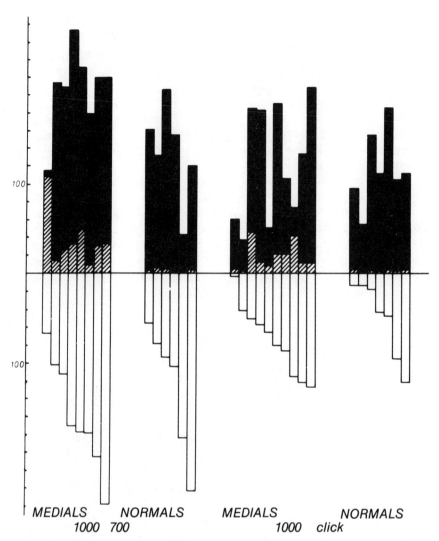

FIG.12.11.   The number of errors made by operated and normal dogs during each phase of the experiment. The symbols are as in Fig. 12.10.

pattern, which will be termed "no-response' strategy," consisted of an initial disappearance of responding to both stimuli during reversal. These patterns occurred with approximately equal frequency in all of the groups. Thus, 4 of the 8 dogs in Group I, 4 out of 6 in Group II, 6 out of 10 in Group III, and 2 out of 7 in Group IV produced the "response-strategy" pattern during reversal learning.

The upper part of Fig. 12.12 compares reversal learning in those operated and normal dogs that showed the "response strategy." These functions were obtained by the Vincent method by dividing the number of trials to criterion for each dog into five equal blocks. Each point represents the percentage of trials within each block on which a response was made. It can be seen that there is no difference between the course of reversal learning in normal and in operated dogs. Removal of the medial prefrontal cortex does not appear to affect the course of reversal learning, regardless of whether similar (upper left) or dissimilar stimuli (upper right) are used. However, the development of differential control by the stimuli is faster when dissimilar rather than similar stimuli are reversed. This applies to both operated and normal animals.

The graphs in the lower part of Fig. 12.12 represent the course of reversal learning for dogs displaying the no-response strategy. These were subjects that failed to respond for two consecutive sessions, and hence their normal training was interrupted by introduction of the aforementioned procedure designed to restore responding. The first block of trials in these lower graphs represents performance before this interruption. Once regular reversal training was reinstated, the remaining trials for each subject were divided into four blocks. Performance during the restoration procedure, between

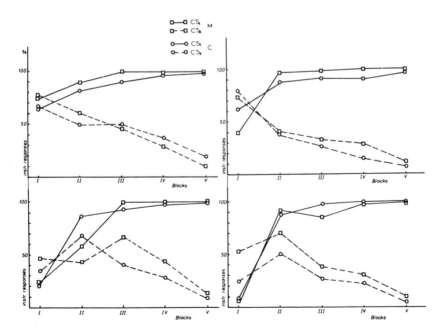

FIG.12.12. Percentage of instrumental responses made by control (C) and operated (M) dogs to similar stimuli (left side) and to dissimilar stimuli (right side). Performance of dogs using the "response strategy" is shown in the upper graphs, and those using the "no-response strategy" in the lower graphs. These functions were derived by use of the Vincent method, as described in the text.

Block I and Block II, is not shown in Fig. 12.12. As can be seen from the lower right-hand panel, with dissimilar stimuli the course of transformation of the inhibitory reflex into an excitatory one is similar in normal and operated animals, whereas the course of inhibition of the response to the previously excitatory stimulus is somewhat retarded in operated dogs, when compared with the controls. With similar stimuli (lower left), transformation of the excitatory reflex into an inhibitory one, and vice-versa, is slightly retarded in operated dogs.

V.   CONCLUSION

Neither of the two hypotheses under consideration can fully explain these data. First, slight or moderate impairment of postoperative retention was not observed in all dogs, and the degree of impairment was not related to the extent of the cortical lesion: Retention in 3 out of 10 operated dogs trained with dissimilar stimuli and retention in 1 out of 8 dogs trained with similar stimuli was not impaired at all. Second, impairment of postoperative retention was not related to the extent of stimulus generalization in preoperative training. Third, medial prefrontal lesions did not affect postoperative reversal learning, regardless of whether similar or dissimilar stimuli were used. Fourth, during reversal learning of the asymmetrical go–no go differentiation, both operated and normal dogs used either the response strategy or the no-response strategy.

What kind of functional process is impaired after ablation of the medial prefrontal cortex? Brennan, Kowalska, and Zielinski (1976) found significant enhancement of stimulus generalization when the medial prefrontal cortex was bilaterally removed in dogs. This effect could easily be explained as being secondary to the primary excitatory effect of drive functions. However, the increase of stimulus generalization was a long-lasting effect that could be observed for several months, whereas disinhibition of the instrumental response in the present experiment did not last longer than several weeks. Moreover, these authors showed that postoperative go–no go differentiation with asymmetrical reinforcement is not impaired and may even be solved better than in normal dogs, if various generalization tests with partial reinforcement preceded acquisition of differentiation. They also showed that compensation of the increased generalization effect during postoperative differentiation took place by inhibition of the response to the onset of stimuli and resulted in prolonging the response latencies to the stimuli. Because this process was brief, it might be supposed that some function other than an inhibitory one was impaired in dogs after ablation of the medial prefrontal cortex.

The medial prefrontal cortex in dogs thus cannot be regarded as the region in which the antihunger center is located, as was proposed by Konorski (1972b). Moreover, the hypotheses of both Konorski (1967, 1972b) and Brutkowski (1964) concerning the inhibitory influence of the medial prefrontal cortex on primary alimentary areas (hypothalamus) are too restricted to explain use of the no-response strategy during reversal learning by both normal and operated dogs.

The functional significance of the medial prefrontal cortex in dogs is thus still not clear and requires further research.

REFERENCES

Brennan, J., Kowalska, D., & Zielinski, K. Auditory frequency generalization with differing extinction influences in normal and prefrontal dogs trained in instrumental alimentary reflexes. *Acta Neurobiologiae Experimentalis*, 1976, *36*, 475-516.

Brutkowski, S. Prefrontal cortex and drive inhibition. In W. J. M. Warren & K. Akert (Eds.), *The frontal granular cortex and behavior*. New York: Mc Graw-Hill, 1964.

Brutkowski, S., & Dabrowska, J. Prefrontal cortex control fo differentiation behavior in dogs. *Acta Biologiae Experimentalis*, 1966, *26*, 425-439.

Dabrowska, J. Dissociation of impairment after lateral and medial prefrontal lesions in dogs. *Science*, 1971, *171*, 1037-1038.

Dabrowska, J. On the mechanism of go-no go symmetrically reinforced task in dogs. *Acta Neurobiologiae Experimentalis*, 1972, *32*, 345-359.

Konorksi, J. Some problems concerning the mechanism of instrumental conditioning. *Acta Biologiae Experimentalis*, 1964, *24*, 59-72.

Konorski, J. *Integrative activity of the brain*. Chicago: University of Chicago Press, 1967.

Konorski, J. Some hypotheses concerning the functional organization of prefrontal cortex. *Acta Neurobiologiae Experimentalis*, 1972, *32*, 595-613.(a)

Konorski, J. Some ideas concerning physiological mechanisms of so-called internal inhibition. In R. A. Boakes & M. S. Halliday (Eds.), *Inhibition and learning*. New York: Academic Press, 1972.(b)

Konorski, J., & Szwejkowska, G. Chronic extinction and restoration of conditioned reflexes. I.Extinction against the excitatory background. *Acta Biologiae Experimentalis*, 1950, *15*, 155-170.

Konorski, J., & Szwejkowska, G. Chronic extinction and restoration of conditioned reflexes. IV. The dependence of the course of extinction and restoration of conditioned reflexes on the "history" of the conditioned stimulus. (The principle of the primacy of the first training.) *Acta Biologiae Experimentalis*, 1952, *16*, 95-113.

Kreiner, J. Reconstructions of neocortical lesions within the dog's brain: Instructions. *Acta Biologiae Experimentalis*, 1966, *26*, 221-243.

Szwejkowska, G. The tranformation of differential inhibitory stimuli into positive conditioned stimuli. *Acta Biologiae Experimentalis*, 1959, *19*, 151-159.

13 The Study of Association: Methodology and Basic Phenomena

R. G. Weisman
Peter W. D. Dodd
Queens University, Kingston, Ontario

Association was the central topic in Jerzy Konorski's great monograph, *Integrative Activity of the Brain* (1967). Konorski's thoughts on method, which form the germinal core of the present chapter, have provided an important record of his insights into the nature of association itself. Thus, in this chapter we have occasion to discuss theoretical as well as methodological questions in the study of association. For the most part, the theoretical questions fit neatly into their methodological context, leaving no reason to introduce them here at the outset. One theoretical question, however, is so central to our discussion that its consideration must preface the remainder of the chapter. We ask, what is the structure of associations? Konorski considered two possible answers. The connectionist answer is that association is a matter of one perceptual unit activating another through the establishment of an actual neural connection between the units. The cognitive, or what Konorski termed the convergence, answer is that association is a matter of one perceptual unit activating another through the establishment of "higher-order" representations thought to include something of the relationship between the two percepts. According to the former view the structure of an association is simply a connection between perceptual units, while according to the latter view it includes a representation of relationship between percepts. Konorski (1967, pp. 167–169) considered the cognitive theory tempting, but proposed to follow the connectionist path in his neurological theorizing. Nevertheless, it seems to us that he expressed a more cognitive view in his discussion of the psychological structure of associations.

We, too, find the cognitive approach to the structure of associations tempting, and we will not seek to avoid it as Konorski did, but rather adopt it and explore its implications, which we hope will prove testable. The implications of a cognitive approach are that animals use relational representations in the course of associative learning to successfully discriminate between compounds of events, and to successfully anticipate elements in a compound of events in time and space. Discrimination refers to the use of a relational representation to distinguish between event sequences already presented. Anticipation refers to expectations concerning events to come, based on a relational representation. For example, the representation "A never follows B" allows an animal to discriminate instances of that sequence from other sequences and to anticipate the absence of A given B. The cognitive view of association asserts that animals may represent a broad range of relationships between events; just how broad is not a matter of debate but of experimentation. In reviewing research for our survey of methods, we discuss the degree to which various kinds of research yield evidence consistent with the connectionist and the cognitive position on the structure of associations.

The research methods adopted by a science define the range of concepts the science may produce. In *Integrative Activity of the Brain,* Konorski, by extending and interpreting the application of scientific method to the study of association, broadened our conceptual horizons. In this chapter we continue his consideration of the evidence from introspection, from the direct application of conditioning procedures, and from multistage studies that evaluate associations learned during an early stage of the experiment by the application of conditioning procedures during the later stages of the experiment. Also, we discuss converging operations, particularly some proposed by Konorski, as definitional tools in the study of association.

I. INTROSPECTION

Introspection is an everyday sort of observational process in which we look within our own minds at ideas and the associations between ideas. The results of such introspection are rarely very startling to ourselves or of much interest to others. Nonetheless the introspective method has a long and honorable history in the study of association. Philosophical associationists from Aristotle to James Mill have trusted their case to the evidence from introspection. Watson (1924, p. 39) rejected introspection as a genuine psychological method because he thought it only another name for an obscure bodily function. Strangely enough, Konorski (1967, p. 187) found the method useful for much the same reason. As introspective evidence in favor of the S-S association in classical conditioning, for example, he noted (1967, p. 267) that in humans a CS reinforced by acid produces an image or even a sour taste in the mouth.

The charm of the method is that one can replicate another researcher's finding almost while reading about it. For example, we can confirm the results of Konorski's human-conditioning experiment simply by observing that thoughts about sour tastes and sometimes even the sensations themselves follow thoughts about biting into a lemon. It would appear that introspection is both a satisfying and a valid method of verifying some specific facts of association. As a general method, however, introspection has been found wanting, for some associations do not find ready verbal representations, and introspection sometimes fails to detect associations that clearly have been made. Introspection is less reliable in the study of association than in perceptual research, mainly because recent common experience is much more important in the former than in the latter research. Furthermore, the difficulty in measuring awareness of associations and reliability of verbal reports make introspection a sometimes cumbersome method. Finally, of course, introspection fails utterly to allow measurement of associations in nonverbal organisms. Other methods of study are proposed by Konorski and in the present chapter to assist us in providing more complete information about all of the associations that are formed in us or in others.

II. ASSOCIATION BETWEEN THE CS AND THE US

A. The Direct Evidence from Conditioning

Psychologists, over the past 50 years or so, have placed considerable confidence in classical conditioning and instrumental learning as methods of studying associative learning, particularly in infrahuman animals. Pavlov, and certainly Watson, hoped that the conditioned reflex would prove to be the basis of all association. This view led to the use of the observable conditioned response, elicited in relatively simple experiments, as a direct measure of association. If the results of associative processes were always clearly revealed in observable behavior, then the direct equation of learned associations with conditioned responses would seem to cause little harm. Indeed, the direct behavioral approach has brought into psychology much important evidence concerning both association and performance in animals. Moreover, it has spawned the development of many useful laboratory procedures.

In the learning laboratory, at least, positive conditioning — that is, a simple forward relationship between the CS and the US — results in measurable conditioned responses and thus direct evidence concerning the nature of the association between the CS and the US. One recent successful application of the direct approach concerned the insufficiency of temporal contiguity as a condition of association. In that work, now well known as blocking, Kamin (1968, 1969) used the conditioned emotional response

(CER) procedure. In CER procedures the CS is positively correlated with shock; suppression of ongoing instrumental behavior in the presence of the CS usually results. Kamin found that training with one stimulus, CS_1, before the introduction of a compound, $CS_1 + CS_2$, may result in conditioned responses to CS_1, but not to CS_2. Thus direct evidence from blocking experiments suggests that animals may associate one stimulus rather than other with a US, even though for several trials both stimuli were contiguous with US onset. (See also Dickinson & Dearing, Chapter 8.)

A variant of the positive conditioning procedure known as autocontingency learning has also yielded some interesting direct evidence concerning the nature of association. In that work Davis, Memmott, and Hurwitz (1975) presented a shock US without a predictive CS. However, for some rats information about the distribution and frequency of the US was available in the temporal schedule of US presentation. In this group there was a minimum fixed time of 3 min between shocks, and the number of shocks per session fixed at 3. Thus, shock was least probable just after a shock and after the third shock of the session. Evidence that rats are somehow able to represent these temporal and numerical relationships is directly available from the pattern of responding engendered by the shock schedule. The terminal performance of 1 rat in this group, shown in Fig. 13.1, makes the point: Responding increased dramatically after each shock, gradually declined with time to the next shock, and remained steady after the final shock of the session. The inference that rats are able to develop representations of the subtle relationships between the US and itself receives compelling support from this direct behavioral evidence.

B. Difficulties with the Direct-Evidence Approach: Silent Associations

The disadvantage of the direct-evidence approach is that one may conclude that an association is absent when it is only behaviorally silent. We include as silent, first, associations that do not elicit the particular CRs an experimenter has observed and, second, associations that often cannot be detected in a single-stage, direct-measurement procedure, but require a multistage procedure.

In the first category of silent associations is the instance in which investigators rest their claims concerning the absence or slow acquisition of an association between a given CS and US on changes in a phlegmatic response system. The nictitating-membrane response system is considered impressively stable, while the heart-rate response system is not. Nonetheless, Van Dercar and Schneiderman (1967), measuring both responses in the same rabbits, found reliable heart-rate CRs, but virtually no nictitating-membrane CRs at a CS–US interval of 6.75 sec, and detectable heart-rate CRs at shorter CS–US intervals many trials before nictitating-membrane CRs. Ex-

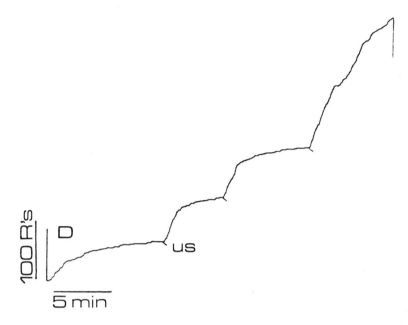

FIG.13.1. Cumulative response records obtained after 170 sessions under conditions of 3 shock deliveries per sessions distributed with a minimum of 3-min intershock interval. From Davis et al., 1975.

perimenters with access only to nictitating-membrane data might conclude, quite erroneously, that rabbits fail to learn a simple forward association when CS and US onset are separated by as little as 6.75 sec. These hypothetical misguided experimenters might also conclude that association, at least in the rabbit, occurs only after many trials of conditioning.

The second and more important category of associations is silent to direct observation. The best-known example of this category is inhibitory—or, as we prefer to call it, negative — conditioning. Often presentation of a CS negatively correlated with the US has no directly measurable effect on behavior. Our hypothetical misguided experimenters might conclude that no negative association has been learned because the negative CS presented in isolation failed to elicit any observable CR. However, Konorski (1967, p. 323) and, in more detail, Rescorla (Chapter 4) have pointed out that, although often silent in direct measurement, negative associations are measured quite easily by indirect techniques. Two techniques now routinely used to measure negative conditioning are inhibition of a CR to a positive CS by a simultaneously presented negative CS (the summation method) and slowed acquisition of a CR to a currently positive but previously negative CS (the retardation method). Both methods are indirect in that they require two stages. Negative conditioning is established in one stage; then its influence on positive conditioning is used to infer negative association. Konorski has provided an excellent example. If in an alimentary situation— one in which the animal receives food—a stimulus never accompanied by food is repeatedly presented, a special sort of inhibitory training occurs in which this stimulus becomes a signal for no food. The result of this training is recognized only indirectly in two ways. First, if later reinforced by food, the stimulus exhibits a great resistance to being conditioned; second, if it is combined with a positive alimentary CS, the effect of the latter is diminished (1967, p. 323).

Multistage research techniques have become the mainstay of experimenters wishing to study the indirect and subtle effects of association on behavior. Many associations judged absent just a few years ago are now seen as only silent to direct measurement. Konorski reminded us that associative learning takes place, after all, in animals' brains, not in their salivary glands or nictitating membranes. Our task, he further reminded us, is to tease from our behavioral experiments the answer to a great and ancient mystery: the nature of association. This task means that we must change our past emphasis on performance and return to the study of learning, and in particular of the representational products of learning.

C. The Indirect Evidence from Conditioning

1. The Logic of the Indirect Approach to Silent Associations

We have been writing as if, of the approaches to the study of association, some methods were direct and some were indirect. In another, more careful, use of words, association is never observed directly; instead, like perception and memory, association is an inference from converging operations and behavioral effects (Garner, Hake, & Eriksen, 1956). Our point has been simply that some associative learning, particularly about positive relationships between a CS and a US, tends to result in an observable behavioral change, whereas some—for example learning about negative relationships between a CS and a US—is often behaviorally silent. In these latter instances conventional wisdom has proposed that associative processing is absent. Only recently have we begun to infer from our experiments that animals do learn about simultaneous, backward, and random, as well as about negative, relationships between a CS and a US, even though none of these relationships regularly results in a conditioned response during initial training. So, it has begun to appear that associative learning is most often behaviorally silent and that it only infrequently, if importantly, results in behavior change.

Accordingly we must develop a methodology and an experimental logic consistent with the subtle hidden nature of associative learning. Konorski clearly foresaw the nature of the problem and realized that converging experimental results and some system of strong inference would be necessary to future researchers in associative learning. The procedures he suggested in the instance of inhibition, summation, and retardation are examples of the sort of multistage experimental methods required to uncover the nature of associative learning.

2. Learned Irrelevance and Simultaneous Conditioning as Examples of Silent Associations

To add some substance to our methodological discussion and to help further define the nature of association, we will examine and compare some recent research concerned with learned irrelevance and simultaneous conditioning. Both studies are multistage experiments, examining variations in the temporal relationship between events, specifically variations considered behaviorally inactive or inert. In the initial stage, shocks and stimuli were presented according to schedules alleged to establish the silent associations. In the test stage a CER procedure gave voice to the silent associations.

Mackintosh's (1973) experiment examined learned irrelevance by presenting a CS and a shock randomly with respect to one another in the first stage. Two other groups received either CS-only presentations or control exposure to the apparatus during the first stage. The 3 groups received identical CER training in the second stage; the 20-sec CS preceded shock while the animals licked for water. The acquisition of conditioned suppression over the first 4 days is shown in Fig. 13.2. The CS-only group showed some slight retardation of acquisition—a typical CS habituation effect—whereas the CS/Shock group was retarded over all 4 days compared with the CS-only and control groups. It is hard to avoid the conclusion that in the first stage, animals learned that the CS was not a reliable predictor of shock.

Simultaneous presentation of a CS and shock is rarely seen to produce a CR. In a unique demonstration of the power of this variation in the temporal parameters of conditioning, Rescorla (1973) investigated simultaneous second-order CER conditioning. The first stage presented CS_1–shock pairings to 3 groups (Groups F,S,U), and unpaired CS_1 and shock to a fourth group (Group C). In the second stage, Groups C and S received presentations of a 60-sec CS_1 and a 30-sec CS_2 that began simultaneously with the onset of CS_1. Group U received CS_1 and CS_2 explicitly unpaired, and Group F received nonoverlapping forward pairings of CS_2 and CS_1. Fig. 13.3 shows suppression ratios for a single 30-sec test presentation of CS_2 while all 4 groups were lever pressing. Groups U and C showed no suppression, and

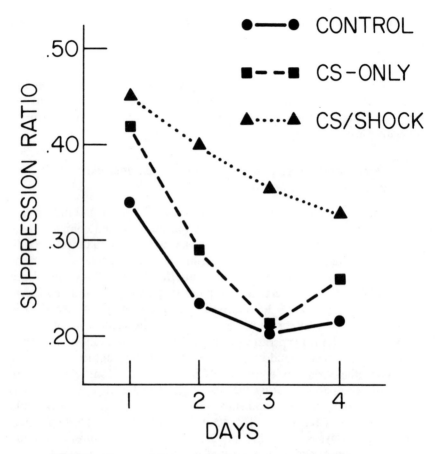

FIG.13.2. The acquisition of conditioned suppression (group mean suppression ratios) following preexposure to the CS alone (CS-only), random CS and shock presentations (CS/Shock), or no preexposure. After Mackintosh (1973).

Group F considerable suppression. Group S suppressed more than Groups U and C, but less than Group F. Clearly, associations between the simultaneously presented CS_2 and CS_1 were revealed in the test stage of Rescorla's higher-order conditioning experiment.

In both Mackintosh's (1973) and Rescorla's (1973) experiments, the CS had no behavioral effect during the initial stage; yet in the test stage there was a clear indication that animals had indeed learned about the random and the simultaneous relationships, respectively. The silent associations underlying stimultaneous conditioning and learned irrelevance continue to be exciting topics of research. One important task is to demonstrate the behavioral effects for these associations in a greater diversity of paradigms —in particular, in some work involving negative conditioning and in some work not involving the retardation of learning test. More than any other research tactic, diverse but converging lines of evidence succeed. A second important task will be establishing procedures to answer the question of what animals learn about negative, random, and simultaneous correlations between CS and US. The structure of these relational representations and the rules for their use in later associative learning remain quite unknown.

Of course, theorists, among them Konorski himself, have proposed connectionist models of the silent associations. Understandably the connectionist concern has been for "what" is connected to the CS percept when it has been negatively or randomly correlated with a reinforcer. One suggestion has been that a "no reinforcer center" becomes connected with the CS percept; another suggestion has been that a negative connection between the

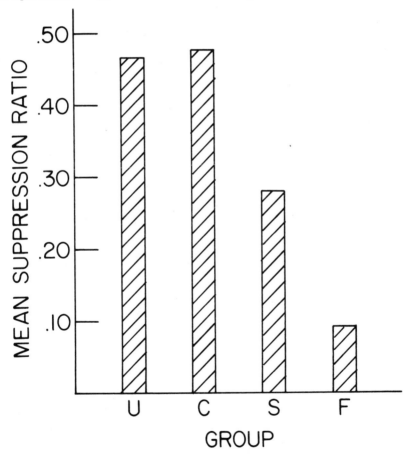

FIG.13.3. Suppression ratios to a second-order CS on a single test trial. Groups F and S received forward and simultaneous second-order conditioning, respectively. Groups U and C received control treatments. After Rescorla (1973).

CS percept and the US center is established. A later suggestion is that excitatory connections between the situation and the reinforcer center may block connections with the CS percept. Left undisclosed are (1) how the system knows which connections to form, given the great possible variation in input; (2) the nature of the connection in the case of learned irrelevance; and (3) the mechanism underlying the fine temporal detail of an animal's anticipation of a period free from the reinforcer, as in the aforementioned study by Davis et al. We find the relational representation simpler and less fanciful than the connectionist alternative because it requires no brain centers of dubious anatomy, no dependence on possibly ephemeral background conditioning, and no additional system to account for the fine detail of expectant behavior.

III.   ASSOCIATIONS BETWEEN NEUTRAL EVENTS

Unlike most other animal-learning theorists except for Tolman (1932), Konorski was perfectly willing to consider and to accept evidence demonstrating the existence of associations between neutral events in infrahuman animals. Indeed he concluded that, although not systematically collected, the evidence supporting the existence of such associations was "indubitable (1967, p. 287)." The CR was seen only as a special case of association in which one of the stimuli regularly produced an overt response (1967, p. 266); presumably the general case includes instances in which neither stimulus regularly produces an overt response. As in the study of inhibition, the methodology suggested in the study of neutral associations uses the CR as a tracer for the otherwise silent association in a multistage experiment. A negative result in this indirect research does not necessarily imply the absence of an association. But surely negative results are never good evidence for the absence of an association; we noted earlier, for example, that failure to obtain an overt nictitating-membrane CR at a CS–US interval of 6.75 sec was hardly proof that rabbits fail to associate a CS with suborbital electric shock in that interval.

A.   Preconditioning Associations between Neutral Events

Quite naturally, cognitive processing of neutral events before conditioning tends to be behaviorally silent—for example, behavioral observation during the presentation of two or more neutral events rarely reveals what animals know about the event sequence or that they can anticipate one element from the presentation of another. Multistage experiments most often provide the methodology for the study of associations formed before the introduction of reinforcers during conditioning. The first stage entails presentation of two or more neutral events in some defined relationship to one another in one or more experimental groups. Presentation of neither event, of only one event, or of both events in a different relationship occurs in the control groups. Representation of relationships between the neutral events is demonstrated when differences between experimental and control groups are observed in the various later conditioning and testing stages of the experiment.

1. Sensory Preconditioning Research

A particularly interesting sensory-preconditioning experiment, conducted here at Queen's University by Tait, Black, Katz, and Suboski (1972), serves as an example of the use of the method to uncover something of the nature of association between neutral events. The unique feature of the study by Tait et al. is the use of a discrimination procedure during the preconditioning stage of the experiment. In the preconditioning stage a tone of 1 frequency (S_1+) preceded a light (S_2), and a tone of another frequency (S_1-) was never paired with the light. There were 7, 14, 28, or 56 presentations each of S_1+ and S_1-. In the next stage during CER training, 10 S_2 presentations coterminated with shock. The results of the test stage, in which S_1+ and S_1- were presented while the rats were licking for water, are shown in Fig. 13.4. With increasing numbers of preconditioning trials the suppression ratios to S_1+ decreased while those to S_1- increased. Thus it appears that the animals had learned that S_1+ was followed by light, and S_1- was not. These results admit the possibility that the preconditioning stage established representations of both the relationship between S_1+ and S_2 and that between S_1- and S_2. Left unresolved, as yet, is whether increased exposure to S_1- during preconditioning simply habituated its unconditioned effects or whether animals learned about the negative relationship between S_1- and S_2.

2. CS Habituation of Correlated Neutral Events

Retardation of the acquisition of a classically conditioned response to a CS presented repeatedly before training, which we will term CS habituation, is now a well-established phenomenon (Lubow & Moore, 1959; Rescorla, 1971; Siegel, 1969). Explanations based on the notion that animals represent the habituated stimulus as irrelevant (Mackintosh, 1973) seem, as a first approximation, particularly apt. Recent research suggests that the procedures of the CS-habituation experiment may be used with profit in the study of the structure of preconditioning associations between neutral events. Our main example is an experiment reported by Mackintosh (1973). In the first stage of that experiment, two groups of rats received several presentations of light before conditioning; one of the groups had the light alone, the other group had the light in a simultaneous compound with noise. A control group received no preconditioning exposure to either light or noise. In the

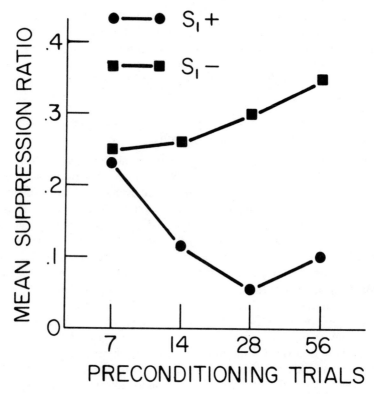

FIG.13.4.   Mean suppression ratios to S_1+ and S_1- as a function of the number of preconditioning trials. After Tati et al., 1972.

second stage all of the rats had CER training, in which the light was paired with a shock. During both stages of the experiment the rats lever pressed for food.

As shown in Fig. 13.5 the light-alone group showed a typical CS habituation effect: retarded acquisition of the CER compared with the control group. The important finding, however, was that the light-plus-noise group was only slightly retarded compared with the contol group.

In an experiment of similar design, Lubow, Schnur, and Rifkin (1976) preexposed one group to a tone–light sequence, a second group to the tone alone, and a control group to neither tone nor light. The tone was paired with shock during subsequent CER training. The suppressive effect of the tone was measured while the thirsty rats licked water from a tube. As in Mackintosh's experiment, preexposure to the CS in compound with another stimulus retarded conditioning much less than preexposure to the CS alone. The subjects of these experiments appear to have represented both serial and simultaneous compounds of the CS and another stimulus as different from the CS alone simply as a function of the repeated presentation of the compound before conditioning. These results suggest the routine formation of relational representations of spatiotemporal compounds of neutral events. None of these results suggest that animals learn to ignore neutral events; rather, it would appear that the events and their spatiotemporal relationships remain represented in fine detail even after repeated exposure.

B. Explicit Correlation between Stimulus Compounds and a Reinforcer

As Konorski pointed out, all conditioning stimuli are compounds. For example, a brief tone originating from a speaker to the left of an experimental subject is a unique spatiotemporal acoustic-compound CS discriminable

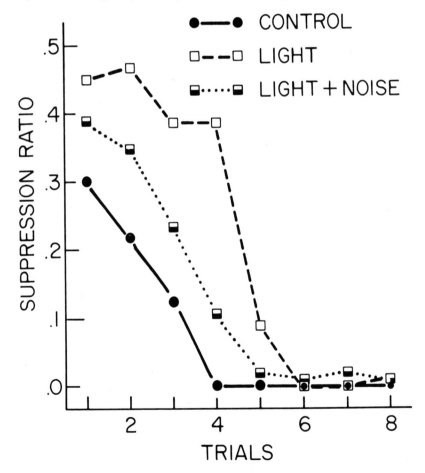

FIG.13.5. Acquisition of conditioned suppression to a light following preexposure to a light alone or to a light-plus-noise compound. After Mackintosh (1973).

from acoustic compounds different in location, duration, frequency, or amplitude because of differences in the sort of relational representations these compounds produce. Our survey is limited to a few demonstrations of relational representations revealed by the various methods used in the literature on stimulus compounding.

1. Generalization Procedures

A useful extension of the familiar stimulus-generalization test is the presentation of the elements of compound stimuli alone and in various combinations in extinction after training. Thomas, Berman, Serednesky, and Lyons (1968) provide an interesting example of the method applied to serial compounds of keylight stimuli paired with food in the pigeon. During the first stage of one of their experiments, a sequence of two line angles, S_1 and S_2, preceded food presentation. Then, in the second stage of the experiment pigeons' keypecks produced S_1 alone, S_2 alone, S_1 followed by a novel keylight, S_2 followed by S_1, and the training sequence, S_1 followed by S_2, in a conditioned-reinforcement generalization test in extinction. Pigeons responded about equally to produce the alternative sequences, but more to produce the training sequence. Thomas et al. comment that the control conditions they employed allow the conclusion that the pigeons learned more than simply that there were two stimuli in a sequence or even the identity of those two stimuli. Rather, the pigeons learned which two stimuli were used and in what order the stimuli appeared.

Interestingly enough, Thomas et al. (1968) and Konorski appear to have come upon the substitution method of demonstrating an association quite independently. Konorski noted that (1967) "by presenting a given stimulus-object in combination with a new stimulus-object instead of that which usually accompanies it, we may detect the existence of associations (p. 188)." The method allows us to infer knowledge of the CS compound from generalization decrement and startle responses produced by the substitute compounds. Thomas et al. of course used the substitution method in tests that followed their S_1 with a novel keylight, but their use of the backward-combination and single-element tests greatly strengthened their experiment.

In a similar variant of the generalization method, which Konorski termed expectation, (1967, p. 188), CS_1 was correlated with a reinforcer in situation I, and CS_2 was correlated with the same reinforcer in situation II. Presentation of CS_1 in situation II or vice-versa produced orientation responses and generalization decrement in the CR. It would appear that the animals had associated CS_1 with situation I, and thus its presentation in situation II was not expected. Much the same logic may be applied to the study by Thomas et al; the birds had associated S_1 with S_2, so that the presentation of a novel stimulus instead of S_2 was not expected.

2. Discrimination Procedures

When some compound stimuli are correlated with a reinforcer and other similar compounds are not, it is possible to study differentiation between responding to the compounds. Further information concerning representations of the reinforced and nonreinforced compounds may be obtained by conducting generalization tests of the sort described in the previous section. Konorski (1967, p. 287) provided an example of this procedure: S_1, S_2, and S_3 are positive conditioned stimuli, whereas the $S_0 S_1$ sequential compound is presented without the reinforcer and established as a negative CS. Such a procedure is said to establish S_0 as a negative CS, but in fact it has further consequences. Substitution of S_2 for S_1 in the successive compound during a generalization test trial resulted in an orientation reaction and generalization decrement in inhibition. The animals knew that S_0 should be followed by S_1, not by S_2.

Discrimination procedures can ensure that responses are made only on the basis of a relational representation of compound conditioned stimuli. In an autoshaping experiment, Looney, Cohen, Brady, and Cohen (1977) presented two sequences that were positively correlated with food — a red keylight followed by a horizontal white line, RH +, and a green keylight followed by a vertical white line, GV +; two other sequences, RV − and GH −, were negatively correlated with food. Keypecking increased to RH + and GV + and declined to RV − and GH −. Because each stimulus was equally often positively and negatively correlated with food, differential responding to the various sequences required that birds acquire relational representations of the conditioned stimuli.

3. Psychophysical Methods

By combining the procedures of generalization and discrimination, it is possible to generate a methodology similar to that employed in animal psychophysics (Blough, 1966; Blough & Blough, 1977). Classical psychophysics sought psychological representation of relatively simple physical events, while the proposed application of animal psychophysical methods explores representations of correlations between events in time and space. In general terms, the procedures establish some relationships between two or more events as positive discriminative stimuli, while other relationships involving the same events are established as negative discriminative stimuli, by correlating the former with reinforcement and the latter with nonreinforcement. Differential responding to the various event combinations, then, provides the empirical basis for inferences concerning the structure of resulting associatve representations. There is not a single psychophysical method, but rather many psychophysical methods, useful in the study of associatve structure. These methods can include simultaneous (yes–no) or successive (rating) discrimination procedures, multiple positive or negative stimulus presentations, and any of the modern measurement systems (e.g., signal detection or multidimensional scaling).

The representation of temporal sequences of events may have important implications for our understanding of association; yet little research has sought to study these representations independent of their inferred role in conditioning experiments. Recently we have turned our laboratory toward the application of psychophysical methods to study the representation of temporal order in sequences of events. Our aim has been to determine the detailed structure of the representation of simple two-event sequences, because such sequences are so common in conditioning experiments. We chose to begin our investigations with an examination of sequences of neutral events; later we expect to extend our work to sequences that include reinforcers.

We have adapted an operant conditional discrimination procedure to explore the representation of temporal order in the pigeon. Our choice was a two-component chained schedule, developed by Reynolds and Catania (1962) in the study of the discrimination of temporal duration. There were two basic modifications to their schedule: In the first component, two-event sequences instead of time-out durations were presented; in the second component, if a specific two-event sequence, instead of a time-out duration, had occurred in the first component, then responding was reinforced. The repeated presentation of many two-event sequences, only one of which was followed by reinforced responding, may be seen as a maintained generalization test in which the events and their temporal order were varied systematically. Such a maintained test provides a very sensitive measure of the discriminability of events and their temporal orders, because responding is continuously constrained by differential reinforcement. The logic of inference was that differential responding in the second component would reflect perceived differences between the animals's representations of reinforced and nonreinforced sequences.

a. The Representation of Temporal Order: Experiment I. In our initial experiment the stimuli used in the first component were 5-sec presentations of green and red lights, which we designated "A" and "B", respectively. For 4 birds the lights were presented on the response key, while for 4 others diffuse green and red overhead lights were used. The absence of A or B for 5 sec in any sequential position was called 'X.' During a trial, after a 20-sec intertrial interval, a 2-stimulus sequence was presented, then the white keylight was illuminated. The sequence was drawn from the nine possible 2-stimulus sequences in the matrix:

AA	AB	AX
BA	BB	BX
XA	XB	XX

The matrix is arranged so that all of the elements ending with the same event are in the same column, and all the elements beginning with the same event are in the same row. On "AB+" trials, after the presentation of the AB sequence, responding to the white keylight during the second component was reinforced by a 2.5-sec access to mixed grain after a variable interval of 15 sec had elapsed. On other trials the white keylight was preceded by a sequence chosen from the matrix and terminated after a fixed time of 15 sec without the reinforcer. In discrimination training, sessions consisted of 12 blocks of 8 trials each. In each block were 3 AB+, 1 AB, and 4 XX trials presented at random without replacement. Each bird was shifted from discrimination training to a maintained generalization test after the ratio of keypecks on XX trials to keypecks on AB trials was less than or equal to 0.25 for, 2 successive sessions. Maintained generalization test sessions consisted of 10 blocks of 13 trials each. In each block were 4 AB+ trials and the 9 test-sequence trials shown in the matrix, presented in random order without replacement. Reinforced and nonreinforced trials had the same parameters as in earlier training. The maintained generalization test continued for 27 sessions.

A discrimination ratio expressed responding during a test-sequence trial as a proportion of responding during AB+ trials; for example, if an animal responded at the same rate during a test trial as during AB+ trials, the discrimination ratio for that trial was 1.0. Fig. 13.6 shows mean discrimination ratios for the 9-test sequences over 9 blocks of 3 sessions each. Results for the overhead-light and keylight groups are plotted separately to illustrate some minor but interesting differences between the effects of the 2 stimulus sources. The evidence obtained in the first block of the maintained generalization test was analyzed to determine how A and B were represented initially. Both A and B appear to have been represented as individual events: BB generated reliably higher discrimination ratios than AA, which in turn generated reliably higher ratios than XX. The sequential location of B also appears to have been represented: AB and XB generated reliably higher discrimination ratios than BA and BX. The sequential location of A, however, does not appear to have been represented: Sequences that began with A resulted in ratios not reliably higher than sequences that began with X, and AX generated a reliably lower discrimination ratio than XA. Thus, prior discrimination of the AB sequence from XX appears to have established a representation of AB that included, at the least, both A and B and the sequential location of B. Over the course of the maintained generalization test, evidence accumulated that the sequential location of A was represented, first, as the reversed sequence, BA, after which the two sequences containing B in the last position, XB and BB, were differentiated reliably from AB. More rapid differentiation of BA than XB or BB further demonstrated the importance of the location of B in the sequence.

Results for the overhead-light and the keylight groups differed in two ways. First, in general, the discrimination ratios of the keylight group declined reliably more rapidly than those of the overhead-light group. Second, discrimination ratios for the XB sequence declined reliably more rapidly than those for the BB sequence in the keylight group, but the discrimination ratios for those sequences declined at the same rate in the overhead-light group. Apparently the keylight group represented the XB

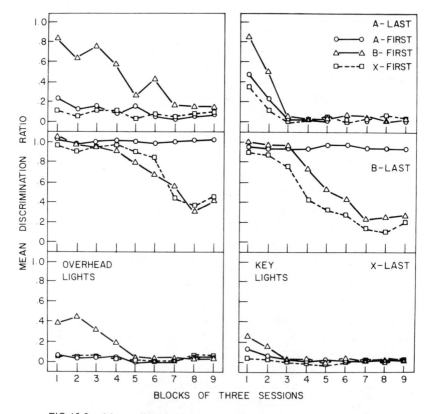

FIG.13.6. Mean discrimination ratios for each 2-stimulus sequence presented in Experiment I during maintained generalization testing over 9 3-day blocks in the overhead-light group (left panel) and in the keylight group (right panel). Ratios of 1.0 indicate that responding after the test sequence equalled responding after the reinforced AB sequence.

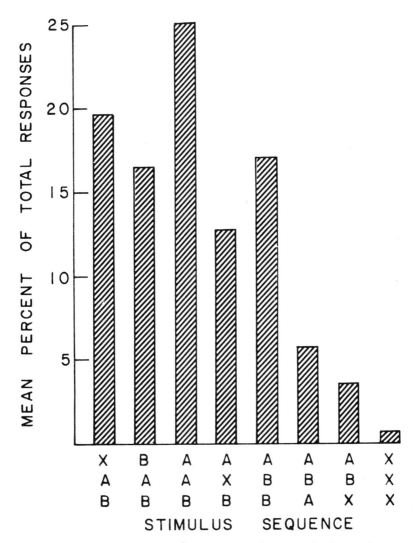

FIG.13.7. Mean percentage of total responses in a generalization test in extinction following each of the three-stimulus sequences shown along the abscissa (Experiment I).

and BB sequences differently, whereas the overhead-light group did not.

After the maintained test, 4 birds — 2 from each of the keylight and overhead-light groups—were given a generalization test in extinction to further explore the representation of the AB sequence. The 2-stimulus sequences were replaced with 3-stimulus sequences that always contained A and B in the correct order, but contained an extra stimulus, A, B, or X, before, between, or after A and B. The 8 3-stimulus sequences are shown along the abscissa of Fig. 13.7, and the mean percentage of total responses is shown along the ordinate. There was very little responding to sequences that did not terminate with B, and the most responding was to the AAB sequence. The broken sequence, AXB, produced less responding than the training sequence, XAB, and the 2 sequences in which B preceded or followed AB produced only slightly less responding than XAB. The intrusion of a different third element reduced responding most when it occurred after AB (see ABA and ABX) and least when it preceded AB (see BAB). The intrusion of a same third element may have reduced responding slightly when it was presented after AB (see ABB) but enhanced responding when it was presented before AB (see AAB). These data are consistent with the structure of the representation of AB we inferred from the maintained test: Both elements are represented in their correct sequential location.

b. *The Representation of Temporal Order: Experiment II.* In our initial experiment, just described, the influence of differential responding to one color over the other and of stimulus generalization between the colors used as A and B was not controlled. In a systematic replication of Experiment I, these factors received special attention. The sequence of colors designated AB was counterbalanced between groups to allow assessment of the role of color independent of serial order. A novel color, C, was introduced as a generalization stimulus during testing. If responding to A were due to generalization from B, then considerable responding to C might be observed as well. The inclusion of C in sequences with A and B is also an application of the substitution method suggested by Konorski and discussed already in the context of the experiment by Thomas et al.

There were several changes in procedure, including those previously suggested, between our first and second experiments: (1) More intense, 40-W instead of 25-W, colored lights were used; (2) the color sequence was counterbalanced — 3 birds had red-yellow and 3 birds had yellow-red overhead lights as the AB sequence during discrimination training; (3) a stricter discrimination criterion — a discrimination ratio equal to less than 0.10 for 3 successive sessions — was used; (4) the novel stimulus, C, a blue overhead light, was substituted for X in the matrix of sequences presented during the generalization tests, and XX as well as CC was presented; (5) a 2-session generalization test in extinction preceded the 27-session maintained test, and AB + trials were omitted from the former but included during the latter generalization test.

The mean percentage of total responses to each test sequence during the generalization test in extinction is shown in Fig. 13.8. Analysis of these data found reliable differences between test-sequence trials but no differential effects of counterbalancing on responding to the 10 test sequences. Further analysis examined the representation of A and B during the extinction test. Both A and B appear to have been represented: BB generated a reliably

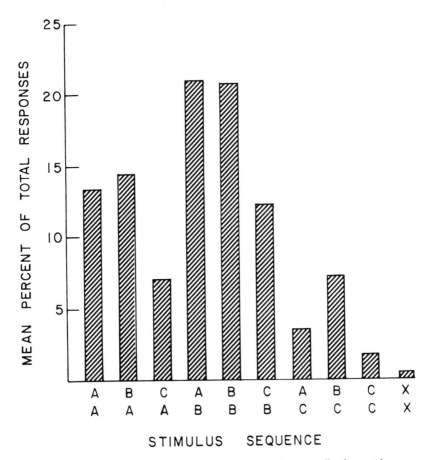

FIG.13.8.   Mean percentage of total responses in the generalization test in extinction following each of the two-stimulus sequences shown along the abscissa (Experiment II).

higher percentage responding than AA, which generated a reliably higher percentage responding than CC. Evidence that B was represented as the last element in the sequence was given by a reliably higher percentage of total responses to test sequences in which B was last than to sequences in which A or C was last. The sequential location of A, however, does not appear to have been preserved; the percentage responding to sequences in which A was first was not different from the percentage responding to sequences in which B was first. These results are in good agreement with those obtained from the first block of the maintained generalization test in our earlier experiment.

The novel overhead light was seen as different from no colored overhead light (X). A separate comparison found that CC (mean = 42 responses) generated reliably more responses than XX (mean $w = 7$ responses). The results also suggest that the birds used the presence of the novel stimulus to discriminate a sequence from AB; the percentage of responding to sequences that included C was reliably lower than to sequences that included only the training stimuli, A and B. The sequential location determined the discriminability of a sequence from AB; sequences that presented C last, AC and BC, generated a lower percentage of responding than sequences that presented C first, CA and CB.

Discrimination ratios, calculated as for the previous experiment, are shown for the maintained generalization test in Fig. 13.9. Because there were no reliable effects of counterbalancing on the discrimination ratio, the average of both the red-yellow and yellow-red groups was plotted. In general the pattern of results obtained in successive blocks of the maintained generalization test is an extension of the findings reported for the extinction test. The sequential location of A was differentiated over the first few blocks of the test: Responding to BB and CB declined reliably from the responding to AB. Responding to other sequences of A and B, AA, and BA, declined as rapidly as responding to CB. Responding to sequences that included C began at a lower level and declined reliably more rapidly than similar sequences without C, giving further evidence that the birds knew that the reinforced sequence did not include C.

A summary of our experimental work seems appropriate. First, discrimination of a sequence of visual events from its absence may establish representations of the elements and of the sequential location of the last, but not the first, element. Second, differentiation of AB from similar sequences appears to follow from the pattern of generalization to those sequences obtained before differentiation. Third, the introduction of a novel stimulus into a sequence, termed substitution by Konorski, can result in generalization decrement even when animals have not fully represented the reinforced sequence.

Those among us who still hope for simple answers might wish to search out an alternative to the sorts of relational representation we have discussed. Spence's (1937) theory, itself designed as a denial of some forms of relational learning, would deem a tempting alternative. One could assign a value to the excitatory potential of each generalization sequence, based on relative delays of reward and on decay of excitatory stimulus traces, positively related to the proximity of each individual stimulus to the reinforcer. The only drawback is that, given the refusal to consider the representation of sequential relationships, this sort of model fails to account for the present data. We can find no estimates of the excitatory strength and temporal decay of A and B that predict initial responding to the generalization sequences, changes during the maintained tests, and responding to the three-stimulus generalization sequences. To predict equal responding to AB and

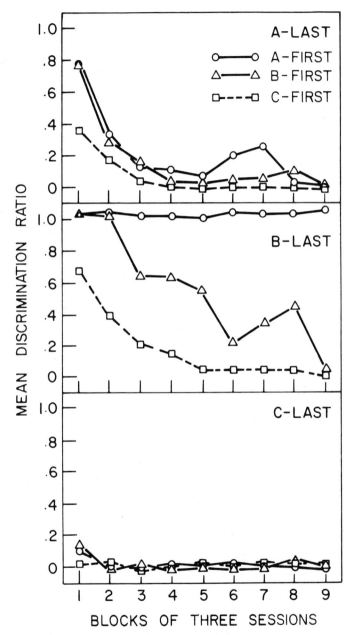

FIG. 13.9.   Mean discrimination ratios for each 2-stimulus sequence presented in Experiment II during maintained generalization testing over 9 3-day blocks.

BB, an additive summation model must attribute equal excitatory strength to A and B in the first sequential location. This approach succeeds in predicting equal responding to BA and AA and to BC and AC in our second experiment (Fig. 13.9), but fails to predict reliably increased responding to BA and BX over AA and AX in our first experiment (Fig. 13.6). Then, of course, equal strength to A and B in the first position cannot account for the differentiation of BB from AB during the maintained tests in both experiments. The meager difference between XB and BB in the maintained test and between AB and ABB in the three-stimulus test is also difficult to predict from additive summation of individual excitatory strengths. Finally, a cross-experiment comparison of responding to XB and CB yielded a result opposite to that predicted from additive summation. In the second experiment, C had excitatory strength; it generated more responses than X. However, the inclusion of C in the CB sequence in our second experiment resulted in fewer responses and more rapid differentiation than its exclusion in the XB sequence in our first experiment. It simply does not appear fruitful to apply Spence's theory to these sequential compounds.

Research concerned with the formation of relational representations of neutral stimuli is hardly complete. Both the sensory preconditioning and CS habituation paradigms may yet provide interesting opportunities to learn more about the representations that are made prior to conditioning. Generalization, discrimination and psychophysical methods have begun to show the complexity of the relationships that animals can represent and in what detail these relationships are retained and used.

IV. THE STUDY OF ASSOCIATION: SUMMARY AND CONCLUSIONS

A. Silent Associations

Full realization that association is not a behavioral phenomenon has been a long time in coming. Resistance stems from a reluctance to accept the notion that such an important determinant of animal behavior is not directly observable. Of course, it is more than a happy accident that associative learning can eventually be inferred from behavioral experiments. The evolution of an associative process without important behavioral effects must be considered unlikely. One need not deny the adaptive significance of association to insist that associative structure is revealed only indirectly in behavior. Specific genetic structure, that most basic of all evolutionary concepts, is often only inferred from phenotypic expression in the anatomy, physiology, and behavior of living things.

Surely whether an association produces an observable CR depends on the circumstances, including, trivially, the sensitivity and range of the experimenter's detectors, and is not a fundamental characteristic of an association. Because associations are not directly observable, the discovery of their existence and substance is entirely a matter of skillful inference and not simply a matter of measurement. In the present context skillful inference is the ability to find and to use diverse behavioral techniques in combinatons that yield unique behavioral effects for each form of associative structure. The effort is especially satisfying if the combination of effects attributed to an associative structure is consistent with the kinds of operations seen as necessary to its establishment.

B. Relational Representations

The discrimination of spatiotemporal compounds of events from one another and from their elements is demonstrated in the differentiation of responding observed in many of the experiments discussed by Konorski and in the present chapter. We shall allow ourselves the not-unwarranted stipulation that animals retain some representation of events successfully discriminated from one another. Now, because the "events" discriminated in these experiments have been spatial and sequential compounds of stimuli, it would seem entirely sensible to suppose that their representations also include some knowledge of the relationship between the stimuli. For example, there seems little objection to statements that the representation of a tone originating from the left includes spatial information, or that the representation of a sequence of neutral events includes temporal information. Resistance is encountered, however, to the assertion that representations formed as the result of conditioning prcedures are relational in character. If one is prepared to admit that animals may represent the order of neutral events, one would be hard put to deny that animals may also represent the various orderings of CS and US employed in conditioning. According to our cognitive view, animals use the same relational representations in anticipating events in a sequence as they do in discriminating one order of events from another. This is the basis of our insistence that associations, too, are relational in character.

Because we agree that associative structure may include representation of relationships, it remains for us to detail the nature of the relational representation. If we believe, for example, that animals learn about negative and random relationships, what is it we believe they learn? It is certainly difficult to determine the kinds and amounts of relational information available from the environment that animals incorporate into their associative structures. One possible representational system would distinguish between *before, during,* and *after* an event (Anderson & Bower, 1973). Even this system, however, might require some temporal structure in order to represent both positive and negative relationships with the detail reflected in behavior. Progress in discerning the structure of relational representation will require further development of both our models of structure and our methods of inferring structure from behavior.

Konorski distinguished between associations that do and do not incude representation of a stimulus that generates an overt response: associations between CS and US and between neutral events. We have distinguished between the silent associations and associations that result in a CR to the CS during original training. One might wish to distinguish between associations that are primarily spatial and those that are primarily temporal. In turn, one could distinguish among the various temporal associations animals might represent. Research in animal learning has focused on the behaviorally active associations. Until more work with the behaviorally silent associations is done, we simply do not know the correct strategy for attempting to classify associations. As our review has shown, the range of methods open to the student of associative learning is much wider than generally realized.

C. Beyond Representation

The representation of knowledge concerning relationships between events—the area traditionally called association—has been the topic of this chapter. We have outlined a methodology and a theory of the representation of association. A weakness of our position is that it has failed to consider the rules of association. We have discussed the representation of association as if it were quite static. The dynamic aspects of associative learning—for example, the formation and application of relational representations — we have termed the rules of association. Information scientists are accustomed to distinguishing between data structures and algorithms. The equivalent distinction in the study of cognition would be between representations and rules. Such a distinction has much to recommend it. It would be convenient if associative representations could be studied independently of associative rules and vice-versa. Another possiblity is that representations and rules should be studied together. The logic of the latter view is that the proper evaluation of an associative theory requires that its statements about the representation and use of knowledge be jointly as well as independently testable.

ACKNOWLEDGMENTS

The authors thank M. D. Suboski, E. Zamble, M. Rilling, R. A. Boakes, and A. Dickinson for their thoughtful criticism of an earlier version of this chapter. The preparation of this chapter was supported by a grant from the National Research Council of Canada.

REFERENCES

Anderson, J. R., & Bower, G. H. *Human associative memory.* New York: Wiley, 1973.

Blough, D. S. The study of animal sensory processes by operant methods. In W. K. Honig (Ed.), *Operant behavior: Areas of research and application.* New York: Appleton-Century-Crofts, 1966.

Blough, D., & Blough, P. Animal psychophysics. In W. K. Honig & J. E. R. Staddon (Eds.), *Handbook of operant behavior.* Englewood Cliffs, N. J.: Prentice-Hall, 1977.

Davis, H., Memmott, J., & Hurwitz, H. M. B. Autocontingencies: A model for subtle behavioral control. *Journal of Experimental Psychology: General,* 1975, *104,* 169–188.

Garner, W. R., Hake, H. W., & Eriksen, C. W. Operationism and the concept of perception. *Psychological Review,* 1956, *63,* 149–159.

Kamin, L. J. "Attention-like" processes in classical conditioning. In M. R. Jones (Ed.), *Miami Symposium on the Prediction of Behavior: Aversive Stimulation.* Miami: University of Miami Press, 1968.

Kamin, L. J. Predictability, surprise, attention and conditioning. In B. A. Campbell & R. M. Church (Eds.), *Punishment and aversive behavior.* New York: Appleton-Century-Crofts, 1969.

Konorski, J. *Integrative activity of the brain: An interdisciplinary approach.* Chicago: University of Chicago Press, 1967.

Looney, T. A., Cohen, L. R., Brady, J. H., & Cohen, P. S. Conditional discrimination performance by pigeons on a response-independent procedure. *Journal of the Experimental Analysis of Behavior,* 1977, *27,* 363–370.

Lubow, R. E., & Moore, A. U. Latent inhibition: The effect of non-reinforced pre-exposure to the conditioned stimulus. *Journal of Comparative and Physiological Psychology,* 1959, *52,* 415–419.

Lubow, R. E., Schnur, P., & Rifkin, B. Latent inhibition and conditioned attention theory. *Journal of Experimental Psychology: Animal Behavior Processes,* 1976, *2,* 163–174.

Mackintosh, N. J. Stimulus selection: Learning to ignore stimuli that predict no change in reinforcement. In R. A. Hinde & J. S. Hinde (Eds.), *Constraints on learning: Limitations and predispositions.* Cambridge: Academic Press, 1973.

Rescorla, R. A. Summation and retardation tests of latent inhibition. *Journal of Comparative and Physiological Psychology,* 1971, *75,* 77–81.

Rescorla, R. A. Second-order conditioning: Implications for theories of learning. In F. J. McGuigan and D. B. Lumsden (Eds.), *Contemporary approaches to conditioning and learning.* Washington, D. C.: Winston, 1973.

Reynolds, G. S., & Catania, A. C. Temporal discrimation in pigeons. *Science,* 1962, *135,* 314–315.

Siegel, S. Effect of CS habituation on eyelid conditioning. *Journal of Comparative and Physiological Psychology,* 1969, *68,* 245–248.

Spence, K. W. Experimental studies of learning and the higher mental processes in infra-human primates. *Psychological Bulletin,* 1937, *34,* 806–850.

Tait, R. W., Black, M., Katz, M., & Suboski, M. D. Discriminative sensory preconditioning. *Canadian Journal of Psychology,* 1972, *26,* 201–205.

Thomas, D. R., Berman, D. L. Serednesky, G. E., & Lyons, J. Information value and stimulus configuring as factors in conditioned reinforcement. *Journal of Experimental Psychology,* 1968, *76,* 181–189.

Tolman. E. C. *Purposive behavior in animals and men.* New York: Century, 1932.

Van Dercar, D. H., & Schneiderman, N. Interstimulus interval functions in different response systems during classical discrimination conditioning of rabbits. *Psychonomic Science,* 1967, *9,* 9–10.

Watson, J. *Behaviorism.* New York: Norton, 1924.

14 Selective Associations

Vincent M. LoLordo
Dalhousie University

I. CONSTRAINTS ON
LEARNING AND PERFORMANCE

Research on learning has concentrated on the search for general laws, which do not depend on the specific choice of stimulus, response, and reinforcer made by an experimenter. It has generally been assumed that the researcher could arbitrarily select the stimuli, responses, and reinforcers to be used in his classical conditioning and instrumental training experiments without markedly affecting the success of the experiments. Of course there have always been researchers who have taken exception to this claim. For example, Mowrer (1947) asserted that autonomic responses can be classically conditioned but not instrumentally trained, whereas skeletal responses can be instrumentally trained but not classically conditioned. In a similar vein Turner and Solomon (1962) suggested that the degree of reflexiveness of a response is an important determinant of its conditionability.

Recently the number of exceptions seems to have multiplied. A variety of data from diverse experimental settings has forced many investigators to conclude that the choice of stimulus, response, and reinforcer is not a matter of indifference, but may be one of the most critical determinants of the success of an experiment on learning. Demonstrations that performance is markedly affected by the particular combination of events chosen by an experimenter have led some researchers (e.g., Rozin & Kalat, 1971; cf. Seligman, 1970) to conclude that the laws of learning will be different for different classes of events, although this conclusion is not necessitated by the demonstrations. Research that has caused the current reevaluation of the generalist position has been reviewed in varying degrees of depth and scope by several investigators (Bolles, 1970, 1972; Garcia, McGowan & Green, 1972; Hinde & Stevenson-Hinde, 1973; Rozin & Kalat, 1971; Segal, 1972; Seligman, 1970; Seligman & Hager, 1972; Shettleworth, 1972a; and others).

In a comprehensive review Shettleworth (1972a) described those experiments whose success was determined by the specific choice of stimulus, response, reinforcer, or combination of these elements as examples of constraints on learning or performance. She also proposed a scheme for classifying experiments in terms of the source of the constraints that they demonstrated—that is, whether they arose from the researcher's choice of stimulus, response, reinforcer, or of some combination of these elements. In Chapter 15 of this volume Shettleworth discusses Konorski's account of constraints resulting from: (1) the choice of a particular class of responses to be reinforced in an instrumental training experiment—for example, in terms of whether elicited responses such as a leg flexion in response to shock or scratching the ear can be instrumentalized—and (2) the choice of a particular combination of discriminative stimulus and instrumental response—for example, the change in the relative effectiveness of qualitative versus directional cues in left-right as compared with "go-no go" discriminations.

This chapter focuses on classical conditioning and instrumental training experiments whose outcomes are constrained by the experimenter's choice of particular combinations of conditioned or discriminative stimuli and USs or reinforcers. Such outcomes are called stimulus-reinforcer interactions.

Two basic experimental designs have been used in recent demonstrations of stimulus-reinforcer interactions. In the single-cue design, four groups of subjects receive the factorial combinations of the two CSs and the two USs (Fig. 14.1). Some measure of performance in acquisition is taken, and if acquisition of a CR occurs more rapidly (or if performance reaches a higher asymptote) with CS1 than with CS2 when US1 is the reinforcer, but the reverse is true when US2 is the reinforcer (as in Fig. 14.1), the outcome is called a stimulus-reinforcer interaction, or crossover effect. The occurrence of such an effect cannot be attributed solely to a difference between the CSs, or solely to a difference between the USs (e.g., differences in learning-rate parameters), but must reflect some selective effect on learning or performance.

Stimulus-reinforcer interactions have also been demonstrated in a compound-cue design, which is illustrated in the top half of Fig. 14.2. In an example of this design using a classical conditioning procedure, two groups of subjects receive a simultaneous compound of the two CSs during conditioning, but the compound CS is paired with different USs in the two groups. The response to the separate elements is assessed in a series of test trials following conditioning. Fig. 14.2 illustrates a stimulus-reinforcer interaction like that in Fig. 14.1.

GROUP 1. CS1 – US1

2. CS2 – US1

3. CS1 – US2

4. CS2 – US2

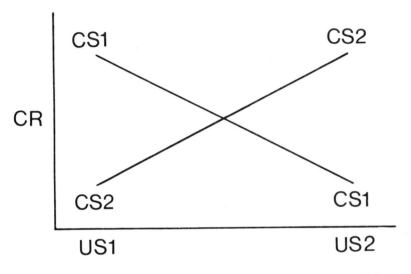

FIG. 14.1.   Single-cue experimental design used to demonstrate stimulus-reinforcer interactions. Four groups of subjects receive the factorial combinations of the two CSs and the two USs. If acquisition of a CR occurs more rapidly (or performance reaches a higher asymptote) with CS1 than with CS2 when US1 is the reinforcer, but the reverse is true when US2 is the reinforcer (as in this hypothetical example), the outcome is called a stimulus-reinforcer interaction.

The single-cue design permits more direct assessment of the relative conditionability of the two cues than does the compound-cue design, in which differences in test responding to the two cues could result from some emergent property of the compound conditioning procedure. However, the compound-cue design has appeal because it permits within-subject comparisons. Examples of each design are presented in the following pages.

Given the associationist bias that has long prevailed in learning theory, stimulus-reinforcer interactions are intriguing, for they suggest that some associations may be selective — that is, that connections between certain antecedent and consequent events (CS1-US1, CS2-US2) within a set of

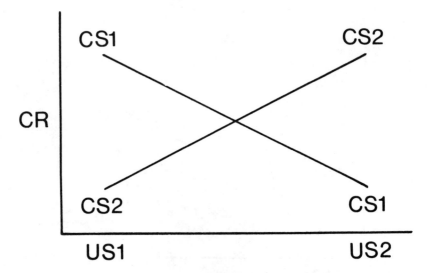

TRAIN | TEST

GROUP 1. (CS1 CS2)-US1   CS1, CS2

2. (CS1 CS2)-US2   CS1, CS2

FIG. 14.2. Compound-cue experimental design used to demonstrate stimulus-reinforcer interactions. Two groups of subjects receive a simultaneous compound of the two CSs during conditioning, but the compound is paired with different USs in the two groups. The response to the separate stimulus elements is assessed in a series of test trials following conditioning. If there is more test responding to CS1 than to CS2 when US1 is the reinforcer, but the reverse is true when US2 is the reinforcer (as in this hypothetical example), the outcome is called a stimulus-reinforcer interaction.

events (CS1, CS2, US1, US2) may be formed very easily, whereas connections between other antecedents and consequents (CS1-US2, CS2-US1) may not be formed at all, or only with great difficulty. To state this possibility another way, consider the sorts of learning-rate parameters that might describe the rate of growth of associations between CSs and USs. In Rescorla and Wagner's recent model (Rescorla & Wagner, 1972), two separate learning rate parameters, α and β, describe the contributions of the CS and US, respectively, to the growth of the association. In these terms, selective associations would be said to occur whenever the rate of growth of a CS-US association can only be characterized by a single parameter, unique to the CS-US combination, that cannot be reduced to the separate parameters for CS and US.

Konorski did not write extensively about stimulus-reinforcer interactions and the possibility of selective associations. However, he did suggest several interpretations of a few sample stimulus-reinforcer interactions, which are presented in Section II of this paper. In Section III, various procedures that can be used to help researchers decide between an account of stimulus-reinforcer interactions in terms of selective associations and several alternative accounts are discussed. These methodological suggestions are used as the basis for the discussion and evaluation in Section IV of several alternative accounts of some recently demonstrated stimulus-reinforcer interactions. The discussion focuses on the question, "Do these stimulus-reinforcer interactions force us to conclude that the relative growth rates of associations between several stimuli and a reinforcer depend on the nature of that reinforcer — that is, that there are selective associations?" Finally, the implications of selective associations for the generality of the laws of learning are discussed in Section V.

II.   KONORSKI'S DISCUSSION OF STIMULUS-REINFORCER INTERACTIONS

In Konorski's 1948 book *Conditioned Reflexes and Neuron Organization,* the possibility of stimulus-reinforcer interactions emerged during a discussion of classical defensive and alimentary conditioning to so-called supramaximal CSs, or stimuli so intense that they elicit defensive behaviors. Konorski noted, as had Pavlov earlier, that it was difficult to condition alimentary CRs to supramaximal CSs. Konorski attributed this difficulty to the presence of inborn, mutually inhibitory connections between defensive and alimentary centers (see Dickinson and Dearing, Ch. 8). Such connections would have to be outweighed or overcome by excitatory, conditioned connections before the supramaximal CS would reliably elicit an alimentary CR. Thus, alimentary conditioning should occur more slowly to a supramaximal CS than to a moderately intense one, which would not elicit defensive behaviors. On the other hand, Konorski (1948, p. 122) suggested that defensive CRs might become conditioned more rapidly to supramaximal CSs than to moderately intense CSs, because the defensive centers stimulated by the supramaximal CS and the US would be "in a state of concealed alliance (p. 123)" — that is, already linked by some inborn, mutually excitatory connections.

Konorski's discussion suggests a straightforward way to demonstrate that the relative conditionability of moderately intense and supramaximal CS depends on the alimentary versus defensive nature of the US. Simply combine the two CSs factorially with the two USs, and observe the acquisition of alimentary and defensive CRs in the various groups. A stimulus-reinforcer interaction of the form shown in Fig. 14.1 would be produced by faster acquisition of the alimentary CR with the moderately intense CS than with the supramaximal one, combined with faster acquisition of the defensive CR with the supramaximal CS than with the moderately intense one.

Would Konorski have interpreted such a stimulus-reinforcer interaction as an example of selective association? Not necessarily. In the aforementioned example, the observed CRs developed more rapidly in two groups given defensive conditioning to a supramaximal CS and alimentary conditioning to a moderately intense CS than in groups given the other CS-US combinations. Konorski would have attributed this outcome to the presence of inborn inhibitory or excitatory connections between CS and US centers, arguing that the number and "direction," or "sign," of these connections determine how many new excitatory connections must be established in the various groups before the threshold for performance of the CR is crossed.

Given the inborn connections proposed by Konorski, the assumption that the association between CS and US grows at different rates in the various groups, thus producing the "observed" stimulus-reinforcer interaction, is unnecessary. The interaction would occur even if the rates of growth of new connections were equal in all four groups — that is, if selective association did not occur. Extending Konorski's reasoning, one must always consider the possibility that a stimulus-reinforcer interaction occurs simply because some of the CSs initially evoke responses allied to (or antagonistic to) the to-be-conditioned effector responses, and thus begin the experiment closer to (or farther from) the threshold for evocation of those conditioned responses (see Henderson, 1973; Jacobs & LoLordo, 1977). In the most extreme case CS1 initially elicits the to-be-CR1 itself, and CS2 initially elicits the to-be-CR2; in other words, the "alliance" in Konorski's statement is overt (above the threshold). Of course, this case is relatively uninteresting, for even if the experiment were continued after it was observed that the CSs were not neutral stimuli, the outcome would not be interpreted by anyone as a demonstration of selective assocation.

Konorski was not proposing precisely this extreme interpretation of our example, but qualified the interpretation in several ways. First, he maintained that there is an "alliance" between CS and US centers whenever the CS and US both initially evoke defensive reactions, even though the defensive reaction initially elicited by the CS is not morphologically like the unconditioned response to the US. That is, Konorski proposed that defensive CRs of various sorts might be conditioned very rapidly to supramaximal stimuli because there was some element common to the diverse defensive reactions elicited by the CSs and the USs. This point was clarified in Konorski's 1967 book and is further developed later in this section.

Second, Konorski considered the possibility that the critical responses initially elicited by the CSs (in our example, the defensive responses to supramaximal stimuli) were elicited only in subthreshold form — in other words, that the "alliance" between CS and US centers was concealed. We have just seen that such a concealed alliance could yield a stimulus-reinforcer interaction, given nonselective facilitation of the initial, subthreshold responses to the CSs by the USs (either associatively or otherwise). Thompson (1976) has also suggested that the sorts of stimulus-reinforcer interactions discussed in this paper may depend on differences in the neural responses unconditionally elicited by behaviorally neutral CSs.

Several additional mechanisms could account for stimulus–reinforcer interactions if the facilitative effect of USs on initial, subthreshold responses to CSs were selective. For instance, Kandel (1976) has recently suggested that a CS may initially elicit in subthreshold form all the responses that can subsequently be conditioned to it, and that various USs may facilitate some of these responses more than others, even leading to a motor CR. This view implies that stimulus-reinforcer interactions should occur, given variations in the strength of the neural responses elicited by diverse CSs, along with variations in the magnitude of the facilitative effects of various USs on those neural responses. The foregoing considerations suggest two distinct possibilities. These stimulus-reinforcer interactions should be interpreted as selective associations if the selective, facilitative effects of USs on the initial responses to the CSs depend on CS-US pairings, or on a positive contingency between CS and US. On the other hand, in some cases the selective, facilitative effect may not depend on CS-US pairings, and thus should be called selective sensitization. Selective sensitization is a reasonably attractive explanation of the hypothesized extraordinarily rapid conditioning of defensive CRs to supramaximal CSs and, in the absence of further data, is a plausible alternative to selective association whenever a CR develops very rapidly.

For any stimulus-reinforcer interaction, control groups must be run before we can decide among the various accounts. These controls are discussed in the next section of this paper.

Following the discussion of conditioning to supramaximal CSs, Konorski (1948) noted another area in which a stimulus-reinforcer interaction would be expected. In a 1911 dissertation from Pavlov's laboratory, Cytovich (in Konorski, 1948; p. 123) apparently demonstrated that the sight and smell of meat did not unconditionally elicit salivation in dogs that had been reared on a milk diet. Nonetheless, when these cues preceded consumption of the meat, they came to evoke a salivary CR after relatively few trials, compared with other indifferent CSs. Presumably, the sight and smell of meat would not have been especially effective CSs when a US other than food was used; so we may assume that a stimulus-reinforcer interaction would occur in a design that combined, for example, the smell of meat versus a visual stimulus (CSs) with meat in the mouth versus electric shock to the paw (USs). Konorski's interpretation of the extraordinary effectiveness of the sight and smell of meat as alimentary CSs, which was based largely on his observation of the immediate attractiveness of the smell of meat to the dogs, was essentially the same as his interpretation of rapid defensive conditioning to supramaximal CSs. He concluded (1948):

> In our view this fact must be considered as showing that between the inborn reflex to the smell of meat (contrary to the inborn reflexes to the smell of some repellent substances) and the alimentary reflex a concealed alliance exists, which is manifested by the rapid and easy formation of the corresponding conditioned reflex and by its strength and stability [p. 124]

In this case Konorski's interpretation of "concealed alliance" is somewhat obscure, but perhaps he meant that both reflexes are appetitive. The smell of meat was an especially effective CS in alimentary conditioning because it elicited a motivational state that facilitated salivary conditioning, although it did not initially elicit salivation. Again Konorski did not seem to be offering selective association as an account of a stimulus-reinforcer interaction. Nor did he offer such an account of the especially slow alimentary conditioning to supramaximal CSs, noted earlier. He attributed that effect to a "concealed antagonism", or inhibitory connection between the defensive motivational state elicited by the CS and the appetitive motivational state presumably underlying salivary conditioning.

In his 1967 book *The Integrative Activity of the Brain*, Konorski classified the varieties of classically conditioned responses. His classification enables us to better understand the account of the stimulus-reinforcer interactions described in his earlier book. Briefly, and ignoring many complexities, Konorski's 1967 account maintained that a supramaximal CS should be especially effective in defensive conditioning (e.g., conditioning of limb flexion with electric shock) because it initially elicits the preparatory or drive CR (fear), which must first be conditioned before the consummatory CR (limb flexion) can be conditioned. If the CS were of moderate intensity, flexion conditioning would occur more slowly because the fear CR would develop more slowly.

According to Konorski's 1967 account, a supramaximal CS should be a very poor CS+ for alimentary conditioning, because the fear response it initially elicits has an inborn, inhibitory effect on performance of the preparatory, alimentary drive CR (hunger). Formation of the hunger CR is a necessary condition for the formation of the consummatory, food CR, which is manifested by salivation.

Parenthetically, Konorski's discussion raises the interesting question of whether supramaximal CSs should be effective inhibitory stimuli in defensive situations — or, to use Konorski's terms, whether they should easily become elicitors of the fear antidrive CR and of the "no-shock CR." Recent experiments by Jacobs and LoLordo (1977) suggest that the answer to this question would be negative. They found that for rats the relative ease of conditioning a response to the onset of white noise and to changes in the level of illumination differed for excitatory and inhibitory conditioning. When the cues were used as a warning signal in a shock-avoidance experiment, white noise became the stronger elicitor of fear. Conversely, when the cues were used as a safety signal, changes in illumination became the stronger inhibitors of fear. Neither cue had overt unconditioned effects on the rate of ongoing shock-avoidance behavior. Nonetheless, the possibility remains that subthreshold defensive reactions to the cues contributed to the observed stimulus-reinforcer interaction, as a Konorskian account would assert (see also Hendersen, 1973; Zielinski, 1965a, 1965b).

Finally, and this prediction is less clear, acccording to Konorski's account the smell of food should be an especially effective CS + for salivary conditioning because it initially elicits the hunger CR, which is a necessary condition for formation of the food CR, and hence of the occurrence of salivation.

To summarize the discussion, Konorski was aware of the likelihood of stimulus-reinforcer interactions in conditioning, and suggested accounts of several examples of such interactions (see Shettleworth, Chapter 15, for a related discussion). Konorski's 1967 account of the stimulus–reinforcer interactions discussed thus far would focus on inborn variations in the development of the preparatory CRs that support conditioning of the consummatory CRs. In other words, facilitation of subthreshold preparatory CRS that are motivationally compatible with the to-be-conditioned preparatory CRs would presumably account for the cases of very rapid conditioning; and facilitation of subthreshold preparatory CRs that are motivationally antagonistic to (or that cancel) the to-be-conditioned preparatory CRs would account for cases of especially slow conditioning (see Dickinson & Dearing, Chapter 8). Such an account is associative, but nonselective, for it does not assert that associations between certain antecedent and consequent events gain strength more rapidly over trials than do associations between other antecedents and consequents. This assertion is the essence of a selective, associative account, regardless of the postulation of differential initial responses to the CSs.

Thus, given his connectionist viewpoint, Konorski asked the right questions about stimulus-reinforcer interactions, and his discussion of various accounts of these interactions suggests that certain experimental procedures will play an important role in the analysis of any stimulus-reinforcer interaction. These include assessment of the unconditioned effects of the various CSs, as well as assessment of the necessity of pairings (or a positive contingency) of CS and US for the observed stimulus-reinforcer interaction. Concern with "control" procedures, which can be used to differentiate between associative and nonassociative effects, has been prominent in recent discussions of stimulus-reinforcer interactions (e.g., Bitterman, 1976; Mitchell, Scott, & Mitchell, 1977; Rescorla & Holland, 1976; Revusky, 1978). Control procedures are discussed in the next section of this chapter.

III.   SOME METHODOLOGICAL CONCERNS

Suppose that a stimulus-reinforcer interaction has been demonstrated in either of the experimental designs illustrated in Figs. 14.1 and 14.2. As Konorski's discussion of stimulus-reinforcer interactions suggested, there are several reasonable alternatives to an interpretation of any such interaction in terms of selective associations. To simplify the argument, let us consider the case in which only CS1 elicits the measured response when US1 is the reinforcer, whereas only CS2 elicits the response when US2 is the reinforcer. Using a label coined by Capretta (1961; see also Revusky & Garcia, 1970), CS1 is called relevant for US1, and CS2 is called relevant for US2.

The first question a connectionist would ask about a stimulus-reinforcer interaction is whether the responding elicited by the relevant stimuli might be nonassociative—that is, attributable to factors other than association by contiguity. Such an outcome would be called selective sensitization — in Konorski's example, selective sensitization of preparatory or drive CRs—if it could be demonstrated that the relevant stimuli initially elicited the to-be-conditioned responses in subthreshold form. Otherwise, such an outcome would be called pseudoconditioning.

What are the control procedures that would enable reasearchers to assess the possibility that the responses evoked by the relevant stimuli were attributable to selective sensitization or pseudoconditioning? One approach would be to include some of the standard controls for these nonassociative factors in one of the experimental designs described earlier. For example, to the four-group, single-cue design shown in Fig. 14.1 could be added four groups, each receiving one CS-US combination, but with CS and US unpaired. In some situations the zero-contingency or random-control procedure (Rescorla, 1967) would seem to be appropriate, whereas in others— for example, in response systems for which there may be little conditioning to background cues, and hence little inhibitory conditioning when the CS is simply negatively correlated with the US—a negative contingency between CS and US would be more appropriate. In cases where the relevant stimulus evokes a CR after as few as one or two trials, neither of these procedures may be as useful as a demonstration that the capacity of the relevant CS to evoke a CR depends on the CS-US interval (i.e., that there is an orderly delay of reinforcement gradient).

Recently, Rescorla and Holland (1976; see also Rescorla, 1978) have proposed a control procedure specifically designed to assess alternatives to selective association as an interpretation of stimulus-reinforcer interactions. In their design all subjects receive both USs, but groups differ in the US that is paired with the CS. For example, in the compound-cue design, Group 1 would receive the CS1-CS2 compound paired with US1, and US2 would occur between trials. Group 2 would receive the compound CS paired with US2, and US1 would occur between trials. If the addition (to the design in Fig. 14.2) of the second US during the intertrial period has no effect on the responses evoked by CS1 and CS2 in the test (i.e., the same stimulus-reinforcer interaction occurs), then it would seem that selective associations have been formed, because any selective, nonassociative effects of the two USs should be equivalent for the two groups. The design proposed by Rescorla and Holland also permits evaluation of the possibility that the occurrence of a given US determines which CS the organism will pay attention to on subsequent trials. Such a phenomenon would yield a stimulus-reinforcer interaction with the design shown in Fig. 14.2, but not with one in which all subjects receive both USs. Using the latter sort of design, Rescorla and his colleagues (Rescorla, 1978; Rescorla & Furrow, 1977) have shown that excitatory associations are formed more rapidly between physically similar stimuli that are paired than between physically dissimilar ones. This result is discussed in detail in the next section.

None of the aforementioned control procedures bears on the possibility that a given stimulus-reinforcer interaction results from nonselective associations superimposed on different initial responses to the two CSs, as in Konorski's discussion of conditioning with moderately intense and supramaximal CSs. In principle, this possibility can be easily assessed. However, it becomes technically difficult to observe the initial responses to the CSs if we concern ourselves with "concealed alliances," which would be manifested by differential elicitation of some neural precursor of the to-be-conditioned effector responses by the two CSs (see Kandel, 1976; Thompson, 1976).

One more methodological point must be made before several recent examples of stimulus-reinforcer interactions can be analyzed. Consider the case in which there is no evidence of conditioning to the less-relevant cue in some demonstration of a stimulus-reinforcer interaction. In such cases the distinction between learning and performance must be made, and it must be recognized that animals might fail to make some response during a CS even though they had learned to associate CS and US (e.g., Holland, 1977). Holland has suggested several generally useful procedures for determining whether some less-relevant CS has been associated with a US. For example, the less-relevant CS could be used as the US in a second-order conditioning procedure, or as the preconditioned "blocking" stimulus in Kamin's (1969) blocking procedure. If the less-relevant CS promotes second-order conditioning, or if prior conditioning to the less-relevant CS successfully blocks conditioning to another, added CS, then the less-relevant CS can be said to have been associated with the US. The critical question remains, however, whether this association was formed more slowly than the association between the relevant CS and the US.

Given these methodological considerations, several recent examples of stimulus-reinforcer interactions can be described and analyzed.

IV.   RECENT EXAMPLES OF STIMULUS-REINFORCER INTERACTIONS

In this section three stimulus-reinforcer interactions that have recently been observed are described. In each case, analysis focuses on the question, "Have selective associations been demonstrated, according to the criteria discussed in the preceding sections of this chapter?"

A. Stimulus-reinforcer Interactions in the Taste-Aversion Paradigm

A 1966 experiment by Garcia and Koelling was largely responsible for arousing interest in stimulus-reinforcer interactions and in the possibility of selective associations. Garcia and Koelling found that when a compound conditioned stimulus consisting of a gustatory stimulus and audiovisual stimulation was paired with visceral upset, only the gustatory stimulus subsequently strongly controlled passive avoidance of drinking. On the other hand, when the compound stimulus was paired with electric shock, only the audiovisual stimulation controlled avoidance of drinking.

In an extension of this research, Domjan and Wilson (1972) minimized differences in the reception of audiovisual and gustatory cues by rapidly infusing flavored solutions directly into the mouths of water-satiated rats, thereby virtually eliminating cues associated with ingestion. Domjan and Wilson extended the single-cue design illustrated in Fig. 14.1 to 6 groups. Three groups received 3 daily presentations of a 35-sec buzzer CS, and 3 other groups received equally long oral infusions of saccharin. The taste CS was terminated by the oral infusion of tap water. Immediately after presentation of the CSs, 1 group from each CS condition was given an intraperitoneal injection of the toxin lithium chloride (LiCl); another was given a brief electric shock and a sham injection; and a third was given only the sham injection. In order to equate contact with the two CSs, 90 to 150 min after each conditioning trial rats were exposed to the CS that they had not received on the conditioning trial. Beginning on the day after the last conditioning trial, all rats were given 2 preference tests. One 15-min, 2-bottle test pitted saccharin against tap water, and the other pitted tap water against tap water plus the sound of the buzzer contingent on licking.

Fig. 14.3 illustrates the results of these tests, expressed as groups mean

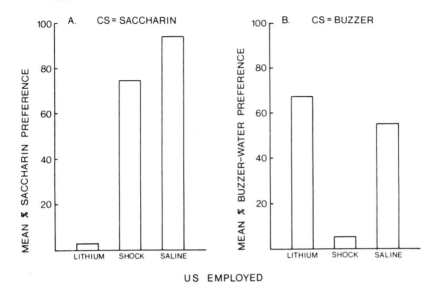

FIG. 14.3. Panel A illustrates saccharin preferences of rats exposed to saccharin, followed by lithium chloride injections, electric shock, or saline injections during conditioning. Panel B illustrates buzzer-water preferences of rats exposed to the buzzer, followed by lithium chloride injections, electric shock, or saline injections during conditioning. Redrawn from Domjan and Wilson, 1972.

preferences for the saccharin (panel A), and for tap water plus buzzer (panel B). Panel A shows that rats that had received saccharin-LiCl pairings showed a marked aversion to saccharin, relative to saline controls. On the other hand, rats that had received 3 pairings of the taste of saccharin and electric shock did not avoid saccharin. These groups did not differ in preference for the buzzer, which they had received 90 to 150 min after each conditioning trial (not shown in Fig. 14.3). Panel B indicates that the rats that had received pairings of the buzzer and electric shock drank at the tube that activated the buzzer much less than saline controls. On the other hand, rats that had received buzzer LiCl pairings produced the buzzer as much as controls. These groups did not differ in saccharin preference. The results of this experiment indicate that the sort of stimulus-reinforcer interaction observed by Garcia and Koelling does not depend on either the use of the compound-cue design or on differences in the reception of audiovisual and gustatory cues.

Garcia and Koelling offered two possible accounts of their effect. First, they suggested that common elements in the temporal-intensity patterns of stimulation may have facilitated cross-modal generalization from reinforcer to cue in two of the four groups, but not in the other two. Audiovisual stimulation and pain presumably had relatively sudden onsets and terminations, whereas taste and illness built up and diminished gradually. Thus there would be considerable stimulus generalization of the response to shock to auditory and visual cues, and of illness to gustatory cues, but little generalization in the other two combinations. This hypothesis, which implies that pairings of CSs and USs are not a necessary condition of the stimulus-reinforcer interaction, was not favored by Garcia and Koelling, who preferred a selective, associative interpretation.

Garcia and Koelling preferred to attribute their stimulus-reinforcer interaction to associations, maintaining that (1966) "natural selection may have favored mechanisms which associate gustatory and olfactory cues with internal discomfort since the chemical receptors sample the materials soon to be incorporated into the internal environment (p. 124)." This view was based on the notion of adaptive specializations in learning — namely, that event-specific associative mechanisms evolve to solve particular problems for a species. Rozin and Kalat (1971) also advanced such a view and expanded it to suggest that, across species, feeding-related cues, rather than taste cues per se, would be most readily associated with the consequences of ingestion. If this were the case, the particular stimulus-reinforcer interaction obtained by Garcia and Koelling (1966) would not be expected to occur with species that depend on cues other than taste in selecting foods. Wilcoxon, Dragoin, and Kral (1971), using quail, have obtained evidence supporting this claim (see Braveman, 1977, for a relevant review).

Domjan and Wilson (1972) also seemed to view their result and that of Garcia and Koelling as evidence for selective associations, as have Schwartz (1974) and Revusky and Garcia (1970). However, Bitterman (1975, 1976) has maintained that the experimental designs used by Garcia and Koelling and by Domjan and Wilson cannot demonstrate selective associations, because they lack controls for pseudoconditioning or sensitization. Certainly such a criticism of the basic designs diagrammed in Figs. 14.1 and 14.2 is valid. However, Domjan and Wilson used a more complex design; all rats received equal exposure to the two CSs, 1 occurring before the US, and the other 90 to 150 min after each conditioning trial. Thus there were groups that received unpaired presentations of saccharin and LiCl, and of buzzer and electric shock. These groups did not drink less saccharin and less buzzer water, respectively, than controls that received no US, suggesting that the aversions to a buzzer paired with shock and to a taste paired with LiCl reflect the formation of associations, rather than sensitization or pseudoconditioning (see also Garcia, McGowan, Ervin, & Koelling, 1968).

Additional evidence that LiCl-based taste aversions reflect the formation of associations between gustatory cues and "illness" can be found in experiments that present all groups with the same number of exposures to CS and US, but vary the temporal contiguity of CS and US. In such studies of the delay-of-reinforcement gradient, failures to obtain taste aversions with the longest CS-US interval are used as controls for aversions obtained with shorter intervals. Given a relatively long interval, say several days, between poisoning and the assessment of the aversion, any general effects of illness-induced neophobia or punishment of ingestion should be minimized. Moreover, the possibility of illness-induced sensitization should be equivalent for all groups. Consequently, the observation of progressively weaker aversions to the taste cue with increases in the CS-US interval would constitute strong evidence for the associative basis of conditioned taste aversions, even if one wanted to propose some form of peripheral reinstatement of the taste cue during a long CS-US interval and thus deny long-delay learning per se (see Bitterman, 1975, 1976; Revusky, 1978). Such an orderly delay gradient has been obtained in several studies of illness-based taste aversions (e.g., Barker & Smith, 1974; Garcia, Ervin, & Koelling, 1966; Revusky, 1968; Smith & Roll, 1967).

Finally, there is little reason to suppose that long-lasting, illness-based sensitization of responding to taste cues occurs, if neophobia is the response assumed to be sensitized. For example, Best and Batson (1977) have recently demonstrated, at least with lithium chloride, that the administration of un-signaled poison does not enhance the neophobic reaction to a novel taste presented several days later, if water intake has been allowed to recover to baseline levels on the intervening days. Furthermore, very strong conditioned taste aversions can be produced even when the neophobic reaction to the taste has been eliminated by means of frequent exposures before conditioning (e.g., Elkins, 1974).

Thus, illness-based taste aversions do reflect associations between gustatory cues and some aspect of the reaction to LiCl. Equally, the aversion to audiovisual cues paired with electric shock is an associative effect (Rescorla, 1968). There is every reason to believe that these processes are still associative when taste-poison and tone-shock pairings are imbedded in one of the designs that have been used to demonstrate stimulus-reinforcer interactions. The question remains whether these associative effects are *selective*, in the sense that taste-illness and tone-shock associations grow more rapidly (given pairings) than taste-shock and tone-illness associations.

This question has several facets, which have been described previously. First, none of the evidence from the control conditions that have just been described rules out the possibility, noted by Rescorla and Holland (1976), that the first US presented determines which stimulus modality will be most strongly attended to on subsequent trials, and hence most strongly associated with whatever US is presented on those trials. The design proposed by Rescorla and Holland, in which all groups receive both USs, would permit evaluation of this possibility.

A second possibility is that a nonselective associative mechanism produces rapid acquisition of aversion to tastes paired with illness and tones paired with shock (and not in the other two CS-US pairs) because tastes (but not tones) initially elicit, in subthreshold form, the same aversion reaction evoked by LiCl, whereas tones (but not tastes) initially elicit the same aversion reaction initially evoked by electric shock. Thus, as in Konorski's discussion of defensive conditioning to moderately intense and supramaximal stimuli, even though associations might grow at the same rate in all CS-US pairs, the CSs that began the experiment "closer to threshold" for particular effector CRs would nonetheless evoke those CRs first. The neural responses to LiCl and electric shock that mediate their aversiveness are not well understood. Consequently we do not know where to look for the different initial neural responses to the gustatory and audiovisual CSs. Such responses should form the basis of stimulus-reinforcer interactions like those observed by Garcia and Koelling, according to the sort of nonselective, associative account discussed by Konorski. Thus such an account remains a viable alternative to a selective, associative interpretation.

The final question to be asked about the stimulus-reinforcer interaction shown in Fig. 14.3 concerns the lack of responding to the less-relevant stimuli. Do rats fail to suppress drinking in the presence of those stimuli because they were not associated with the USs, or because the association was somehow not manifested in performance? Because we already know that a variety of exteroceptive cues can be associated with illness (see Best, Best, & Henggeler, 1977, for a recent review of this literature), and that taste cues can be associated with electric shock (Best et al., 1977; Krane & Wagner, 1975), the question becomes, "Were the less-relevant stimuli associated with the USs in the demonstrations of stimuli-reinforcer interactions, and, if so, how rapidly were those associations formed, relative to the associations between the relevant CSs and USs?" As noted in the previous section, Holland (1977) has suggested several indirect procedures for assessing whether an association has been formed (e.g., between buzzer and illness). Whatever procedure is used must have the additional feature of assessing the conditioned responses to tastes and audiovisual cues by means of the same dependent variable, because the relative rate of formation of associations between these cues and various USs, rather than the possibility of association between less-relevant CSs and USs, is at issue.

In any case, the fact that strong taste aversions can be acquired as a result of one or two pairings of taste with illness leaves little room for faster acquisiton of associations between, for example, a buzzer and illness.

To sum up the discussion of stimulus-reinforcer interactions in the taste-aversion paradigm, it seems clear that the responses conditioned to the relevant stimuli do reflect the formation of associations. Whether the stimulus-reinforcer interaction reflects differences in the rate of growth of associations within relevant CS-US pairs and less-relevant pairs, or differences in the initial tendencies of various CSs to evoke the to-be-conditioned responses, or both of these phenomena, is unclear. A similar analysis is now applied to a second area in which stimulus-reinforcer interactions recently have been demonstrated.

B. CS-US Similarity

In the last section, we noted that Garcia and Koelling (1966) proposed, but did not pursue, an account of their stimulus-reinforcer interaction based on the notion of cross-modal stimulus generalization from USs to CSs. Testa (1974) has recently extended this account, arguing that associations based on the similarity of the temporal-intensity patterns of stimulation produced by CS and US are largely responsible for stimulus-reinforcer interactions (see also Testa & Ternes, 1977). This statement can be interpreted in two ways. First, it could be asserted that similarity of CS and US should affect responding to the CS even in the absence of temporal contiguity between CS and US, in other words, that pseudoconditioning is association on the basis of similarity (see Wickens & Wickens, 1942). As we have seen, such an account of the stimulus-reinforcer interaction observed by Domjan and Wilson is incorrect.

Alternatively, Testa maintained that the similarity of CS and US might affect the rate of formation of an association between those events, given that they were paired. Testa (1975) provided data compatible with this assertion. In a single-cue design like that shown in Fig. 14.1, the similarity of the spatial loci and temporal-intensity patterns of visual CSs and air-blast USs significantly affected the rate of acquisition of conditioned suppression in rats when CSs and USs were paired. Acquisition was faster when CS and US were similar than when they were dissimilar. As Testa noted, the lack of control groups in his experiment leaves open the possibility that the outcomes are attributable to pseudoconditioning or stimulus generalization, rather than to modulation of the rate of growth of contiguity-based associations between CSs and USs by the similarity of those events. However, recent experiments by Rescorla and Furrow (1977) are not subject to such interpretations. These studies examined the effects of similarity in the context of second-order conditioning, thereby allowing the experimenters considerable flexibility in selecting events. In two experiments with rats, similar events were defined as those in the same sensory modality, either auditory or visual, whereas in a study with pigeons all stimuli were visual, and similar events were defined as those on the same dimension, either wavelength or linetilt.

All three experiments used variations of the experimental design proposed by Rescorla and Holland (1976), in which all subjects received both USs —in this case, both first-order CSs—during the second-order conditioning phase. The various groups thus differed in terms of which first-order CS was paired with a second-order CS, and which first-order CS was presented alone. (See Fig. 14.4 for a representative experimental design.) Some subjects received pairings of similar events, whereas others received pairings of dissimilar events. Whether the response measured was conditioned suppression of food-reinforced behavior or conditioned approach to a food

PHASE

FIRST-ORDER   SECOND-ORDER

GROUP 1. $T_1 \longrightarrow US, L_1 \longrightarrow US, T_2-$ $T_2 \longrightarrow T_1, L_1-$

2. $T_1 \longrightarrow US, L_1 \longrightarrow US, T_2-$ $T_2 \longrightarrow L_1, T_1-$

3. $T_1 \longrightarrow US, L_1 \longrightarrow US, L_2-$ $L_2 \longrightarrow T_1, L_1-$

4. $T_1 \longrightarrow US, L_1 \longrightarrow US, L_2-$ $L_2 \longrightarrow L_1, T_1-$

FIG. 14.4.   Design of an experiment to demonstrate the effects of stimulus similarity upon the acquisition of a second-order conditioned response. During first-order conditioning, auditory and visual CSs (T_1, L_1) are separately paired with the US, but the stimulus that is to be used as the second-order CS (either T_2 or L_2) is not reinforced. In the second-order conditioning phase, groups 1 and 4 receive pairings of similar second- and first-order stimuli (T_2-T_1 or L_2-L_1) whereas groups 2 and 3 receive pairings of dissimilar events (T_2-L_1, L_2-T_1). All groups receive the other first-order CS between trials, so that all receive equal exposure to both first-order CSs during the second-order conditioning phase. Redrawn from Rescorla and Furrow, 1977.

magazine in rats, or conditioned pecking in pigeons, it was acquired more rapidly when the first- and second-order CSs were similar than when they were dissimilar. As was suggested earlier, the experimental design used in these studies rules out the possibility that the stimulus-reinforcer interactions are attributable to nonassociative effects, such as selective sensitization. It also rules out the possibility that presentation of a particular first-order CS simply increases attention to similar events, so that there would be rapid acquisition of conditioned responding to those events regardless of which reinforcer followed them. Such a notion would predict equal conditioned responding in groups given similar and dissimilar CS-US pairs. Furthermore, because the response to the two first-order CSs (e.g., green and horizontal line) was the same (e.g., pecking a key) at the start of second-order conditioning, the more rapid conditioning with similar first- and second-order CSs than with dissimilar ones would not follow from differences in the initial probability of evocation of a subthreshold form of the to-be-conditioned response by the two CSs. Such a mechanism would result in better second-order conditioning with one second-order CS than with the other, regardless of the first-order CS used.

Thus, these experiments demonstrate that the rate of formation of a contiguity-based association between CS and US can be a function of the similarity of those two events—that is, that there can be selective associations in the sense used throughout this paper. Rescorla and Furrow (1977) have noted that little is understood about the mechanism(s) of such selective associations—for example, whether similarity affects what is associated, or only the rate of formation of associations.

C. Attention in the Pigeon: Differential Effects of Food and Shock

In the course of developing procedures for instrumental avoidance training of pigeons, Foree and I observed that the auditory element of an audiovisual compound discriminative stimulus gained stronger control of the shock-avoidance response than did the visual stimulus. This outcome was discrepant with "common knowledge" that pigeons were primarily visual animals, and the discrepancy led to an experiment designed to ascertain whether the relative dominance of visual and auditory cues depended on the reinforcer (Foree & LoLordo, 1973).

AVOIDANCE

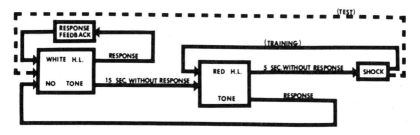

APPETITIVE

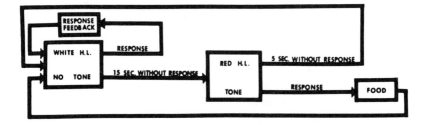

FIG. 14.5. Diagramatic representations of the avoidance (top) and appetitive training procedures used by Foree and LoLordo (1973). Note the similar consequences of responding in the presence of "white houselight-no tone" in the two procedures. Also note the different consequences of failing to respond in the presence of "red houselight-tone" during test versus training in the avoidance condition. (Copyright (1973) by the American Psychological Association. Reprinted by permission.)

Groups of food-deprived pigeons were trained to depress a foot treadle in the presence of a compound stimulus consisting of tone and red houselight to avoid electric shock or obtain grain. Responding in the absence of the compound stimulus postponed its next appearance. These procedures are diagramed in Fig. 14.5. Briefly, the final training schedule included an intertrial period of at least 15 sec, during which white houselights and no tone were on. Each response on the foot treadle in the intertrial period extended it for 15 sec from the time of the response, at the end of which time a 5-sec trial period began. During the trial red houselights and a 440-Hz tone were on. In the appetitive condition the first response produced 5-sec access to grain. During reinforcement the tone and all houselights were off, and the grain magazine was illuminated. If no treadle response occurred within 5 sec, the intertrial stimuli were reinstated. In the avoidance condition, a response during the trial reinstated the intertrial conditions. Failure to respond within 5 sec after onset of the compound stimulus resulted in a brief shock. The trial stimuli remained on, and a brief shock was presented every 5 sec, until a response was made.

When the birds were responding on at least 75% of the compound trials, but responding infrequently between trials, the degree to which the compound and each element controlled treadle pressing was determined. Twenty test trials each of the compound training stimulus, the tone alone, and the red houselights alone were presented in random order. Responses to the compound or to either element were reinforced. Thus, this experimental design was like that diagramed in Fig. 14.2. The right-hand panel of Fig. 14.6 shows that, in the appetitive test, the compound and the red light exerted strong control over treadle pressing, whereas the tone exerted very little. On the other hand, in the test that followed avoidance training the compound and the tone controlled much more responding than the light.

The feedback for the treadle-press response varied between reinforcement conditions in the first study. Food-reinforced responses resulted in blackout of the chamber lights and elevation of an illuminated feeder, whereas the avoidance response merely reinstated the intertrial conditions of white houselight and no tone. To more nearly equate feedback in the two conditions, LoLordo and Furrow (1976) replicated the first study, but in both conditions simply raised a darkened grain magazine for 5 sec whenever the treadle was pressed during a trial. The grain magazine was filled in the appetitive condition, but empty in the avoidance condition. Intertrial conditions were reinstated when the grain magazine was lowered.

The left-hand panel of Fig. 14.6 shows that the stimulus-reinforcer interaction was just as strong as in the first experiment. Thus, the greater visual feedback in the appetitive condition of the first study than in the avoidance condition was not responsible for the observed stimulus-reinforcer interaction.

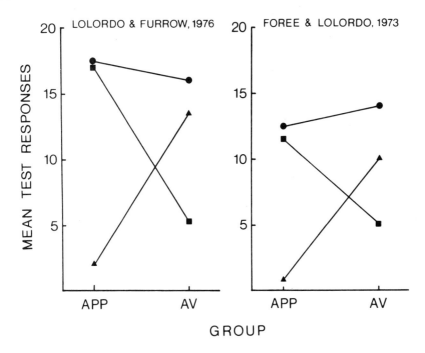

FIG. 14.6. Mean test responses to the compound (filled circles), the red houselight (filled squares), and the tone (filled triangles) for groups of pigeons given appetitive (App) or avoidance (AV) training in studies by Foree and LoLordo (1973, right panel) and LoLordo and Furrow (1976, left panel). (Copyright (1973) by the American Psychological Association. Reprinted by permission.)

Jacobs, Foree, and I have obtained the same stimulus-reinforcer interaction in the single-cue design. Briefly, when food was the reinforcer, birds that received the red light discriminative stimulus acquired the treadle-press response in fewer sessions than birds that received the auditory stimulus. The reverse relationship held when shock avoidance was the reinforcer. Most of the subjects in all 4 groups did acquire the discriminated treadle-press response within 20 sessions. The results indicate that the stimulus-reinforcer interaction obtained by Foree and LoLordo does not depend on some special property of compound-cue designs, although in the appetitive condition the relative extent of visual dominance was enhanced in the compound-cue design.

Until recently we had not performed any experiments to ascertain that the stimulus–reinforcer interaction we observed was an associative effect, largely because alternative, nonassociative accounts seemed implausible in the case of discriminated, instrumental behavior. In any case, Shapiro, Jacobs, and I have recently extended our research with pigeons to the classical conditioning paradigm, in an attempt to control for sensitization and other nonassociative effects.

Food-deprived pigeons implanted with stainless steel electrodes attached to the pubis bones received daily 30-min sessions. The basic experimental design was like that in Fig. 14.2. Two groups of pigeons received pairings of an audiovisual compound CS with a US. For 1 group the US was food, whereas the second group received electric shock. Two additional groups of birds, run subsequently, received uncorrelated presentations of the CS and either food or shock.

For the experimental groups each trial consisted of a 5-sec presentation of the compound CS — onset of a red houselight and a 440-Hz tone — followed immediately by the US—either a 3-sec access to mixed grain or a brief electric shock. Trials were presented at 20-sec intervals, roughly the same density used in the study by Foree and LoLordo. Birds in the control groups received the same number of CSs and USs per session as the experimental groups, but the 2 events were independent. For some control birds the CS occurred at fixed times, for others the US occurred at fixed times, and for still others both CS and US were presented aperiodically. Three independent observers rated the birds throughout the sessions, and characterized any changes in behavior that occurred during the CSs.

Pigeons in the experimental groups acquired a distinctive response to the CS by the fifth day of conditioning. This response was stereotyped for all birds within a given reinforcement condition. During the CS all food birds typically pecked either around the food hopper or on the wall above the hopper. Two birds began pecking only when the CS was presented, but 1 occasionally pecked between trials. In all cases the rate of pecking increased during the CS. Birds in the shock group typically pranced vigorously and made elaborate head-bobbing movements, in response to the compound CS. Because all birds made a distinctive CR on at least 75% of the trials on day 5, the tendency of the separate elements to evoke the CR was assessed on the following day. In the test, birds received presentations of red light, tone, and no signal (blank trials) interspersed among presentations of the compound CS. Reinforcement occurred on all but the blank trials, which were included to assess control by time-correlated stimuli.

Fig. 14.7 illustrates the percentage of trials of each kind on which a CR occurred for each of three birds in the two experimental groups. The three food birds responded on nearly all presentations of the compound and the red light, but responsed infrequently to the tone or on blank trials. Thus, the strong visual dominance observed by Foree and LoLordo (1973) when birds were required to press a treadle to obtain food also occurred when no particular instrumental response was required. The three shock birds responded most frequently on presentations of the compound, somewhat less frequently on presentations of the tone, and even less frequently on

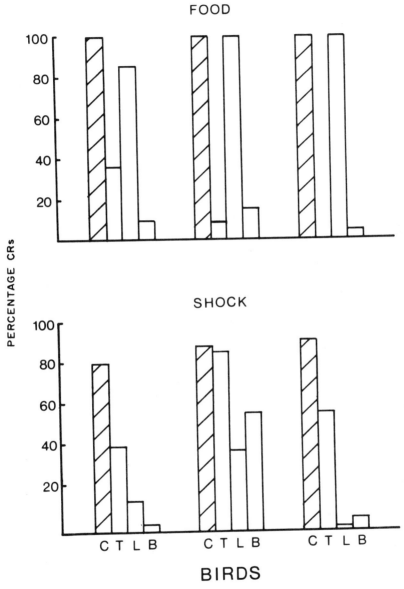

FIG. 14.7. Percentage of conditioned responses to a compound of the tone and red houselights (C), to the tone (T), to the red houselight (L), and on blank trials (B) following classical conditioning in which the compound had been paired with a US. The upper panel illustrates the percentage of pecking CRs of three birds that had received pairings of the compound CS and food. The lower panel illustrates the percentage of "prancing" CRs of three pigeons that had received pairings of the compound with electric shock.

presentations of the red light or on blank trials (see Shettleworth, 1972b, for a related result). Apparently the auditory dominance of treadle pressing to avoid shock, as observed by Foree and LoLordo (1973), depends on signal-shock pairings, not on the shock-avoidance contingency.

None of the birds in the control groups exhibited any differential responding to the CSs. Although individual birds in the appetitive groups (N = 3 in each subgroup) pecked at the wall containing the hopper, or at the floor, ceiling, or in the air, intermittently during the sessions, such pecking was no more likely during the CS than at other times. Similarly, pigeons in the aversive conditioning group tended to prance intermittently, but no more frequently during the CS than between trials.

The data from both the experimental and control groups have been replicated in a second experiment in which trials were widely spaced, rather than massed. The occurrence of a stimulus-reinforcer interaction in the paired groups of these two experiments, along with the virtual absence of responding in the control groups, suggests that the stimulus-reinforcer interaction has an associative basis. As in the earlier discussion of stimulus-reinforcer interactions in the taste-aversion paradigm, in this case it is possible that associations were formed between the red light element and electric shock and between the auditory element and food, even though the measures of pecking and prancing presented in Fig. 14.7 yielded no strong evidence for such associations, and though no consistent changes in skeletal-motor behavior were observed during test presentations of these less-relevant stimuli. The development of discriminative performance in all groups when the single-cue design was applied to the treadle-press procedure also suggests that associations can be formed between red light and shock, and between tone and food. However, the stimulus-reinforcer interaction observed in that very experiment suggests that the association between red light and shock is weaker than the association between tone and shock, and that the association between tone and food is weaker than the association between red light and food.

Although the stimulus–reinforcer interaction observed in our laboratory reflects differences in the strengths of associations between various CS–US pairs, it is unclear whether selective association has occurred. There are two plausible alternative accounts, which have been presented during the discussion of earlier examples. The first alternative maintains that the presentation of food increases attention to visual cues, so that those cues are more strongly associated with whatever US follows them. A parallel argument could be made for shock and auditory cues. This account could be tested easily if the experimental design proposed by Rescorla and Holland (1976), in which each subject receives both USs during a session, were applied to the classical conditioning procedure that yielded the data of Fig. 14.7. This has not been done, but Foree and LoLordo (1975) have obtained some data that bear on account that has just been proposed.

Six additional pigeons were trained to press a treadle in the presence of the compound disciminative stimulus for food reinforcement. Once they were responding and being reinforced on 75% of the trials, a punishment contingency was added. Each intertrial response produced a brief electric shock in addition to postponing the onset of the next trial. Shock intensity was gradually increased until the birds showed signs of being shocked, but continued to press the treadle during the compound-trial stimulus. In the test, all 6 pigeons pressed the treadle on most presentations of the compound and the red light, but responded infrequently to tone. The presence of shock neither attenuated visual control, nor increased auditory control, of the treadle-press response, relative to the outcome of the appetitive procedure used by Foree and LoLordo (1973). This result suggests that the auditory dominance observed when the compound CS was paired with shock did not simply result from shock-produced increases in attention to auditory cues, which would have resulted in auditory control of food-reinforced treadle pressing.

On the other hand, data obtained by Foree (1974) are compatible with the assertion that presentation of food simply increases attention to visual cues. Pigeons were trained to peck a white key for food; then the discriminated, treadle-press avoidance schedule was added, so that the avoidance and appetitive schedules were in effect concurrently. After the pigeons learned to depress the treadle to avoid shock on at least 75% of the presentations of the compound stimulus, they were tested as in the earlier experiments. Most of the pigeons depressed the treadle on an equal, and large, proportion of both auditory trials and visual trials. This outcome is compatible with the assertion that the presence of food in the situation increases the pigeon's attention to red light, resulting in a stronger association between red light and shock than would have occurred otherwise. Clearly, further tests must be made of the hypothesis that presentation of a US biases the subject's attention toward certain modalities, and that this bias accounts for the stimulus-reinforcer interaction observed by Foree and LoLordo (1973).

Finally, as in the taste-aversion example discussed earlier, it is not clear whether the stimulus-reinforcer interaction being discussed stems from a greater initial tendency of tone than red light to evoke some subthreshold aversive reaction, and/or a greater initial tendency of red light than tone to elicit some subthreshold appetitive reaction. If such were the case, as in Konorski's discussion of defensive versus appetitive conditioning to moderately intense versus supramaximal CSs, the observed stimulus-reinforcer interaction would follow, even if the (acquired) associative strength superimposed on this base grew nonselectively. In a sense this Konorskian account could be construed as explaining one case of the effects of CS-US similarity—similarity on an affective dimension—upon association by contiguity.

The unconditioned effects of the red houselight and tone CSs have not been thoroughly assessed in our laboratory; thus, the possibility that the observed stimulus-reinforcer interaction depends on differential initial reactions to tone and red light cannot be ruled out.

To sum up, the stimulus-reinforcer interaction observed by Foree and LoLordo (1973) and later replicated in a classicial conditioning paradigm is an associative effect. Whether this associative effect is selective — that is, reflects faster growth of a tone-shock than of a red light-shock association, and faster growth of a red light-food than of a tone-food association—is yet to be determined.

V. IMPLICATIONS OF SELECTIVE ASSOCIATIONS

The study of stimulus-reinforcer interactions is just beginning. Analysis of such interactions has suggested a variety of associative interpretations, as well as some nonassociative ones. The three sample stimulus-reinforcer interactions described in Section IV all appear to be associative effects, although only Rescorla and Furrow's (1977) experiments on similarity conclusively demonstrate selective associations — or, dependence of the growth rate of an association between a CS and a US on the particular CS-US combination.

Little has been said about the implications of selective associations for the laws of learning, for the very good reason that it is not clear what those implications are. Possible implications may be considered in terms of the question, "Do relevant stimuli obey special laws?" In the taste-aversion literature there are many experiments that bear on this question, and they have been recently reviewed by Revusky (1977). He concluded that there is nothing in the literature on conditioned taste aversions that demands the postulation of more than parametric differences between taste-illness associations and those formed in more conventional classical conditioning paradigms. Even long-delay learning poses no particular problems, for the form of the delay gradient is the same in taste-aversion learning and in other paradigms (e.g., Revusky, 1968).

The same question asked about associations between taste and illness can be asked about associations between red houselight and food, or between tone and shock, and in the procedure used by Foree and LoLordo (1973). LoLordo, Jacobs, and Foree (1975) have performed an experiment that asks whether a red houselight paired with the availability of food is affected like an arbitrarily chosen stimulus in a standard paradigm for the study of stimulus control, the blocking paradigm.

Blocking refers to the prevention or attentuation of conditioning to a

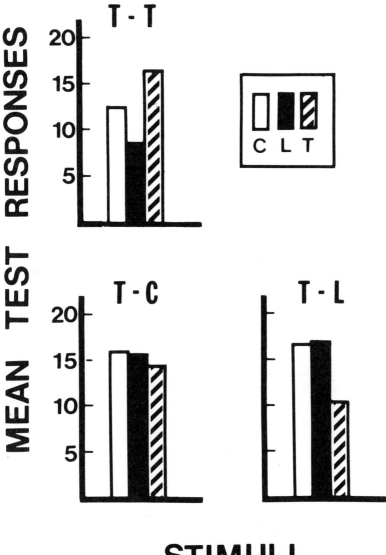

FIG. 14.8. Mean number of test responses to the compound, the tone alone, and the red light alone, by pigeons that had been trained to depress a treadle in the presence of a tone for food reinforcement. Following this training, red houselight replaced tone for groups T-L, whereas red houselight was added to tone for group T-C. Group T-T received additional training with the tone.

stimulus, B, when it is added to an already conditioned stimulus, A, with reinforcement unchanged. Blocking has been demonstrated in a wide variety of procedures, including both classical (e.g., Kamin, 1969) and instrumental conditioning (e.g., Miles, 1970). We asked the question whether prior conditioning to the less-relevant tone would block conditioning to an added red houselight in the appetitive procedure used by Foree and LoLordo. Specifically, a group of pigeons was trained to press the treadle for food in the presence of a tone. After these birds responded on at least 75% of the trials in a session, they were given 5 days of overtraining. Then the red light was added to the tone, and the pigeons received 5 additional days of training with the compound discriminative stimulus. Finally, responding to the compound and to each of the elements was assessed in a test session. Fig. 14.8 reveals that the birds in the experimental group (Group T-C) responded as frequently to the red light in the test as did birds that had received only the red light in the second phase of the study (Group T-L). Thus, there was no evidence of blocking. Furthermore, because a group (Group T-T) that received only the auditory stimulus during both phases of training responded significantly less frequently to red light than did Groups T-C and T-L, test responding to light in the latter groups was in large part acquired during the second phase of the study, although stimulus generalization of the effects of reinforced responding to tone, or acquisition of responding to light during the test, or both, occurred as well.

Parallel results were obtained in the avoidance training situation, where tone is the relevant stimulus. Prior asymptotic conditioning of avoidance responding to the red light stimulus failed to block conditioning to the added tone. These results suggested that blocking might not apply to relevant stimuli, at least in our situation. However, perhaps some stimuli other than relevant stimuli may be impervious to blocking. Hall, Mackintosh, Goodall, and dal Martello (1977) found that prior conditioning to a weak tone CS only slightly attenuated conditioning to a bright light, which was added to the tone in a CER procedure with rats (see also Feldman, 1975). The authors attributed these results to the much greater salience of the light than the tone. Extending these results, perhaps there would have been a complete failure of blocking had the added light CS been even brighter, which would suggest that relevant stimuli are not the only ones that can acquire control when used as the added stimulus in a blocking paradigm. Thus, there is no firm evidence that the tone-shock or red light-food associations in our experiments obey special laws.

Perhaps the product of learning is different for selective associations and other associations. Garcia, Hankins, and Rusiniak (1974) have suggested that, whereas tone-shock pairings cause the rat to treat the tone as a signal for shock, taste-illness pairings make the taste unpleasant, as can be inferred when (1974) "the rat rubs its chin against the floor as in a sign of disgust and grooms itself vigorously (p. 828)." It is not clear that this difference is essential, because rejection of a taste that has been paired with illness is a form of withdrawal from a localized cue, and rats and pigeons have been shown to withdraw from localized visual signals for shock (Karpicke, Christoph, Peterson, & Hearst, 1977; see also Biederman, D'Amato, & Keller, 1964), as well as from signals for the absence of food (Wasserman, Franklin, & Hearst, 1974). Thus, if a taste paired with illness has become unpleasant, it seems reasonable to call a localized visual cue paired with shock "unpleasant" as well. Hence, the implications of selective associations for the product of learning remain unclear. (See also Rescorla & Furrow, 1977.)

One concern of researchers in learning has been that the study of "constraints," including stimulus-reinforcer interactions, would result only in the accumulation of isolated entries in a "catalogue," with each entry describing some exceptional case, about which predictions that failed to take into account the particular nature of the events would be incorrect. In a sense, research on the stimulus-reinforcer interactions is still at this stage of development. However, signs of hope can be found in the broadened scope of learning theorists' vision that has accompanied the study of these exceptional cases. A variety of provocative, potentially general accounts of stimulus-reinforcer interactions have been proposed, and in the course of evaluating these accounts our understanding of the mechanisms of conditioning will almost certainly be enhanced.

ACKNOWLEDGMENTS

The author thanks W. J. Jacobs, A. Randich, A. L. Riley, K. Shapiro, W. Stanley, P. Urcuioli, and J. Willner for their many helpful criticisms of an earlier draft of this chapter. Research reported here was supported by grants from the National Science Foundation and the National Research Council of Canada. This chapter was largely prepared while the author was on sabbatical leave at the University of Colorado, Boulder.

REFERENCES

Barker, L. M., & Smith, J. C. A comparison of taste aversions induced by radiation and lithium chloride in CS-US and US-CS paradigms. *Journal of Comparative and Physiological Psychology,* 1974, *87,* 644-654.

Best, M. R., & Batson, J. D. Enhancing the expression of flavor neophobia: Some effects of the illness-ingestion contingency. *Journal of Experimental Psychology: Animal Behavior Processes,* 1977, *3,* 132-143.

Best, P. J., Best, M. R., & Henggeler, S. The contribution of environmental non-ingestive cues in conditioning with aversive internal consequences. In L. M. Barker, M. R. Best & M. Domjan (Eds.), *Learning mechanisms in food selection.* Waco, Texas: Baylor University Press, 1977.

Biederman, G. B., D'Amato, M. R., & Keller, D. M. Facilitation of discriminated avoidance learning by dissociation of CS and manipulandum. *Psychonomic Science,* 1964, *1,* 229-230.

Bitterman, M. E. The comparative analysis of learning. *Science,* 1975, *188,* 699-709.

Bitterman, M. E. Flavor aversion studies. *Science,* 1976, *192,* 266-267.

Bolles, R. C. Species-specific defense reactions and avoidance learning. *Psychological Review,* 1970, *77,* 32-48.

Bolles, R. C. Reinforcement, expectancy, and learning. *Psychological Review,* 1972, *79,* 394-409.

Braveman, N. S. Visually guided avoidance of poisonous foods in mammals. In L. M. Barker, M. R. Best, & M. Domjan (Eds.), *Learning mechanisms in food selection.* Waco, Texas: Baylor University Press, 1977.

Capretta, P. An experimental modification of food preferences in chickens. *Journal of Comparative and Physiological Psychology,* 1961, *54,* 238-242.

Cytovich, I. S. In Konorski, J. *Conditioned reflexes and neuron organization.* Cambridge: Cambridge University Press, 1948.

Domjan, M., & Wilson, N. E. Specificity of cue to consequence in aversion learning in rats. *Psychonomic Science,* 1972, *26,* 143-145.

Elkins, R. L. Conditioned flavor aversions to familiar tap water in rats: An adjustment with implications for aversion therapy treatment of alcoholism and obesity. *Journal of Abnormal Psychology,* 1974, *83,* 411-417.

Feldman, J. M. Blocking as a function of added cue intensity. *Animal Learning and Behavior,* 1975, *3,* 98-102.

Foree, D. D. *Stimulus-reinforcer interactions in the pigeon.* Unplublished doctoral dissertation, University of North Carolina, Chapel Hill, 1974.

Foree, D. D., & LoLordo, V. M. Attention in the pigeon: The differential effects of food-getting vs. shock avoidance procedures. *Journal of Comparative and Physiological Psychology,* 1973, *85,* 551-558.

Foree, D. D. & LoLordo, V. M. Stimulus-reinforcer interactions in the pigeon: The role of electric shock and the avoidance contingency. *Journal of Experimental Psychology: Animal Behavior Processes,* 1975, *1,* 39-46.

Garcia, J., Ervin, F. R., & Koelling, R. A. Learning with prolonged delay of reinforcement. *Psychonomic Science,* 1966, *5,* 121-122.

Garcia, J., Hankins, W. G., & Rusiniak, K. W. Behavioral regulation of the *milieu interne* in man and rat. *Science,* 1974, *185,* 823-831.

Garcia, J., & Koelling, R. A. Relation of cue to consquence in avoidance learning. *Psychonomic Science,* 1966, *4,* 123-124.

Garcia, J., McGowan, B. K., Ervin, F. R., & Koelling, R. A. Cues: Their relative effectiveness as a function of the reinforcer. *Science,* 1968, *160,* 794-795.

Garcia, J., McGowan, B. K., & Green, K. F. Biological constraints on conditioning. In A. H. Black & W. F. Prokasy (Eds.), *Classical conditioning II: Current research and theory.* New York: Appleton-Century-Crofts, 1972.

Hall, G., Mackintosh, N. J., Goodall, G., & dal Martello, M. Loss of control by a less valid or less salient stimulus compounded with a better predictor of reinforcement. *Learning and Motivation,* 1977, *8,* 145-158.

Henderson, R. W. Conditioned and unconditioned fear inhibition in rats. *Journal of Comparative and Physiological Psychology,* 1973, *84,* 554-561.

Hinde, R. A., & Stevenson-Hinde, J. *Constraints on learning.* London: Academic Press, 1973.

Holland, P. C. Conditioned stimulus as a determinant of the form of the Pavlovian condition-ed response. *Journal of Experimental Psychology: Animal Behavior Processes.* 1977, *3,* 77-104.

Jacobs, W. J., & LoLordo, V. M. The sensory basis of avoidance responding in the rat. *Learning and Motivation,* 1977, *8,* 448-466.

Kamin, L. J. Predictability, surprise, attention, and conditioning. In B. A. Campbell & R. M. Church (Eds.), *Punishment and aversive behavior.* New York: Appleton-Century-Crofts, 1969.

Kandel, E. R. *Cellular basis of behavior.* San Francisco: W. H. Freeman, 1976.

Karpicke, J., Christoph, G., Peterson, G., & Hearst, E. Signal location and positive and neg-ative conditioned suppression in the rat. *Journal of Experimental Psychology: Animal Behavior Processes,* 1977, *3,* 105-118.

Konorski, J. *Conditioned reflexes and neuron organization.* Cambridge: Cambridge University Press, 1948.

Konorski, J. *Integrative activity of the brain.* Chicago: University of Chicago Press, 1967.

Krane, R. V., & Wagner, A. R. Taste aversion learning with delayed shock US: Implications for the "generality of the laws of learning." *Journal of Comparative and Physiological Psychology,* 1975, *88,* 882-889.

LoLordo, V. M., & Furrow, D. R. Control by the auditory or the visual element of a com-pound discriminative stimulus: Effects of feedback. *Journal of the Experimental Analysis of Behavior,* 1976, *25,* 251-256.

LoLordo, V. M., Jacobs, W. J., & Foree, D. D. *Failure to block control by a relevant stimulus.* Paper presented at the 1975 meeting of the Psychonomic Society, Denver, Colorado.

Miles, C. G. Blocking the acquisition of control by an auditory stimulus with pretraining on brightness. *Psychonomic Science,* 1970, *19,* 133-134.

Mitchell, D., Scott, D. W. & Mitchell, L. K. Attenuated and enhanced neophobia in the taste-aversion "delay of reinforcement" effect. *Animal Learning and Behavior,* 1977, *5,* 99-102.

Mowrer, O. H. On the dual nature of learning — A reinterpretation of "conditioning" and "problem-solving." *Harvard Educational Review,* 1947, *17,* 102-148.

Rescorla, R. A. Pavlovian conditioning and its proper control procedures. *Psychological Review,* 1967, *74,* 71-80.

Rescorla, R. A. Probability of shock in the presence and absence of CS in fear conditioning. *Journal of Comparative and Physiological Psychology,* 1968, *66,* 1-5.

Rescorla, R. A. Some implications of a congitive perspective on Pavlovian conditioning. In S. H. Hulse, H. Fowler, & W. K. Honig (Eds.) *Cognitive processes in animal behavior.* Hillsdale, N.J.: Lawrence Erlbaum Associates, 1978.

Rescorla, R. A., & Furrow, D. R. Stimulus similarity as a determinant of Pavlovian condition-ing. *Journal of Experimental Psychology: Animal Behavior Processes,* 1977, *3,* 203-215.

Rescorla, R. A., & Holland, P. C. Some behavioral approaches to the study of learning. In M. R. Rosenzweig & E. L. Bennett (Eds.), *Neural mechanisms of learning and memory.* Cambridge, Mass.: MIT Press, 1976.

Rescorla, R. A., & Wagner, A. R. A theory of Pavlovian conditioning: Variations in the effec-tiveness of reinforcement and non-reinforcement. In A. H. Black & W. F. Prokasy (Eds.), *Classical conditioning II: Current research and theory.* New York: Appleton-Century-Crofts, 1972.

Revusky, S. Aversion to sucrose produced by contingent X-irradiation: Temporal and dosage parameters. *Journal of Comparative and Physiological Psychology,* 1968, *65,* 17-22.

Revusky, S. Learning as a general process with an emphasis on data from feeding experiments. In N. W. Milgram, L. Krames, & T. M. Alloway (Eds.), *Food aversion learning.* New York: Plenum Press, 1977.

Revusky, S. Reply to Mitchell. *Animal Learning and Behavior,* 1978, *6,* 119-120.

Revusky, S. H. & Garcia, J. Learned associations over long delays. In G. H. Bower (Ed.), *The psychology of learning and motivation* (Vol. 4). New York: Academic Press, 1970.

Rozin, P., & Kalat, J. W. Specific hungers and poison avoidance as adaptive specializations of learning. *Psychological Review,* 1971, *78,* 459-486.

Schwartz, B. On going back to nature: A review of Seligman and Hager's *Biological boundaries of learning. Journal of the Experimental Analysis of Behavior,* 1974, *21,* 183-198.

Segal, E. F. Induction and the provenance of operants. In R. M. Gilbert & J. R. Millenson (Eds.), *Reinforcement: Behavioral analyses.* New York: Academic Press, 1972.

Seligman, M. E. P. On the generality of the laws of learning. *Psychological Review,* 1970, *77,* 406-418.

Seligman, M. E. P., & Hager, J. L. *Biological boundaries of learning.* New York: Appleton-Century-Crofts, 1972.

Shettleworth, S. J. Constraints on learning. *Advances in the Study of Behavior, 1972, 4,* 1-68.(a)

Shettleworth, S. J. Stimulus relevance in the control of drinking and conditioned fear responses in domestic chicks. *Journal of Comparative and Physiological Psychology,* 1972, *80,* 175-198.(b)

Smith, J. C., & Roll, D. L. Trace conditioning with X-rays as an aversive stimulus. *Psychonomic Science,* 1967, *9,* 11-12.

Testa, T. J. Causal relationships and the acquisition of avoidance responses. *Psychological Review,* 1974, *81,* 491-505.

Testa, T. J. Effects of similarity of location and temporal intensity pattern of conditioned and unconditioned stimuli on the acquisition of conditioned suppression in rats. *Journal of Experimental Psychology: Animal Behavior Processes,* 1975, *1,* 114-121.

Testa, T. J., & Ternes, J. W. Specificity of conditioning mechanisms in the modification of food preferences. In L. M. Barker, M. R. Best, & M. Domjan (Eds.), *Learning mechanisms in food selection.* Waco, Texas: Baylor University Press, 1977.

Thompson, R. F. The search for the engram. *American Psychologist,* 1976, *31,* 209-227.

Turner, L. H., & Solomon, R. L. Human traumatic avoidance learning: Theory and experiments on the operant-respondent distinction and failures to learn. *Psychological Monographs,* 1962, *76* (Whole No. 559).

Wasserman, E. A., Franklin, S. R. & Hearst, E. Pavlovian appetitive contingencies and approach versus withdrawal to conditioned stimuli in pigeons. *Journal of Comparative and Physiological Psychology,* 1974, *86,* 616-627.

Wickens, D. D., & Wickens, C. D. Some factors related to psuedoconditioning. *Journal of Experimental Psychology,* 1942, *31,* 518-526.

Wilcoxon, H. C., Dragoin, W. B., & Kral, P. A. Illness-induced aversions in rat and quail: Relative salience of visual and gustatory cues. *Science,* 1971, *171,* 826-828.

Zielinski, K. The influence of stimulus intensity on the efficacy of reinforcement in differentiation training. *Acta Biologiae Experimentalis,* 1965, *25,* 317-335.(a)

Zielinski, K. The direction of change versus the absolute level of noise intensity as a cue in the CER situation. *Acta Biologiae Experimentalis,* 1965, *25,* 337-357.(b)

15

Constraints on Conditioning in the Writings of Konorski

Sara J. Shettleworth
University of Toronto

I. INTRODUCTION

A. Constraints on Conditioning: the Issues

The term "constraints on learning"—or, equivalently, "constraints on conditioning"—has come to refer to a diverse collection of demonstrations that qualitatively different events in conditioning lead to different results (see Shettleworth, 1972, for a review). A number of cases in which stimuli and reinforcers seem to interact have been discussed by LoLordo (Ch. 14).

In reviewing examples of constraints on conditioning, Seligman (1970), Rozin and Kalat (1971), Bolles (1970), and others who discussed the subject in the early 1970s drew attention to two distinct classes of issues. The one that has had greater attention subsequently is the question of mechanism: Are there constraints on conditioning that are truly associative? Alternatively, is a given phenomenon due to associative variables that are already well-understood, like stimulus similarity or effective duration, or is it due to performance factors of some kind? LoLordo's chapter illustrates the kind of detailed experimental analysis these questions call for.

The second, orthogonal, aspect of the "constraints on conditioning" problem is what can be referred to as the biological one. It is based on the claim, made most explicitly by Rozin and Kalat (1971), that to fully understand learning we must view it as part of the animal's whole adaptation to its environment. This general statement has at least two implications. First, knowledge about the function of learning for a particular species in its natural habitat is important. Animals are likely to be adapted to acquire most readily those modifications of behavior that are useful in nature, and

they will acquire them in the way most suited to the conditions they en-
counter in nature. This kind of consideration helps make sense of such
things as the long delays in poison-avoidance learning, because poisons do
not usually make their effects felt immediately upon consumption. A second, separate implication of viewing learning as part of behavioral biology
is that behavior of animals in learning experiments may profitably be viewed
in terms of concepts from ethology. For example, Timberlake and Grant
(1975) have suggested that the ethological notion of drive system may help
predict the form of classically conditioned responses in autoshaping and
other conditioning preparations.

The "biological" approach to constraints requires more of a reorienta-
tion of theory and experiment by learning theorists than does the analysis of
mechanism. Perhaps for this reason it has received less attention than the
latter approach. However, even though the "biological" aspect of
constraints on conditioning is orthogonal to the problem of mechanism, if
there turn out to be constraints on conditioning that cannot be traced to
familiar variables such as stimulus similarity (cf. Testa, 1974), one way to
do more than catalogue them will probably be to draw on considerations
about the function of learning in its natural context. The evolutionary explanations of social structure in sociobiology show how ecologically based,
functional accounts of behavior provide an organizational principle for
diversity of the sort we might expect to find in learning if we begin to look
for it. The idea that viewing learning in terms of its function for the species
can only result in an endless catalogue of special cases (Revusky, 1977;
Rozin & Kalat, 1971) is far too pessimistic.

B. Konorski's Views on "Constraints"

Konorski was well aware of both aspects of the constraints on conditioning
problem, but he had more to say about mechanism. In this chapter I first
briefly describe Konorski's general notions about how conditioning might
be constrained. Then I review in some detail three constraints on instrumental conditioning that he analyzed. Konorski's analysis of at least these examples differed in important ways from the approach to similar problems
taken in more recent literature, and the final section develops this point.

Konorski's conceptualization of the physiology of associations provides a
natural basis for constraints on conditioning, because he assumed that functional connections could be formed only where potential connections were
already developed in ontogeny (1948, p.87).[1] In one sense it is trivially true
that something can only happen when there is the potential for it. However,

---

[1]Page references are to the original, 1948, edition of *Conditioned Reflexes and Neuron
Organization* and to the Second Edition (1970) of *Integrative Activity of the Brain*.

in Konorski's framework the potential connections form a nontrivial basis for constraints on conditioning, in that not all possible potential connections exist. According to Konorski (1948),"the cerebral cortex cannot be considered as a uniform structure capable with equal facility of putting through any intercentral connexions, like an automatic telephone exchange, but that it possesses a definite organization, which makes the formation of certain connexions an easy and rapid process and others, on the other hand, difficult and complex (p. 124)."

It is all too easy to use the notion of absent potential connections in a circular way to explain differences in conditioning due to qualitative features of events. For example, the failure of auditory quality to control left–right differentiation discussed in Section III is attributed to lack of potential auditory–kinesthetic connections without any independent verification that such connections are lacking. This is a common problem in the constraints literature. Even where conditioning phenomena could in principle be tied to independent observations, as in relating species-specific defense reactions to easily acquired avoidance responses (Bolles, 1970), this is usually not done. It must be remembered, however, that what may appear as a circular explanation on a behavioral level is not necessarily circular in a framework like Konorski's, where behavioral observations are a means of revealing the way the brain works.

In attributing differences in conditioning to differences in potential connections, Konorski made explicit an assumption of many more recent discussions of constraints on learning. "Preparedness" (Seligman, 1970) certainly lends itself to a connectionistic interpretation, as does Thorndike's (1911) earlier statement of the notion. Garcia and others have interpreted the stimulus relevance in taste-aversion learning in terms of anatomical proximity of centers for the events involved (Garcia, McGowan, & Green, 1972). And Thompson (1976) has proposed a physiological basis for specificity in conditioning that is closely allied to some of Konorski's views (see Section II).

Konorski's overall view of the organization of behavior is also compatible in a more general way with some current approaches to constraints. His conception of the control of learned behavior in terms of drive centers with mutually excitatory and/or inhibitory relationships (see Dickinson & Dearing, Chapter 8) and of the determination of instrumental responding jointly by internal (drive) and external (CS) factors (see Morgan, Chapter 7) is virtually identical with the scheme many ethologists use to analyze behavior (cf. Hinde, 1970). Such notions may prove useful if not essential in analyzing such problems as constraints on instrumental responding by response and reinforcer (Shettleworth, 1975) and differences among reinforcers (Hogan & Roper, 1978).

The foregoing notions about constraints seem applicable to any type of conditioning, but the three cases that Konorski considered in detail and that are discussed here were from instrumental conditioning. Konorski analyzed them all in terms of his 1967 model of instrumental conditioning. According to this model, instrumental performance required two connections between the explicit CS or situation and the instrumental response. One was a direct CS–response connection, as evidenced by the fact that different responses could be trained to different CSs using the same drive and reinforcement. The CS also controlled the response indirectly, by arousing the appropriate drive, which was in turn connected to the response. The necessity for this connection is evidenced by such things as the effects of satiation. Although this model may seem unduly complicated compared with the 1948 model (see Mackintosh & Dickinson, Chapter 6) or with other models of instrumental performance, it permits (as will be seen) a novel and potentially fruitful way of accommodating a variety of constraints on instrumental learning.

II.   SPECIAL PROPERTIES OF SPECIFIC TACTILE STIMULI

When the CS for instrumental, food-reinforced movement of a dog's paw is a tactile stimulus to the leg involved, the CS is referred to as a specific tactile stimulus (STS). The association between paw movement and the STS is especially strong compared with that between paw movement and an auditory stimulus or a tactile stimulus to another part of the body. It is more resistant to satiation and to extinction (Dobrzecka & Wyrwicka, 1960) and more quickly restored after extinction than the same paw movement to a buzzer. No nonassociative controls were reported for this situation, but the STS is not just a generally prepotent stimulus, because a buzzer is a better CS for salivation than the STS (Dobrzecka & Konorski, 1962). The STS also has special properties in transfer paradigms. After training to the STS, transfer of food-reinforced paw movement to a buzzer is slow or nonexistent, whereas transfer is quick from buzzer to STS (Dobrzecka & Konorski, 1962). When STSs to the two forepaws are used in left-leg–right-leg differentiation, acquisition is much faster than with tactile stimuli to the two sides of the body (Dobrzecka, Konorski, Stepièn, & Sychowa, 1972).

Konorski's interpretation of these results is that preexisting sensorimotor connections involving a given leg facilitate conditioning to the STS. Dobrzecka, Sychowa, and Konorski (1965) cut the presumed connections in the cortex and found that indeed the special character of the STS was destroyed. It became no different from a buzzer in extinction and no longer interfered with transfer of the conditioned paw movement to another CS. These connections could normally function in various ways to facilitate elicitation of paw movement by the STS. The STS might have an original "associative strength" greater than zero, or it might increase learning rate or performance in some way. There is some evidence that a non-zero initial associative strength is involved even if it is not the only factor. The STS is described (Dobrzecka & Konorski, 1962) as having powerful "motorigenic properties," as evidenced by the fact that it sometimes elicits an innate placing reaction before conditioning. This seems to mean that some actual connections already exist. However, the exceptional strength of conditioning to the STS is explained in the same paper in terms of the presence of strong potential connections.

A similar lack of clarity in interpretation is found in a later paper by Dobrzecka (1975) on a related phenomenon. Here tactile stimuli to the hind legs were instrumental CSs for left-leg–right-leg differentiation involving the forelegs. The STS signaled movement of either the ipsilateral (Group 1) or contralateral foreleg (Group 2). Unlike the case in which the STS is applied to the forelegs, Group 1 had great difficulty in acquiring the differentiation. The relative ease with which Group 2 learned (1975) "was based on the readiness for a definite motor reaction which potentially exists in the form of postural reflexes (p. 367)." This was due in part to the fact that the dogs turned to orient to the site of application of the STS and in doing so freed the ipsilateral foreleg, leaving the contralateral leg supporting them. However, Dobrzecka states that the orienting reflex wanes fairly soon, and that further difficulty in learning beyond this point is due to the Sherringtonian crossed reflexes involved in locomotion — that is, the coupling of hindleg and opposite foreleg somehow participates in the conditioning.

The results with specific tactile stimuli could perhaps be interpreted as an example of a more general principle of facilitation of association by similarity of location (Testa, 1974) — in this case, location of CS and response. However, it seems more likely that the special features of the STS are due not to its supporting an especially high rate of learning but to its already facilitating the to-be-conditioned movement at a subthreshold level. There is little evidence from other preparations that similarity of CS location and response location has much effect unless the animal's responding is biased by orienting responses. Harrison and his colleagues (e.g., Downey & Harrison, 1972) have compared acquisition of control of responding by location of sounds in monkeys and rats. The rats learned equally well whether or not the sound was adjacent to the lever they were to press. Monkeys, however, learned much faster in adjacent than nonadjacent conditions because they tended to look in the direction of the sound and then respond to the manipulandum they were looking at.

More important, although some of the special features of the STS do not follow directly from its supporting an especially fast rate of learning, Konorski was able to interpret them quite naturally in other ways. The asymmetry of transfer between STS and auditory stimuli was attributed by

Dobrzecka and Konorski (1962) to the fact that in training with the STS CS-movement connections predominate while CS-drive-movement connections are relatively weak, as evidenced by the relatively small effect of satiation. Thus drive – movement connections cannot facilitate transfer. With original training to the sound, drive–movement connections are established and, together with the preexisting facilitatory effect of the STS on movement, they ensure quick transfer to the STS.

Suggestions that some CSs may already have special relationships to the USs or to performance of classically or instrumentally conditioned behaviors are not unique to Konorski's interpretation of the effects of specific tactile stimuli. LoLordo (Chapter 14) discusses some other examples from Konorski's work. Elsewhere, Frontali and Bignami (1973) attribute differences in difficulty of passive avoidance learning to noise rather than light to "motorigenic" properties of the noise. Hendersen (1973) presents evidence that certain CSs have relatively permanent unconditioned fear inhibitory properties that are simply overlaid by the effects of conditioning procedures. Rozin and Kalat (1971) attribute species differences in the stimuli that most readily control avoidance of noxious food to differences in "eating related cues." It is hard to see what an "eating related cue" could be, except one that has some preexisting influence on feeding in the way specific tactile stimuli have a special relationship to certain movements.

Perhaps most encouraging to Konorski's interpretation of the effect of STSs, Thompson (1976) notes that for classical conditioning of the rabbit nictitating-membrane response, an unconditioned tone temporarily increases the excitability of the final common path for the behavioral response. He further suggests that this preexisting influence on excitability is necessary for conditioning to occur (1976): "Such a requirement could easily account for the differing degrees of effectiveness of various stimuli as CSs and perhaps for the specificity of stimuli in situations such as conditioned taste aversion with a sickness UCR (p. 217)." This conjecture suggests a way to look for "constraints" independently of the results of conditioning. At the same time, however, such constaints represent a mechanism that some would consider less interesting theoretically than differences in rates of learning that depend on interactions of CSs and USs (cf. LoLordo, Chapter 14).

III. CUE SPECIFICITY IN LEFT–RIGHT DIFFERENTIATION

The best-known example of a constraint on conditioning from Konorski's laboratory, and some of the most compelling evidence for a true constraint on associability, is the difference in learning left–right versus go–no go discriminations with qualitative versus directional cues. In a number of experiments with compound and single stimuli, left-right differentiation (either leg movement or approach to one of two feeders) was learned more readily to directional than quality cues, while the reverse was true of go–no go differentiation (Dobrzecka & Konorski, 1967, 1968; Dobrzecka, Szwejkowska, & Konorski, 1966; Lawicka, 1964, 1969; Szwejkowska, 1967a, 1967b). Konorski's (1964) interpretation of this "crossover" interaction is that the two types of differentiation tasks are controlled in different ways. With directional cues the dog associates the different movements with kinesthetic cues from differential orienting responses to cues in different locations. Qualitatively different stimuli from the same place do not produce different orienting reactions unless they differ greatly in intensity or are from different modalities, and pure acoustic-kinesthetic connections are not potentially available — although there seems to be no independent evidence of this. In go – no go tasks, on the other hand, the different qualities control excitation versus inhibition of the hunger drive and hence paw movement; in other words, acoustic – kinesthetic connections are not necessary.

Whether or not one accepts Konorski's analysis of this example of the "cue specificity principle" (Konorski, 1964), it is pretty clearly associative. At least it is by the criterion outlined by Schwartz (1974), in that a crossover is demonstrated. It does not satisfy the more rigorous criterion suggested by Rescorla and Holland (1976). They require that all the events involved in a crossover design be experienced by each subject, and here no dog was simultaneously trained on left-right and go–no go discriminations. It seems possible that when responses are to be differentiated as to location (i.e., left-right differentiation), animals might attend especially readily to the location of CSs. This kind of effect could be interpreted as basically different from an associative constraint; Rescorla and Holland's design would reveal it. However, it is hard to see how this reasoning would predict the effects in go–no go differentiation. Moreover, the dogs in these experiments apparently do not fail to attend to the qualities of the stimuli. They show orienting responses on the first few trials when a CS is presented in a new location (Dobrzecka & Konorski, 1967). Therefore, they have associated the quality and location of the CS, but apparently this association cannot mediate control of the differentiation by quality cues.

This example of an apparently associative constraint on conditioning is unlike any other in the literature. The "belongingness" of location cues to spatially differentiated responses might seem to support the case for association by spatial similarity (Testa, 1974). In fact, however, the stimuli were generally located above and below or anterior and posterior to the animals. Overmier, Bull, and Trapold (1971) describe a situation in which a stimulus on one side or the other of a dog signals a response on that side to avoid shock to the ipsilateral or contralateral hind leg. Unlike the situation described by Dobrzecka (1975; see also Section II), learning was equally fast in both these conditions. It was significantly impeded only when there was no consistent relationship between signal and shock locations.

Konorski indicated that this constraint on conditioning is not predictable from some generalized category of associative specificity, such as association by similarity. He noted (1967, p. 439) that in analogous visual tasks, rats can associate qualitatively different visual cues with response location. Because this constraint is already quite well documented with dogs, research with other species and other stimulus modalities might be worthwhile to see where it fits in some overall characterization of constraints on conditioning.

IV.   INSTRUMENTALIZATION OF REFLEXES

A. Instrumental-Response Classification

Beginning with Thorndike (1911) many experimenters have observed that all identifiable responses an animal makes are not equivalent in instrumental training. Reinforceability may depend on the reinforcer used, so that a response readily acquired for one reinforcer is hardly performed for another (e.g., Sevenster, 1973). Also, some responses have long been considered involuntary—that is, they cannot be brought under the control of any reinforcer (Hearst, 1975). Despite the oldness of the problem, there is still no workable response-classification system that predicts the effects of instrumental training. The notion that responses are either voluntary or involuntary or that there is a continuum of voluntariness, however intuitively appealing, is just a restatement of the results of training and fails to capture the special features of instrumentalized "involuntary" activities that we discuss in this section.

Konorski devoted Chapter 11 of his 1967 book to the question of what types of movements can be instrumentally trained, or "instrumentalized." He distinguished four types of movements, "according to their origin": (1) elicited responses, or URs; (2) already-established instrumental CRs (e.g., manipulatory or locomotor responses); (3) passively produced movements; and (4) electrically induced movements. He reviewed evidence that the last two categories — movements performed without the active participation of the animal — cannot be truly instrumentalized; this is not considered here. He claimed that the already-established CRs can be trained easily, but that the reflexes, or URs, are a problematical category. Some URs can be trained fairly readily, but some only with great difficulty or perhaps not at all.

It is questionable (1967) whether movements can be classified unambiguously ·as reflexes (URs) versus "movements produced as a result of already established instrumental CRs (p. 467)." The latter category are more easily trained because they already possess "the food instrumental character (p. 465)," which seems to be a way of saying they have pre-existing, appropriate drive–response connections. However, movements in this second class have eliciting stimuli, and they are trained, like reflexes, by making use of this fact (1967): "The best method of training is that the manipulanda are first baited and when the animal performs a proper movement to get the bait he receives food from another place (p. 465)." Thus, Konorski's classification is virtually the same as that between emitted and elicited responses, and equally difficult to maintain rigorously (cf. Hearst, 1975).

More detailed consideration of the relevant data may suggest a more satisfactory classification scheme. In this section I review the work Konorski describes on "instrumentalized reflexes" in dogs and cats and compare the data and interpretation with that from similar work on golden hamsters from my laboratory. Unless otherwise referenced the results from hamsters are those described in Shettleworth (1975).

B. Peculiarities of Instrumentalized Reflexes

Konorski described "instrumentalization" of the following elicited responses in dogs and/or cats: leg flexion to shock, moving the leg in response to binding with a band, scratching the ear, cleaning the anus, rubbing the face, barking, and yawning. Ear scratching in cats for food illustrates what appears to be a typical procedure for instrumentalizing reflexes (cf. Gorska, Jankowska, & Kozak, 1961; Jankowska, 1959). First, although "spontaneous" scratching is reinforced, initially scratching is also elicited by putting a piece of cotton in the cat's ear. The cotton elicits not only scratching with the hind paw, the desired movement, but also other movements that could function to remove the cotton, such as rubbing the ear with the forepaw and shaking the head. These movements may still appear in later sessions when cotton is no longer used, suggesting that cotton in the ear has become associated with the experimental situation and that the instrumental movement is due in part to this association. This is a different mechanism from the association of drive and situational factors with the motor center that Konorski postulated for conditioning of other types of movements. However, there must also be some specific association of ear scratching, as opposed to rubbing and head shaking, with the situation because the latter movements apparently disappear in a well-trained animal. Similarly, when we train face washing in hamsters, other types of grooming movements generally do not increase in frequency, even though they normally vary in frequency together with face washing (Shettleworth, 1975).

A second peculiarity of "instrumentalized reflexes" is that the trained movements often are systematically simplified versions of the untrained ones. For example, in cats the unconditioned cleaning reflex elicited by wetting the fur around the anus consists of sitting down, raising the hind leg, turning the head, and licking the anus. Food-reinforced anus cleaning consists of sitting and raising the leg (Gorska, Jankowska, & Kozak, 1961). Similarly, when yawning is reinforced in dogs, the frequency of yawning itself changes very little, but "pseudoyawning" appears and increases in frequency. Moreover, from Thorndike (1911) onward, most reports of attempts to train comfort movements in various species note that the trained movements are minimal forms of the original motor patterns (Beninger, Kendall, & Vanderwolf, 1974; Lorge, 1936; Paxinos & Bindra, 1970; Soltysik, personal communication; but cf. Hogan, 1964). Because reduction of the trained movement is such a reliable feature of reinforced grooming, it might be instructive as regards the mechanisms of instrumental learning generally.

C. Theories of how Reflexes are Instrumentalized

An uninteresting account of minimal responses is that they illustrate the "law of least effort." This is a reasonable interpretation of cases in which the experimenter reinforces any recognizable component of a full-blown reflex, as was apparently true in the work to which Konorski (1967) referred: "Since we usually do not require that the animal perform a 'better' movement for food reinforcement this movement is regularly performed in the experimental situation. It is easy, however, to force the cat to perform a better movement by not offering him the food immediately: then the animal will 'correct himself' (p. 464)." Gorska et al. (1961) imply that they followed a similar procedure in training anus cleaning. However, minimal movements are also found in these situations even if only well-developed forms are reinforced (Soltysik, personal communication).

Recent work in my laboratory also suggests that minimal grooming does not develop simply because it is the least-effortful reinforced response. When hamsters are reinforced for one of a number of different motor patterns under identical schedules, only grooming movements become minimal in duration, whereas digging, rearing, and "scrabbling" (scratching and jumping against the walls) become, if anything, more well-developed. There is some evidence that the abnormally short bouts of face washing even develop when only longer bouts are reinforced (Shettleworth, 1973). Similarly, Konorski noted (1967) that, unlike grooming, a movement of the leg elicited by binding with a band and reinforced "remains roughly the same as it was when provoked by the US. (p. 463)." And, as further evidence that the "law of least effort" does not necessarily apply to reinforced "URs," there are a number of reports in the literature of such activities becoming so well-developed that they interfere with consumption of reinforcers (review in Shettleworth, 1973).

A second possibility raised by Konorski (1967, p. 466) is that as the reinforced movement becomes a food CS, it elicits alimentary responses incompatible with some of its components. For example, orienting to the feeder is incompatible with turning the head and licking the anus. Cats performing pseudo–anus cleaning did sit facing the feeder (1967, p. 466). In contrast, shock-elicited leg flexion or movements elicited by binding with a band are not incompatible with orientation to the feeder, and thus these responses would not be expected to be reduced when reinforced.

Again, however, research from my laboratory shows that at least in hamsters differences between grooming and other trained responses can be observed when this peripheral competition is unlikely to be involved. For example, face washing and digging in the substrate are equally incompatible with orientation to a feeder placed on a wall (one could argue that face washing is more compatible than digging), but food-reinforced digging increases in bout length while face washing decreases. Thus, any incompatibility between food CRs and grooming may be more central than is implied by the notion of incompatible responses. It may be, instead, that a state of food anticipation actively inhibits grooming. Observations of the hamsters in various situations outside the food-reinforcement experiments (Shettleworth, 1975, 1978b) show that grooming is indeed suppressed when the hamsters are hungry and/or anticipating food, whereas the action patterns (APs) that readily increase greatly for food are facilitated under the same conditions, or at least are not suppressd. Thus, when face washing or any other response is reinforced it becomes, as Konorski noted, a food CS. But because food CSs facilitate only certain activities and inhibit others, only some APs can be performed at a high rate for food reinforcement. This notion attributes reinforcer-specific constraints on performance of certain responses to an incentive motivational mechanism in which there is some specificity in the activities facilitated by anticipation of given reinforcers (see Boakes, Ch. 9). This account also suggests that the relative modifiability of different responses will vary with the reinforcer in a way related to the independently observed reinforcer-specific motivational system. Findings with punishment (Shettleworth, 1978a) encourage this notion. One of the responses (open rearing) that was easy to train with food was difficult to suppress with contingent shock, while another (scrabbling) was easily suppressed.

In principle the organization of the reinforcer-specific behavior system can be observed in classical conditioning experiments (cf. Timberlake & Grant, 1975), and in fact the effects of food CSs do parallel the instrumental-training results to a certain extent (Shettleworth, 1978b). However, this type of analysis can not be entirely straightforward, because the choice of CS may determine what CRs are exhibited (Holland, 1977).

A final possibility is that perhaps only one component of a motor pattern, such as face washing or yawning, is "voluntary" (i.e., trainable). Because responses such as grooming seem to have a high degree of intrinsic organization (Fentress, 1972), this possibility seems unlikely, but it would account for such observations as "pseudoyawning." Moreover, Soltysik and other Eastern European workers have distinguished two types of minimalized reflexes, both of which could appear in the same experiment (Soltysik, personal communication). These were abortive forms of the true reflex, and sham forms. The abortive forms are like those already described: Full performance of the activity appears to be cut short by an interfering movement, such as turning to the food tray. Sham forms, however, appear to be voluntary movements, for the animals can perform them readily. Sham forms are true instrumental responses and do not depend on activation of the motor center for the true reflex. In principle, then, they might be distinguishable electrophysiologically from the abortive forms of reinforced reflexes.

Konorski's account of what goes on when a reflex or UR is instrumentalized is that a representation of the eliciting stimulus (US) is associated with the experimental situation or explicit CS. In contrast, "normal" instrumental responses occur because a representation of the movement itself is associated with the experimental situation. In effect, the cat trained to scratch does so because it "feels an itch", not because it "feels like scratching." The evidence for this special mechanism comes from cases like that of the aforementioned instrumental ear scratching, in which the whole system of movements elicited by the US for the reflex is performed.

Thus Konorski, following Jankowska and Soltysik (1960; Soltysik, personal communication), proposed that the mechanism of instrumentalized reflexes was basically different from that of other instrumental responses. There are several serious problems with this approach. First, as discussed at the beginning of this section, it is not clear that the different classes of responses, especially "URs" and already-established CRs, are distinguishable.

Second, on a behavioral level it would seem impossible to maintain a distinction between a movement's arising through activation of its "own" center and its being elicited by some stimulus. This is not to say, however, that this is not a tenable distinction on the physiological level, with which Konorski was primarily concerned. Finally, the peculiarities of instrumentalized reflexes—that is, persistent performace of sham and abortive forms —do not seem to follow from this special mechanism. If, for instance, the eliciting stimulus for yawning is well associated with a training situation, why doesn't the dog yawn at a high rate?

In contrast to Konorski's approach, I have tried to account for differences among responses in terms of different outcomes of a single mechanism. In effect, all movements may be equally susceptible to being instrumentalized, and in the same way; but the classical conditioning embedded in the instrumental paradigm can either facilitate or inhibit instrumental performance, depending on some preexisting organization of the behavior. Now, in fact, Konorski's 1967 model of instrumental performance could handle the differences among instrumental responses in a way quite similar to this, by taking into account preexisting facilitatory or inhibitory drive-response connections. In a model quite close to this, Hogan and Roper (1978) have proposed that we think of instrumental training as bringing a response under the control of a drive system (e.g., hunger, in the case of food reinforcement). One need only add to this a more explicit statement about preexisting connections, or potential connections, to suggest some constraints at least on instrumental performance, if not on learning. Some results from my laboratory are consistent with this. The responses that hamsters readily performed at high rates for food reinforcement were also facilitated by hunger. Those responses, such as grooming, that showed little or no rate increase and were performed abortively when reinforced were inhibited in hungry hamsters. Thus, it may be that Konorski's basic model of instrumental performance can encompass the problems in instrumentalizing certain reflexes, even if his specific proposal of a special mechanism for such responses cannot be rigorously maintained.

V.   CONCLUSIONS

A. Konorski's Analyses of Constraints on Conditioning

Two general points emerge from this discussion of Konorski's analyses of three constraints on instrumental conditioning. The first is that Konorski's general approach to these problems contrasts with that which is implicit in more recent discussions of other constraints on conditioning. For example, consider the case described by Foree and LoLordo (1973: see also Chapter 14). Auditory stimuli selectively control responding for shock avoidance, and visual stimuli control responding for food. Because the avoidance and food-reinforcement paradigms are precisely parallel in these experiments, it has seemed necessary to conclude that the relative salience or associability of visual as compared with auditory stimuli is affected by the reinforcer used. Essentially, the animal is assumed to associate exactly those events manipulated by the experimenter in a way that reflects such manipulation fairly precisely. (See Mackintosh & Dickinson, Chapter 6 for a discussion of the advantages of this approach.)

In constrast, Konorski's analysis of the constraints on conditioning discussed here shows that two tasks that are parallel from the experimenter's point of view (e.g., those involving food reinforcement of a movement) can be seen as involving different *kinds* of learning by the subject. Because instrumental performance is the result of several associations — between CS and drive, drive and response, and CS and response — performances controlled by the same paradigm can result from different combinations and degrees of these associations.

A review of the three examples discussed in this chapter may clarify this point. In the case of the special properties of specific tactile stimuli, the CS – paw-movement associations preexist for the STS, and they block acquisition of other associations that are acquired when a buzzer is used. The paw movement to the STS results primarily from CS–response connections, whereas paw movement to the buzzer results from relatively strong CS–drive-response connections. In the crossover design with quality and directional cues for left–right or go–no go differentiation, Konorski's suggestion again is that the two tasks are controlled in different ways. Go–no go differentiation involves control of the hunger drive—and hence the response— by CSs, whereas left–right differentiation involves direct CS–response associations. In addition, it is necessary to conclude that dogs, at least, cannot associate auditory-quality cues with different movements, and this seems to be a true limitation of their associative potential. However, this deficit can be seen not to affect go–no go tasks only by an assumption like Konorski's that in some way these involve a different kind of learning than left–right tasks.

In the final example reviewed in this chapter—the peculiarities of certain instrumentalized reflexes, such as like grooming — Konorski's special approach is most apparent. There, he explicitly assumed that reflexes come to be performed instrumentally through a different mechanism from that responsible for instrumentalization of other movements. This analysis does have some serious shortcomings, which have been discussed. However, the overall notion that constraints on conditioning may lie not so much in the ease with which animals make a fixed kind of association as in exactly what associations they do make is one that might be worth exploring in other cases.

B. The Diversity of "Constraints on Conditioning"

Throughout this chapter and elsewhere (1972), I have emphasized that a diversity of mechanisms could be responsible for constraints on conditioning. For example, in this chapter preexisting connections, failures of association, and different ways of learning superficially similar things have all been suggested. This diversity is to be expected because the classification of phenomena as constraints on learning has arisen through historical accident: Anything that violates the supposedly prevailing assumption of a tabula rasa animal is a "constraint on learning (Revusky, 1977)." Examples of "constraints", "preparedness" or "belongingness" are defined in terms of a rather general class of experimental outcomes— that is, any effects of intrinsic qualities of events. These might be constraints on ease of acquiring associations, on the way learned associations are expressed in performance (Holland, 1977), or on the principles of association formation themselves (Rozin & Kalat, 1971). Of course "constraints" may be defined in terms of a more restricted set of outcomes, so as to exclude nonassociative effects (e.g., LoLordo, Chapter 14; Rescorla & Holland, 1976; Schwartz, 1974). This is entirely appropriate to the first approach to the area outlined in Section I. But in terms of the "biological" approach, it is not so appropriate. If nonequivalence of events in conditioning is viewed as one aspect of the adaptedness of experiential modifications of behavior, then all mechanisms through which this is achieved are of equal interest. This point was made most vividly by Garcia et al. (1972), when they said that it does not matter whether taste-aversion conditioning is classical, instrumental, or pseudo. This statement has been cited (Bitterman, 1975) as exemplifying the nonanalytical character of a functional approach. But the fact is that it may not matter for an animal's survival whether it avoids poisoned foods for associative or nonassociative reasons. (On the other hand, however, more subtle analysis may reveal that it does matter.) Of course the mechanism is of great interest, but there is no logical justification for focusing on associative mechanisms to the exclusion of others. Rescorla and Holland (1976) have made a similar point about the paucity of relationships among events experimenters have studied, although they have done so from the more abstract viewpoint of the nonbiological approach to constraints.

Recent work indicates that the claims of the early commentators on constraints on learning (e.g., Rozin & Kalat, 1971; Seligman, 1970) may not turn out to be justified for the specific cases that inspired them. For example, taste-aversion learning does not seem to follow qualitatively different laws from other conditioning preparations (Revusky, 1977), and autoshaping seems to be acquired as are other forms of classical conditioning (see Hearst, Chapter 2). However, it is still to be hoped that the study of these and other examples of constraints on conditioning will permanently enrich and broaden the experimental study of learning and bring it more into touch with other approaches to animal behavior. That is a development I think Konorski would have applauded.

ACKNOWLEDGMENTS

Preparation of this chapter was supported by a grant from the National Research Council of Canada. Travel to the conference was made possible by a grant to the University of Toronto from the Canada Council. I am very grateful to Dr. S. Soltysik for a description of some of his unpublished results and thoughts on the issues discussed in Section IV and for a partial translation of Jankowska and Soltysik (1960), and to Dr. E. Jankowska for her help. This chapter was prepared in collaboration with Vin LoLordo so that our coverage would not overlap unnecessarily. Author's address: Department of Psychology, University of Toronto, Toronto, Ontario, Canada M5S 1A1.

REFERENCES

Beninger, R. J., Kendall, S. B., & Vanderwolf, C. H. The ability of rats to discriminate their own behaviors. *Canadian Journal of Psychology*, 1974, *28*, 79–91.

Bitterman, M. E. The comparative analysis of learning. *Science*, 1975, *188*, 699–709.

Bolles, R. C. Species-specific defense reactions and avoidance learning. *Psychological Review*, 1970, *77*, 32–48.

Dobrzecka, C. The effect of postural reflexes on the acquisition of the left foreleg–right foreleg differentiation in dogs. *Acta Neurobiologiae Experimentalis*, 1975, *35*, 361–367.

Dobrzecka, C., & Konorski, J. On the peculiar properties of the instrumental conditioned reflexes to "specific tactile stimuli." *Acta Biologiae Experimentalis*, 1962, *22*, 215–226.

Dobrzecka, C. & Konorski, J. Qualitative versus directional cues in differential conditioning. I. Left leg–right leg differentiation to cues of a mixed character. *Acta Biologiae Experimentalis*, 1967, *27*, 163–168.

Dobzecka, C., & Konorski, J. Qualitative versus directional cues in differential conditioning. IV. Right leg - left leg differentiation to non-directional cues. *Acta Biologiae Experimentalis*, 1968, *28*, 61–69.

Dobrzecka, C., Konorski, J., Stepien, L., & Sychowa, B. The effects of removal of the somatosensory areas I and II on left leg–right leg differentiation to tactile stimuli in dogs. *Acta Neurobiologiae Experimentalis*, 1972, *32*, 19–33.

Dobrzecka, C., Sychowa, B., & Konorski, J. The effects of lesions within the sensory-motor cortex upon instrumental response to the "specific tactile stimulus." *Acta Biologiae Experimentalis*, 1965, *25*, 91–106.

Dobrzecka, C., Szwejkowska, G., & Konorski, J. Qualitative vs. directional cues in two forms of differentiation. *Science*, 1966, *153*, 87–89.

Dobrzecka, C., & Wyrwicka, W. On the direct intercentral connections in the alimentary conditioned reflexes type II. *Bulletin of the Polish Academy of Sciences*, 1960, *8*, 373–375.

Downey, P., & Harrison, J. M. Control of responding by location of auditory stimuli: Role of differential and non-differential reinforcement. *Journal of the Experimental Analysis of Behavior*, 1972, *18*, 453–463.

Fentress, J. C. Development and patterning of movement sequences in inbred mice. In J. Kiger (Ed.), *The biology of behavior.* Corvallis: Oregon State University Press, 1972.

Foree, D. D., & LoLordo, V. M. Attention in the pigeon: The differential effects of food-getting vs. shock-avoidance procedures. *Journal of Comparative and Physiological Psychology*, 1973, *85*, 551–558.

Frontali, M., & Bignami, G. Go–no go avoidance discrimination in rats with simple "go" and compound "no go" signals: Stimulus modality and stimulus intensity. *Animal Learning and Behavior*, 1973, *1*, 21–24.

Garcia, J., McGowan, B. K., & Green, K. F. Biological constraints on conditioning. In A. H. Black & W. F. Prokasy (Eds.), *Classical conditioning II.* New York: Appleton-Century-Crofts, 1972.

Gorska, T., Jankowska, E., & Kozak, W. The effect of deafferentation on instrumental (Type II) cleaning reflex in cats. *Acta Biologiae Experimentalis*, 1961, *21*, 207–217.

Hearst, E. The classical–instrumental distinction: Reflexes, voluntary behavior, and categories of associative learning. In W. K. Estes (Ed.), *Handbook of learning and cognitive processes* (Vol. 2). Hillsdale, N. J.: Lawrence Erlbaum Associates, 1975.

Hendersen, R. W. Conditioned and unconditioned fear inhibition in rats. *Journal of Comparative and Physiological Psychology*, 1973, *84*, 554–561.

Hinde, R. A. *Animal Behavior.* New York: McGraw-Hill, 1970.

Hogan, J. A. Operant control of preening in pigeons. *Journal of the Experimental Analysis of Behavior*, 1964, *7*, 351–354.

Hogan, J. A., & Roper, T. J. A comparison of the properties of different reinforcers. *Advances in the Study of Behavior,* 1978, *8,* 155–255.

Holland, P. C. Conditioned stimulus as a determinant of the form of the Pavlovian conditioned response. *Journal of Experimental Psychology: Animal Behavior Processes,* 1977, *3,* 77–104.

Jankowska, E. Instrumental scratch reflex of the deafferented limb in cats and rats. *Acta Biologiae Experimentalis,* 1959, *19,* 233–247.

Jankowska, E., & Soltysik, S. Conditional motor reflexes elaborated from unconditional motor reflexes reinforced by food. In E. A. Asratian (Ed.), *Central and peripheral mechanisms of motor activity in animals.* Moscow: Academy of Sciences (USSR), 1960. (Russian)

Konorski, J. *Conditioned reflexes and neuron organization.* Cambridge: Cambridge University Press, 1948.

Konorski, J. Some problems concerning the mechanism of instrumental conditioning. *Acta Biologiae Experimentalis,* 1964, *24,* 59–72.

Konorski, J. *Integrative activity of the brain.* Chicago: University of Chicago Press, 1967.

Lawicka, W. The role of stimuli modality in successive discrimination and differentiation learning. *Bulletin of the Polish Academy of Sciences,* 1964, *12,* 35–38.

Lawicka, W. Differing effectiveness of auditory quality and location cues in two forms of differentation learning. *Acta Biologiae Experimentalis,* 1969, *29,* 83–92.

Lorge, I. Irrelevant rewards in animal learning. *Journal of Comparative Psychology,* 1936, *21,* 105–128.

Overmier, J. B., Bull, J. A., & Trapold, M. A. Discriminative cue properties of different fears and their role in response selection in dogs. *Journal of Comparative and Physiological Psychology,* 1971, *76,* 478–482.

Paxinos, G., & Bindra, D. Rewarding intracranial stimulation, movement, and the hippocampal theta rhythm. *Physiology and Behavior,* 1970, *5,* 227–231.

Rescorla, R. A., & Holland, P. C. Some behavioral approaches to the study of learning. In M. R. Rosenzweig & E. L. Bennett (Eds.), *Neural mechanisms of learning and memory.* Cambridge, Mass.: MIT Press, 1976.

Revusky, S. Learning as a general process with an emphasis on data from feeding experiments. In N. W. Milgram, L. Krames, & T. H. Alloway (Eds.), *Food aversion learning.* New York: Plenum Press, 1977.

Rozin, P., & Kalat, J. W. Specific hungers and poison avoidance as adaptive specializations of learning. *Psychological Review,* 1971, *78,* 459–486.

Schwartz, B. On going back to nature: A review of Seligman and Hager's *Biological boundaries of learning. Journal of the Experimental Analysis of Behavior,* 1974, *21,* 183–198.

Seligman, M. E. P. On the generality of the laws of learning. *Psychological Review,* 1970, *77* 406–418.

Sevenster, P. Incompatibility of response and reward. In R. A. Hinde & J. Stevenson-Hinde (Eds.), *Constraints on learning: Limitations and predispositions.* New York and London: Academic Press, 1973.

Shettleworth, S. J. Constraints on learning. *Advances in the Study of Behavior,* 1972, *4,* 1–68.

Shettleworth, S. J. Food reinforcement and the organization of behavior in golden hamsters. In R. A. Hinde & J. Stevenson-Hinde (Eds.), *Constraints on learning: Limitations and predispositions.* New York and London: Academic Press, 1973.

Shettleworth, S. J. Reinforcement and the organization of behavior in golden hamsters: Hunger, environment and food reinforcement. *Journal of Experimental Psychology: Animal Behavior Processes,* 1975, *1,* 56–87.

Shettleworth, S. J. Reinforcement and the organization of behavior in golden hamsters: Punishment of three action patterns. *Learning and Motivation,* 1978, *9,* 99–123. (a)

Shettleworth, S. J. Reinforcement and the organization of behavior in golden hamsters: Pavlovian conditioning with food and shock unconditioned stimuli. *Journal of Experimental Psychology: Animal Behavior Processes.* 1978, *4,* 152-169. (b)

Soltysik, S. Personal communication, 1977.

Szwejkowska, G. Qualitative versus directional cues in differential conditioning. II. Go–no go differentiation to cues of a mixed character. *Acta Biologiae Experimentalis,* 1967, *27,* 169–175.(a)

Szwejkowska, G. Qualitative versus directional cues in differential conditioning: 3. the role of qualitative and directional cues in differentiation of salivary conditioned reflexes. *Acta Biologiae Experimentalis,* 1967, *27,* 413–420.(b)

Testa, T. J. Causal relationships and the acquisition of avoidance responses. *Psychological Review,* 1974, *81,* 491–505.

Thompson, R. F. The search for the engram. *American Psychologist,* 1976, *31,* 209–227.

Thorndike, E. L. *Animal intelligence.* New York: Macmillan, 1911.

Timberlake, W., & Grant, D. L. Auto-shaping in rats to the presentation of another rat predicting food. *Science,* 1975, *190,* 690–692.

16 Cognitive Processes in Conditioning

W. K. Estes
Rockefeller University

The notion of "higher cognitive processes" in conditioning raises a number of questions. First, what do we conceive these processes to be? Second, what do the processes have to do with the presumably more basic ones of conditioning? Third, why is it timely to consider the relationship between basic mechanisms of association and higher processes? To set the context for what will have to be very limited treatment of these matters, I shall first comment on each of these questions.

For present purposes I take higher processes to be those of attention, perception, and memory, which are presumed to be basic to the mental life of human beings and perhaps manifest also in animals. But henceforth I refer simply to "cognitive" processes in order not to beg the question concerning their relationship to those of conditioning. These terms come from a quite different intellectual tradition and belong to a rather different scientific universe of discourse from those customarily used in the interpretation of classical and instrumental conditioning. Consequently, considerable groundwork needs to be laid for any useful effort toward bringing out significant relationships between these very disparate types of concepts.

One finds quite different views, even among specialists, with regard to the relationship of conditioning to other topics of psychology. These range from the strict operationalism of some eclectics, for whom conditioning is to be observed only in conditioning experiments, to the monistic theoretical view that conditioning is implicated wherever learning occurs. Hilgard and Marquis (1940) were perhaps the first to examine systematically the place of conditioning in psychology and to distinguish carefully between definitions of conditioning and its subvarieties (classical, instrumental, etc.) in terms of operations and of processes. Many experimentalists seem, implicitly at least, to have been satisfied to proceed on the assumption that classical and instrumental conditioning are defined simply by the operations and procedures used to demonstrate them, and this view continues to be strongly defended by some investigators (e.g., Gormezano & Kehoe, 1975).

More influential, understandably, among those interested in progressing toward general learning theories is the process orientation — the view that the term *conditioning* refers to a process or a collection of processes, most clearly demonstrable to be sure under special experimental arrangements, but nonetheless operative in many if not all forms of learning. The approaches of Hull (1937, 1943) and Spence (1937, 1951) and my own work during the 1950s (Estes, 1950, 1959) certainly fall in this category. From the strict operationalist viewpoint, any connection between conditioning and higher mental processes must exist only in analogies. From the process-oriented viewpoint, deeper relationships are imaginable, and the matter of trying to define them becomes an important theoretical task. This task has in recent years been receiving increasingly explicit attention. (e.g., Bolles, 1975; Estes, 1970; Konorski, 1967; Mackintosh, 1975; and others).

I.   A FRAMEWORK FOR INTERPRETING COGNITIVE ASPECTS OF CONDITIONING AND LEARNING

At this stage it seems to me that, within the process orientation, two principal aspects of the potential relationships between conditioning and cognitive processes need to be distinguished. One of these might be termed a multilayered conception of conditioning and of learning processes and mechanisms, the other a multifaceted conception of the area embracing both conditioning and learning, on the one hand, and the so-called cognitive processes, on the other.

The term *multilayered* is chosen to express the idea that such concepts as conditioned excitation and conditioned inhibition, which grew out of conditioning experiments and are used to interpret them, apply much more generally, but do not necessarily apply evenly across phylogenetic and ontogenetic levels. These conditioning processes may account for all or nearly all of the cognitive life of the lowest organisms. As one considers organisms progressively higher on the phylogenetic scale, however, the proportion of cognitive or intelligent activity so interpretable steadily diminishes. Using the multilayered conception, one does not conceive that the processes of conditioning disappear at the higher levels of the phylogenetic scale, but rather that they generally become subordinate to superimposed systems or processes — in man, especially verbal processes — and can be observed in relatively pure form only within restricted subsystems, such as visceral processes, certain reflexes, and perhaps motivational or emotional reactions. Both the lower and the hgher organisms can be observed to behave quite similarly in conditioning situations because demands of the task and the experimental restraints do not bring into play the additional capacities and more elaborate processing capabilities of higher organisms, just as a master carpenter and a small boy with a toy hammer may perform quite similarly if the task is driving a nail into soft wood.

The more elaborate system characteristic of higher organisms and man preserves many of the functional properties of the simpler ones, but also exhibits properties that appear novel or emergent. In the first class might be included the properties of retention loss in relation to repetitions and spacing of learning experiences — notably similar in form from the simplest forms of conditioning to complex verbal learning—or the functional properties common to stimulus variability in conditioning and to encoding variability in verbal learning. In the second class are, for example, characterisitics of the cognitive operations by which people encode different syntactic elements of sentences. These latter cannot at present be related to terms applicable to conditioning and elementary learning. It may in principal be possible to trace their evolutionay development from more primitive processes—as has been done for other aspects of bodily function in comparative anatomy and physiology—but the task largely remains to be accomplished.

On the multifaceted view of the broad area of learning and cognition, it is not appropriate to speak of conditioning as a lower process and of memory or perception as higher processes. Rather, processes of memory and perception are presumed to occur at all phylogenetic and developmental levels, though with important differences in their overt manifestations, related to differences in information-processing capacity.

According to this view it has been natural not to apply the terminology of memory and perception at all in relation to conditioning experiments, mainly because most of the special capabilities of human memory and perceptual systems are not obviously manifest in conditioning situations. However, one of the by-products of progress in research on conditioning is that conditioning seems to be far more complex than realized in the early Pavlovian period; witness the emergence of aspects of conditioning that resemble phenomena customarily associated with higher processes—for example, the role of information value or "surprise" in current work bearing on stimulus selection and conditioning (e.g., Kamin, 1969; Mackintosh, 1975). Then it becomes less clear whether the terminology should not be changed in accordance with changes in apparent complexity.

No universal agreement can be expected among psychologists on the definition of the terms *perception* and *memory,* but the following serves our present purposes. *Perception* is taken to denote the processing of stimulation resulting in an encoded representation that includes only part of the information in the original stimulus, but with the information retained being sufficient to identify the stimulus as a member of a category that is relevant to the task or situation. For example, when one hears a spoken letter, it is encoded by the auditory-perceptual system in terms of auditory-distinctive features (e.g., place of articulation, aspirated or nonaspirated) that distinguish it from other letters, but with information regarding loudness, the voice of the speaker, and so on discarded. The encoded representation is presumed to become associated with others in the course of learning.

Memory is taken to refer to information retained by an organism as a consequence of an experience in any stimulating situation, with the implication that the information may be reflected in future behavior in numerous different ways. Assumptions or hypotheses regarding perception and memory must ultimately be testable by virtue of implications about overt behavior describable in stimulus–response terms. But neither perception nor memory is assumed to be *equivalent* to any finite set of stimulus–response statements. An account of what has been observed in terms of concepts of perception and memory is, thus, more abstract than one in terms of stimulus–response associations (Estes, 1975).

With these understandings about usage in mind, it is easy to see why there is little or no occasion to apply the vocabulary of perception and memory to the simplest conditioning situations. Typically only one stimulus is involved, and procedures are usually so stereotyped that the way the stimulus is encoded can be assumed to be determined by the sensory system involved, uniformly over different subjects, and uninfluenced by the individual organism's experience. Similarly only one response is usually recorded, and hence no evidence is available that more is remembered (has been learned) than is reflected in the observed changes in stimulus–response relationships.

It has long been assumed by process-oriented investigators that the study of conditioning might ultimately help us to better understand the higher cognitive processes. But, according to the multifaceted viewpoint, it also seems possible that results of research on memory and perception might help to advance and illuminate research on conditioning. Even granting that memory should be assumed to be involved in learning at all phylogenetic and developmental levels, some might feel it most natural to assume that the only workable strategy would be to try to fully understand memory in its more restricted or elementary forms, as manifest in conditioning experiments, and then proceed upward to higher organisms and more complex behavioral situations. But one must consider the possibility that the assumption is not well-founded. To be sure, higher organisms, especially man, present more highly elaborated learning and memory systems and generally must be studied in more complex situations with less rigid experimental control; but, on the other hand, there are many ways of testing or interrogating the memory of a human subject that have no counterpart in situations in which an investigator is attempting to trace changes in the contents of the memory of a paramecium or of a mouse in a conditioning situation. In part because of the availability of many alternative behavioral indices of memory at the human level, a good deal has been accomplished toward refining methods of measurement and analyzing the separable functions of learning and performance variables.

In view of these considerations, the possibility arises that what was once perhaps considered only a one-way commerce may become a two-way interaction between developments in research and theory at the level of conditioning and the study of memory in the higher organisms and man. Although it must be emphasized that no one else should be thought responsible for the specific viewpoint I have set down here, it is of interest to see how many elements of it have begun to appear in the work of some influential investigators of conditioning and learning over a period of time. Of special relevance to the theme of the present volume is the emergence of some of these views in the writings of Jerzy Konorski during a period of some years ending with the publication of his integrative volume in 1967.

II.   CONDITIONING AND HIGHER PROCESSES IN KONORSKI'S SYSTEM

On the matter of relationships between conditioning and higher mental processes, by the early 1960s Konorski's thinking had already gone far beyond that of the behavior theorists. He seems to have been the first to see clearly that we do not need to look beyond or above conditioning experiments for evidence of higher processes, such as those of perception and memory. These are to be found in the conditioning experiment itself, and we can see them if we simply look more deeply than we have been accustomed to.

Konorski's presentation at the Leyden Symposium (1962) was the first occasion on which I had ever heard an experienced and dedicated investigator of conditioning make serious use of the term perception in the interpretation of conditioning phenomena. One instance occurred in a proposed reinterpretation of the observation, first reported by Pavlov, that in the course of forming a conditioned discrimination an animal typically shows an initial increase in responding to the negative stimulus. The customary interpretation has been that the strength of association between the negative stimulus and the conditioned response increases on early trials, because of generalization from reinforcement of the positive stimulus; then the generalization somehow begins to weaken, as does the strength of association between the negative stimulus and the CR. Konorski proposed instead that the animal's state of information concerning the different reinforcement contingencies associated with the positive and negative stimuli grows monotonically over conditioning trials and that from the outset the animal perceives the difference between the two stimuli (rather than having to learn to discriminate them). The reason for the initial increase in responding to the negative stimulus, Konorski proposed, is that the orienting reflex elicited by the negative stimulus when it is novel suppresses the conditioned response; habituation of this orienting reflex over subsequent trials then removes this suppressive factor and allows the animal to reveal the information it has been accruing concerning the different reinforcing events associated with the positive and negative stimuli. After also considering a series of transposition experiments involving responsiveness to absolute versus relational properties of stimuli, Konorski offered the general conclusion that one cannot generally expect to develop satisfactory interpretations of conditioning solely in terms of laws relating response tendencies to the stimuli presented to the animal. The animal extracts different information from the same stimulus, depending on the nature of the task and the contingencies involved (Konorski, 1962).

The International Congress of Psychology in Washington, D. C., the following year was the occasion of Konorski's first presentation to a large audience of psychologists of his expanding conception of theoretical mechanisms in conditioning. In his invited address to the Congress he remarked (1964), "Since in my own country nobody would ever dream of regarding me as a psychologist, I can only conclude from this invitation that the boundary between psychology and brain physiology is becoming more and more illusory (p. 45)." In this lecture, concerned with mechanisms of instrumental conditioning, Konorski presented early versions of the schemata for associative structures arising in various forms of conditioning that were to eventuate in the elaborate system presented in his 1967 book.

Even at that relatively early stage, Konorski's schema for the associative structure of an instrumental conditioned response was rather more complex than the simple S–R connections (perhaps broken by an O, in obeisance to vague organismic factors, or by an H, denoting habit) that had been taken to represent the outcome of a conditioning experience in much of the earlier literature. For example, his first effort toward representing instrumental conditioning, shown in the upper panel of Fig. 16.1, emphasizes that the CS exerts its influence on the learned response through multiple routes and that joint operation of the linkage by way of the drive center and of the direct connection to the instrumental response mechanism are indispensable for performance. Further development of these notions and their research implications are treated by Morgan (Chapter 7).

The ideas put forward in Konorski's papers of the 1960s concerning the role of perception in conditioning and the form of associative structures proved to be only precursors of the comprehensive system presented in his 1967 volume — the first major attempt in nearly two decades (i.e., since

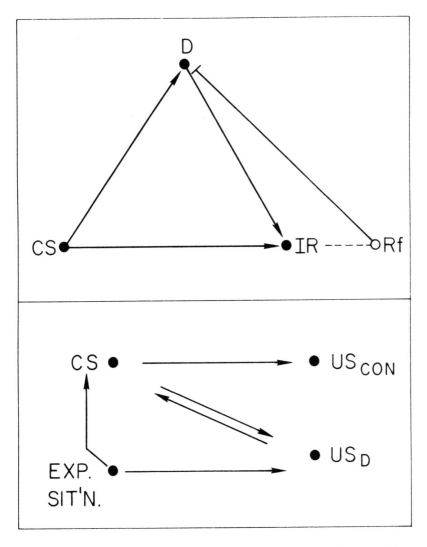

FIG. 16.1. Schematizations of the associative structure established during conditioning. The upper diagram follows Konorski (1964, 1967), the elements associated being a
conditioned stimulus (CS), a drive center (D), an instrumental response (IR), and reinforcement (RF). Solid lines terminating in arrows or bars denote excitatory and inhibitory effects, respectively, and the dashed line denotes temporal succession. The
lower diagram portrays a representation of the corresponding structure established by
classical condtioning, the representation of the US here being subdivided into consummatory (Con) and drive (D) aspects and a representation of the context provided by
the experimental situation (Exp. Sitn.) added. After Konorski (1967).

Hebb, 1949) to develop in broad outline a system that might encompass perception and memory, as well as conditioning and other more elementary forms of learning, in each case with full justice being done to both psychological and physiological aspects.

In this more comprehensive treatise, Konorski clarified his views on the relationship between conditioning and other forms of learning. He assumed specificially that classical conditioning is simply a particular case of the formation of an association between two stimuli (in this instance, a CS and a US) in which the recipient stimulus uniformly produces a specific overt response (the UR). Further, he clearly assumed that the process of formation of associations is the same in classical conditioning as in ordinary human experience. Conditioning is a typical, though experimentally privileged, case of interperceptual association, in which the CR provides a convenient "tracer" of the formation of the association.

However, even in the simplest cases, conditioning involves more than a single association. The associative structure proposed as the outcome of successful classical conditioning, shown in the lower panel of Fig. 16.1, differs in only a few essentials from the one earlier put forward for instrumental conditioning (upper panel of Fig. 16.1). Two aspects of the unconditioned stimulus representation are distinguished—the drive, US_D, whose input must combine with that of the CS and the context in order to produce a response, and the consummatory component, US_{CON}, whose activation whether by external stimulation or by way of the association from the CS leads to occurrence of the UR. An advance is the explicit inclusion of context (nonspecific stimulation from the experimental situation). The one basic respect in which I find Konorski's formulation unsatisfying is that the associative structures are closed networks. The various representations of stimuli and context are connected in pairwise fashion, and there is no natural way for the resultant structure to enter into larger systems as a unit.

Meditation on this problem led me (Estes, 1973a) to a conceptualization that differs from Konorski's in that the representation of CS, US, and context are associated with each other, not directly, but by way of higher-order control elements. The control element, C_1, representing the joint occurrence of a CS and a background context is reactivated when elements of both are reinstated, inputs from CS and context combining multiplicatively. This control element and one, C_2, representing the occurrence of the US in a context comprising the same experimental background plus drive stimuli would be associated by common connections to a higher-order control element, C_{12}. If the initial CS–US presentations of a conditioning experiment were followed by other training in the same situation—for example, via higher-order conditioning procedures—then the control element C_{12} would in turn be associated with others, thus enlarging the associative network to represent memory for the sequence of experiences.

A timely illustration of the importance of including explicit representation of context on a par with experimentally manipulated stimuli is given by a study of the conditioning and extinction of taste aversion by Archer, Sjoden, Nilsson, and Carter (1977). In their study, a saccharine CS and lithium chloride US were paired in a specific background context, X_1. Then extinction of the conditioned aversion was carried out in a distinctively different context, X_2. Tests for saccharine preference in each context following the extinction showed almost complete recovery of the aversion if the test was in X_1, but not if it was in X_2. If, instead, extinction occurred in X_1, the extinction was much slower; but partial recovery of the aversion was still observed if the subsequent test occurred in X_2. Clearly there was strong context specificity even though the contextual cues were entirely exteroceptive. This result is of special interest, for it has been held that the aversion is readily associated only with interoceptive cues (Garcia & Ervin, 1968).

By far the most novel and elaborate extension of Konorski's earlier work in his 1967 volume is his treatment of perception. The perceptual system is conceived to be multilayered, with inputs to the lowest-level units coming from sensory analyzers. The concept of analyzer is basic, yet never fully defined; it evidently refers to the peripheral sensory mechanisms selectively tuned to particular dimensions or properties of stimuli.

Beyond the analyzers, the first principal level of the perceptual system constitutes the afferent, or "transit" field, in which groups of neurons integrate inputs from individual receptors into patterns that are then represented in gnostic units. Output from the transit fields feeds back to mechanisms responsible for the control of receptor input (e.g., targetting reflexes) and also exits upward in the system to the gnostic units in the cortex.

This last conception is treated in more detail by Wagner (Chapter 3). Gnostic units are formed at successively higher levels of organization as a consequence of experience with objects or stimulus patterns. For example, a child's familiarization with letters leads to the formation of gnostic units representing individual letters; continuing experience in learning to read leads to the formation of higher-level units corresponding to syllables or words, which may then be perceived as wholes without being decomposed into constituent elements.

A great deal of evidence bearing on the systematic treatment of perception is presented by Konorski, nearly all coming from a combination of everyday life observations and clinical material, and the same is true of his treatment of associations (with regard to the latter, again see Wagner, Chapter 3). The one major missing ingredient, from my viewpoint, in Konorski's treatments of both perception and association at the psychological level is the absence of any apparent interaction with, or at least any reference to, bodies of relevant research in experimental psychology. To point up some of the possibilities for remedying this lack, I wish in the remainder of this chapter to focus on a single topic in which it appears that experimental results now available could fruitfully be brought to bear on ideas growing out of theoretical developments such as Konorski's.

Of the several major aspects or components of Konorski's system, I propose to consider one aspect — namely, the treatment of memory — in some detail, hoping to elucidate both the existing and the potential interactions between this theory, growing out of a neurophysiological approach to conditioning, and current research and theory on human memory.

The conventional view, of course, is that the relationship between these two disciplines should be one of reduction. Concepts and theories of human memory should prove reducible to concepts and theories of neurophysiology, and the experimental phenomena of the former should be explained, if not indeed explained away, by the latter. However, reduction requires both a clear picture of the phenomena to be explained and a securely anchored theory to do the explaining. Probably few will argue with the idea that we are a long way from satisfying the necessary conditions at either level. But more importantly, I would like to develop the view that progress toward the desired objective will be unnecessarily slow and tortuous if the experimental psychology of memory and the study of the mechanisms responsible for memory in the brain proceed independently.

It is quite widely recognized by psychologists, to be sure, that we can and should take account to the greatest extent possible of current knowledge about the structure and function of the nervous system when formulating theories of psychological processes such as memory. But I think we have also reached a point where information from research on memory can advantageously be brought to bear on theoretical concepts and issues growing out of neurophysiological research. The particular concepts I wish to examine in order to illustrate this thesis are the treatments of short- and long-term memory in Konorski's system, on the one hand, and in current experimental psychology, on the other.

III. THE DUAL-ASPECT CONCEPTION OF MEMORY

It is well known that major theoretical developments often arise nearly simultaneously but independently in widely different research contexts. An interesting case in point is the theoretical distinction between short-term and long-term memory. At the same time that Atkinson and Shiffrin were formulating their information-processing model for short- and long-term memory (Atkinson & Shiffrin, 1968) — first presented at the Moscow Congress in 1966 and quickly to become one of the major influences on psychological research, especially in short-term human memory — Konorski was completing his 1967 volume, in which he developed, almost entirely on the basis of neurophysiological considerations, a conception of memory organized around two principal types or levels — transient memory, the counterpart of Atkinson and Shiffrin's short-term store, and associative memory, the counterpart of Atkinson and Shiffrin's long-term store.

In Konorski's system memory depends on the establishment in the brain of functional entities termed *gnostic units,* as a consequence of sensory experiences. Gnostic units are activated whenever the neural input arising during any sensory experience reaches the sensory-projection areas. The transient memory that persists immediately following any perceptive process is attributed to a continuing state of excitability of gnostic units as a consequence of the activity of self-reexciting chains of neurons, possibly in the form of reverberatory corticothalamic loops. As long as the gnostic units remain in the state of heightened exitability, they can be aroused by input from nonspecific activating systems and lead to partial reinstatement of the sensory experience. The functions of this transient memory system Konorski conceived to be twofold: first, to enable planning, the counterpart of working memory in current psychological theories (e.g., Baddeley & Hitch, 1974); second, to enable the organism to keep track of what it has just done and thus avoid useless repetitions of actions (Konorski, 1967, pp. 494–495).

Long-term memories, in contrast, are assumed by Konorski to be associative. The connections responsible develop after a coincident activity of two gnostic units, an additional necessary condition being the arousal of the recipient unit by input from nonspecific activating systems (associated with biological drives or emotive states). An association becomes functional in a later test situation only if the specific sensory input appropriate to one of the units recurs together with nonspecific activation of both members by the emotive system.

I find the dual-aspect interpretation proposed by Konorski appealing in many ways. In fact I arrived at much the same conception, though by way of a very different intellectual route (Estes, 1972). I was trying to account for detailed patterns of short-term recall data. Konorski, on the other hand, was clearly guided mainly by very general biological considerations, neuroanatomical evidence with regard to possible circuitry, and, on the empirical side, familiarity only with delayed-response data and a few incidental observations from conditioning experiments.

The immediate possibilities for following up and testing the idea that memory comprises a transient nonassociative and a long-term associative component are somewhat asymmetrical. On the one side, there seems to be no hope of much guidance from neurophysiology concerning further development of these ideas in the near future. Furthermore, the methods of conditioning and animal learning experiments do not yet offer any obvious ways of getting incisively at the assumed properties of the transient versus long-term memory distinction, although the growing interest in experimental studies of animal memory (e.g., Medin, Roberts & Davis, 1976; Shimp, 1975; Wagner, Chapter 3) may begin to redress matters within the foreseeable future. On the other hand, the methodology that has developed in the experimental study of human memory, though remote from physiological processes, provides increasingly sharp means of analyzing the data of controlled experiments on memory so as to isolate assumed processes and enable the testing of specific hypotheses concerning the functioning of the system.

It is not to be expected that specific findings of studies on human memory will carry over directly to conditioning, but it is often found that looking at a given empirical area from a fresh viewpoint proves salutory. In experimental psychology this kind of benefit takes the form of stimulating attention to aspects of measurement that have not been fully exploited. In the present instance it seems to me that the stage may be set for a contribution of just this sort, by the great differences between current research efforts in human memory and in conditioning regarding the degree to which distinguishable aspects of the information stored in memory concerning stimulus–response sequences have been incorporated into theories.

Of particular import is the distinction between item and order information—that is, memory for occurrence of certain events in a given situation as distinguished from memory for the order in which they occurred. Taking adequate account of the distinction has proven important in the development of recent theories of human short-term memory (e.g., Estes, 1972; Lee & Estes, 1977; Murdock, 1974). One wonders whether the same might not prove true in the case of theories of conditioning.

In consequence of these considerations, it might be more than an intellectual exercise to review the ways in which current knowledge, methods, and theory of human memory can be brought to bear on the distinction between transient and long-term memory. Because the associative nature of long-term memory is hardly in question, I shall concentrate on the question of specific evidence for the nonassociative nature of transient, short-term memory.

IV.   EXPERIMENTAL EVIDENCE ON THE NON-ASSOCIATIVE CHARACTER OF TRANSIENT MEMORY

A. Clarification of the concept.

Before considering specific findings, we need to be clear about just what we should be looking for. This step requires a clear statement of the alternative models we wish to distinguish at a conceptual level, and a specification of the property or properties of observable data that would distinguish these models. To this end we may refer to Fig. 16.2,, in which (at the left) I have schematized the events occurring on a trial of a typical short-term recall experiment. Following the occurrence of a "ready" or "start" signal, a series of simple stimuli—for example, individual letters—is presented to the experimental subject. In experiments I describe later, the letters appear singly on a display screen, one after another in the same location. The letters that would actually be presented on a trial are a random selection, of course, not the ordered sequence of letters shown in the figure. The presentation rate is relatively rapid, typically 2 or 2-1/2 letters per sec, and the subject is required to say aloud the name of each letter as it appears, thus enabling the experimenter to be sure that all the letters are being perceived and that the subject is not able to rehearse the items. After the to-be-remembered letters have been presented, the subject attempts to recall them, either immediately or following a retention interval; in the latter case the retention interval is filled with a distractor activity—for example, the presentation of a rapid sequence of random digits that the subject must pronounce aloud as they appear, to preclude rehearsal and allow a determination of memory at a given

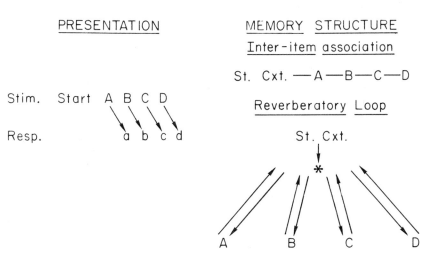

FIG. 16.2.   Schematic representation of the events of a typical trial in a short-term recall experiment on the left, and of two types of models for short-term retention on the right. In the experimental paradigm the capital letters denote stimulus items presented successively to an experimental subject, and small letters denote the subject's response of vocalizing the name of each as it occurs. At the upper right is a schematization of a conception of interitem associations or connections; at the lower right, a schematization the idea of nonassociative short-term memory maintained by reverberatory loops, the cyclically reactivating circuits continuing in a state of activity as long as input continues from the original stimulus context.

time following a single presentation of an item.

The two alternative types of models we wish to distinguish are illustrated at the right of Fig. 16.2. In the interitem-association model, which would represent long-term memory in Konorski's dichotomy, the first item of the sequence is connected by an associative link with the stimulus context, the memory representation of the second letter with that of the first, the third letter with the second, and so on. In contrast, in the nonassociative (reverberatory loop) model, it is assumed that the memory representation of each of the letters that has been presented is maintained independently by a reverberatory neural circuit.

Now with regard to empirical implications, it is clear that according to the interitem-association model the likelihood that a given letter (say C, in the illustration) will be recalled must depend on whether its predecessor, B, was recalled on the same trial, For only if B was recalled will the B–C connection be activated. On the other hand, in the reverberatory-loop model, whether or not C is recalled depends only on its state of activation and not on the state of activation of any of the other letters. Thus, we understand recall data to support the nonassociative conception only if the recall of any two items is statistically independent; that is, for any items A and B, $P(AB) = P(A) \times P(B)$. But if A and B are associated, then $P(AB)$ should be greater than $P(A) \times P(B)$.

B. Earlier studies.

Some of the studies designed to yield evidence on the independence or nonindependence of recall of different items presented on a trial of such an experiment have produced results that are not entirely in agreement. In an experimental paradigm preferred by Wickelgren (1965), the comparison is made between recall of different items that occur at the same serial positions of different sequences but in the two cases accompanied by different items in the other positions, with conditions so arranged that expectations would be different on hypotheses of independence versus dependence. For example, letting capital letters signify items, a comparison might be made between recall of B and D in the sequence ABCD versus their recall in the sequence ABAD. If memory under the given conditions is nonassociative so that independence holds, there should be no difference in the two cases. But if memory is associative, then recall of B and D would be poorer in the sequence ABAD because both would be associated with the same stimulus A, thus leading to interference at recall. In a study employing this design with auditorily presented items, Wickelgren (1965) reached the conclusion that auditory short-term memory is associative, because he obtained a difference in recall between the two conditions: however, in a later study using visually presented items, no difference was observed and the conclusion was suggested that visual short-term memory is nonassociative (Wickelgren & Whitman, 1970).

A rather extensive study by Baddeley (1968) included instances of both visual and auditory presentations and a slightly different design. In one of these experiments, for example, Baddeley used randomly ordered sequences of consonant letters as the stimuli to be remembered. The letters used were of 2 types: letters that were auditorily similar with respect to their vowel phoneme, B, C, and D, and letters that were auditorily dissimilar, J, L, and R. On any trial a random sequence of 6 letters was presented, the letters appearing singly in a display screen at the rate of .75 sec per letter. Presumably the subjects pronounced the letters, at least subvocally, as they appeared.

On reasoning similar to that of Wickelgren's study, we would expect impaired recall of letters that follow auditorily similar letters in the sequence if memory is associative but not if it is nonassociative.

Consider, for example, the sequence JDLCRB (upper panel, Fig. 16.3), illustrating the pattern DSDSDS, where S denotes a member of the auditorily similar and D a member of the dissimilar set. If memory for order were maintained by associations between successive items, then a letter following an S letter should be recalled more poorly than one following a D letter, for in the former case generalization would occur to and from other S letters, thus producing confusion errors. For example, the e sound would be associated with the letter L in view of the successive occurrence of D and L in the sequence, but also with R in view of the successive occurrence of C and R thus providing a source of interference when either D, C, or B is encountered during a recall test.

On the other hand, if memory is nonassociative, there would be no such expectation; in fact we should predict poorer recall of the similar letters themselves, because the memory traces of these would be more difficult to distingush from one another than traces of dissimilar letters at the time of recall. The results of Baddeley's study clearly favor the nonassociative hypothesis, as illustrated in terms of a subset of his data reproduced in Fig. 16.3. It is seen that in neither of the two types of sequences shown are there any cases in which recall is impaired if a dissimilar letter follows a similar one. But if we compare recall of similar versus dissimilar letters presented in the same serial position of the two sequences in either panel, we find poorer recall of the similar letter in every case.

C. Isolating nonassociative, transient memory.

In the course of some research on short-term recall, the idea emerged that the sometimes apparently conflicting results of experiments such as those of Wickelgren and Baddeley might be accounted for if, under some of the conditions studied, recall actually depended on some mixture of nonassociative,

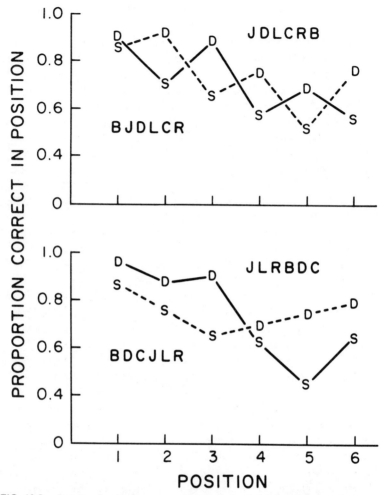

FIG. 16.3. Results of a short-term recall experiment in which auditorily similar (S) letters, B, C, D, and auditorily dissimilar (D) letters, J, L, and R, were intermixed as indicated. It is seen that for each serial position of a list of to-be-remembered items, recall is better for the dissimilar than for the similar items, as anticipated on the basis of a non-associative model; but in no case is recall impaired at the position following a similar as opposed to a position following a dissimilar letter, contrary to expectations from an associative model. after Baddeley (1968, Exper. V).

short-term memory and associative, long-term memory (Lee & Estes, 1977). To check on this possibility, we contrived an experimental arrangement that might be expected to yield separate cases in which the associative component would and would not be expected to be involved. These are illustrated in Fig. 16.4. In both cases the subject was presented on each trial with a sequence of 12 items comprising a mixture of 4 letters and 8 ramdomly selected digits. Only the letters were to be recalled, but they were arranged in 2 different ways with respect to the distractor digits. Under the *spaced* condition, the letters of each pair were separated by 3 digits, whereas in the *clustered* condition the letters were all grouped together. In both conditions the subject was required to pronounce aloud the name of each letter and digit as they appeared successively at the rate of 2-1/2 characters per sec; then, at the end of the trial, the subjects were to recall the letters, also indicating the position among the 12 possible "slots" in which they believed that each letter had occurred.

In the upper line of data are shown the conditional probabilities of recall of the third of the four letters provided that the second was recalled on that trial; and in the lower line are shown the conditional probabilities of recall of the third letter provided that the second letter had not been recalled. If memory is associative, the first value should of course be larger than the second. What we observe is that in the clustered condition, where there might be a mixture of nonassociative and associative memory, there is indeed a large advantage in favor of recall of the third letter when the second had been recalled (thus providing the stimulus for an associative link, if there were one, between the two letters). But in the spaced condition, although recall was nearly as high as in the clustered condition, there was no difference whatever in the direction predicted by an associative hypothesis, and in fact even a slight difference in the "wrong" direction.

To check further on the idea that the clustered condition produces a mixture in which recall depends in part on short- and long-term memory systems, I reanalyzed some data of an earlier experiment (Estes, 1973b), with the result shown in Fig. 16.5. Here again the letters to be remembered

	SPACED	CLUSTERED
	Q725M617J492P	...561QMJP49...
$L_2 \rightarrow L_3$	.70	.87
$\overline{L}_2 \rightarrow L_3$	.75	.71

FIG. 16.4. Sample stimulus sequences for a short-term recall experiment conducted with *spaced* input strings — that is, the to-be-remembered letters are spaced out by distractor digits — as compared with a *clustered condition,* in which the to-be-remembered letters are all adjacent. The conditional probability that the third letter of the sequence is recalled given that the second was recalled on the trial is much larger than the probability of recall of the third letter given nonrecall of the second in the clustered condition but not in the spaced—supporting the idea that recall in the latter case is based on nonassociative memory, but in the former perhaps on a mixture of nonassociative and associative. Data from Lee and Estes (1977).

were randomly selected consonants and in all cases were presented under the clustered conditions. The upper panels show results for the usual procedure, in which letters were presented at a rate of 2-1/2 per sec and were pronounced aloud by the subjects, whereas the lower panel shows results for a variation in which the letters were presented at a rate of 5 per sec and were not (indeed could not) be pronounced by the subjects—thus eliminating any auditory input. In each panel we show the same conditional probabilities that were exhibited in Fig. 16.4 but separately for retention intervals in which the letters were followed by 12, 24, 36 random digits. The significant result from the standpoint of present interests is that the differences between the conditional probabilities of recall of the third letter, given that the second was or was not recalled, are small at the shortest retention interval and

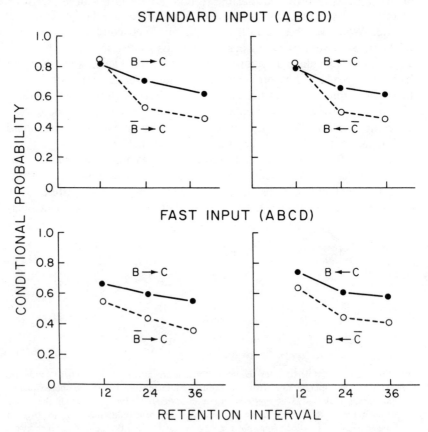

FIG. 16.5.   Comparisons of conditional probabilities of recall of a particular letter of a sequence provided that the preceding letter was or was not recalled, as in Fig. 16.4, but here plotted as a function of retention interval. Data from Estes (1973b).

increase with the duration of the retention interval. This result is, of course, precisely what would be expected if recall at short intervals is a mixture of nonassociative and associative memory, with perhaps the former predominating, but with the short-term, nonassociative component dropping out as the retention interval increases.

These results are just a sample of a rather large assemblage that seems to support the conception of a very short-term memory system that is nonassociative and may well be accounted for in terms of the recycling or reverberatory-loop hypothesis put forward independently by Konorski and by this writer.

The methods of the Lee and Estes (1977) study enable us to carry out another kind of analysis, which is quite instructive with regard to detailed properties of the short-term memory process and without any counterpart yet available to us in the case of conditioning or animal-learning experiments. The idea involved is illustrated in Fig. 16.6. Suppose that the sequence of letters shown at the top has been presented to a subject, the task being to try, at the end of a retention interval, to recall each letter and also to indicate the serial position in which it occurred. If the data for many such trials are pooled together, we can then plot functions of the form labeled "memory position gradients," showing the empirically determined probability with which a letter presented at any given serial position is recalled by the subject in any one of the possible positions. The gradients shown for the middle two letters of the illustrative trial show the regularly obtained result that each letter is most often recalled at its correct position, somewhat less often at the two adjacent positions, and least often at the most remote position.

Now if recall under the given conditions is mediated by a nonassociative, short-term memory process, as illustrated in Fig. 16.2, and if errors in which a letter is recalled but placed in an incorrect position arise as a consequence of random perturbations in the recycle times of the reverberatory loops, resulting in some of the letters being reactivated out of their correct order, then the frequency of errors that involve inversions of the order of any two letters should be predictable from the memory-position gradients. The kind of error referred to is illustrated at the bottom of the figure, the letters Q and R both being recalled but in one case in the correct order and in another case in an inverted order. In terms of the nonassociative model, the probability of an inversion is simply the sum of the probabilities that any point on the memory-position gradient for the letter Q falls to the right of any point on the gradient for the letter R. In the Lee and Estes (1977) study these computations were carried out, and it was shown that for the spaced strings (Fig. 16.4) the frequency of inversion errors was very accurately predicted from the memory-position gradients, as expected on the nonassociative hypothesis; whereas for clustered strings, inversions of order

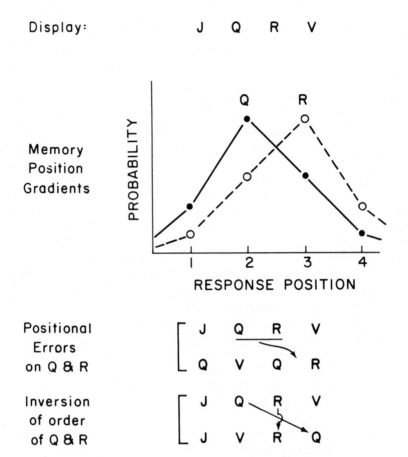

FIG. 16.6. Illustration of memory-position gradients representing the theoretical uncertainty distribution of an individual's memory for positions of particular letters of a to-be-remembered sequence. The x-axis represents the time interval during which the displayed letters were presented, the numerals 1 to 4 denoting the positions of the 4 letters within this interval. The functions are experimentally determined in terms of the observed probabilities with which subjects assign recalled items to particular positions at recall. Because in the example the gradients for the letters Q and R overlap, one must predict that there will be occasions on which both letters are remembered out of their true positions but still in the correct order, as in the first illustration of the bottom, and instances in which they are recalled in the incorrect order, as in the second illustration. The frequencies of these types of errors prove to be predictable from the position gradients, a result strongly supporting the nonassociative model for transient memory. From Lee and Estes (1977).

did not occur as often as expected, indicating that recall in that case depended on a mixture of nonassociative and associative memory.

V. EXTENSION OF THE NONASSOCIATIVE MODEL
TO CONDITIONING

It will be perhaps almost obvious how the concepts of the nonassociative model illustrated in Figs. 16.2 and 16.6 could be applied to a conditioning paradigm. Conditioned and unconditioned stimuli would of course correspond to individual items in the recall experiment, and the representation of each stimulus would be maintained for a short time after its occurrence by means of reverberatory circuits. The memory-position gradients for two adjacent items, Q and R in Fig. 16.6, could be reinterpreted as representing the CS and US respectively, the only change needed being to relabel the x-axis in terms of temporal units rather than discrete response positions.

If the administrations of the two stimuli were too far apart, their representations would never be recycled simultaneously, and if they were too close together the position gradients would overlap to such an extent that the organism could not remember which had come first; and in either case there would be little or no basis for anticipation of the UR upon recurrence of the CR. For intermediate intervals between onsets, the memory gradients would only partially overlap, and there would be information maintained concerning the order of the two stimuli.

It is of interest in this connection to recall the literature on simultaneous and backward conditioning, which has largely resisted effective interpretation in terms of existing theories of learning and conditioning. In general, evidence of conditioning is difficult to obtain when the CS and US are presented simultaneously or in "backward" order. Nonetheless in a number of studies a systematic trend has been observed, with some apparent conditioned responses occurring after the first few presentations of CS and US in a simultaneous or backward arrangement, but with the probability of a CR declining over a subsequent series of paired presentations (e.g., Heth, 1976; Spooner & Kellogg, 1947).

In terms of the extension of the short-term memory model sketched above, we would expect the original response to the US to occur anticipatorily upon the appearance of the CS only if the animal has information in memory that the US has previously followed the CS. On the first few trials of a typical experiment, performance may largely reflect short-term memory, and in terms of the model, if the CS and US have been presented simultaneously or very close together, their temporal uncertainty functions would almost completely overlap. In that case, shortly after a   paired presentation, the organism would have only very ambiguous information in memory as to the temporal order of the stimuli. On some occasions it would remember, incorrectly, that the CS had preceded the US and then might make an anticipatory CR upon a new presentation of the CS. But over a series of trials this tendency would be inhibited when the animal had opportunity to learn that the CS alone was followed instead by absence of the US. The results of Heth (1976), showing that the reduction in frequencies of CRs with successive simultaneous pairings of CS and US is specifically a function of the number of pairings. with other factors controlled, seems in line with this analysis.

Even more striking in this context are some new results of Rescorla and-Furrow (1977) showing that a CS that has been paired simultaneously with a US, and thus demonstrates little or no tendency itself to evoke CRs, may nonetheless provide an adequate basis for higher-order conditioning if it in turn is preceded on a series of trials by another, originally neutral stimulus. It terms of the model, on the simultaneous CS–US trials, the animal is left with the memory that the CS is usually followed by termination or absence of the US, whereas on the higher-order conditioning trials it is left with the memory that the new stimulus has preceded a stimulus that has frequently been experienced in conjunction with the US.

With regard to potential applicability of the item-order distinction in conditioning, we might take note of the familiar fact that different measures of conditioning often yield apparently sharp disagreements (Warner, 1932; Zener, 1937). In particular, after even a single occurrence of a traumatic event, such as an electric shock or a blast of air to the cornea, one can virtually always detect some behavioral symptoms of memory for the event, even though very often the specific unconditioned response to the given US does not occur on presentation of the CS.

Some indices of the presence of memory for the US are much more sensitive than others, a stabilimeter often showing a clear change in activity level (e.g., freezing or hyperactivity) even though no response closely resembling the UR is evoked. One interpretation is that the organism remembers some attributes of the events that occurred, but not enough to specify the unconditioned response completely and perhaps not of the right kind to specify the order of the stimuli and responses that occurred on the conditioning trial. Repeated trials are needed, not to strengthen some one connection or item of information, but rather to provide opportunities for adding additonal information with regard to both the events that occurred and their temporal ordering.

Going beyond the short-term memory model, I would assume, in keeping with the ideas of Anderson (1973, 1977), that when the reverberatory processes responsible for transient memory of the two stimuli are maintained long enough, a trace is left in the assemblage of neural units involved, together with a trace representing the context in which the stimuli occurred, and this memory trace provides the basis for longer-term retention.

VI. CONCLUDING COMMENT

These illustrations from current research and theory may suffice to make the point that some of the ideas concerning relationships between conditioning and other mental processes that Konorski put forward primarily on the basis of neurobiological considerations can make some meaningful contact with concepts and models growing out of research on human memory. It seems reasonable to expect that we may see continuing progress toward a common body of theory, with this progress being by no means automatic but, rather, depending strongly on the degree to which results obtained in each of the disciplines feeds back to catalyze and direct both experimental and theoretical work in the other.

ACKNOWLEDGMENTS

Research reported in this chapter was supported in part by Grant GB41176 from the National Science Foundation and by Grant MH23878 from the National Institute of Mental Health.

REFERENCES

Anderson, J. A. A theory for the recognition of items from short memorized lists. *Psychological Review,* 1973, *80,* 417–438.

Anderson, J. A., Silverstein, J. W., Ritz, S. A., & Jones, R. S. Distinctive features, categorical perception, and probability learning: Some applications of a neural model. *Psychological Review,* 1977, *84,* 413–451.

Archer, T., Sjoden, P. O., Nilsson, L. G. & Carter, N. *Role of exteroceptive context in taste-aversion conditioning and extinction.* Unpublished manuscript, University of Uppsala, 1977.

Atkinson, R. C., & Shiffrin, R. M. Human memory: A proposed system and its control processes. In K. W. Spence & J. T. Spence (Eds.), *The psychology of learning and motivation* (Vol. 2). New York: Academic Press, 1968.

Baddeley, A. D. How does acoustic similarity influence short-term memory? *Quarterly Journal of Experimental Psychology,* 1968, *20,* 249–264.

Baddeley, A. D. & Hitch, G. Working memory. In G. H. Bower (Ed.), *The psychology of learning and motivation* (Vol. 8). New York: Academic Press, 1974.

Bolles, R. C. Learning, motivation, and cognition. In W. K. Estes (Ed.), *Handbook of learning and cognitive processes* (Vol. 1). Hillsdale, N. J. Lawrence Erlbaum Associates, 1975.

Estes, W. K. Toward a statistical theory of learning. *Psychological Review,* 1950, *43,* 94–107.
Estes, W. K. The statistical approach to learning theory. In S. Koch (Ed.), *Psychology: A study of a science* (Vol. 2). New York: McGraw-Hill, 1959.
Estes, W. K. *Learning theory and mental development.* New York: Academic Press, 1970.
Estes, W. K. An associative basis for coding and organization in memory. In A. W. Melton & E. Martin (Eds.), *Coding processes in human memory.* Washington, D. C.: Winston, 1972.
Estes, W. K. Memory and conditioning. In F. J. McGuigan & D. B. Lumsden (Eds.), *Contemporary approaches to conditioning and learning.* Washington, D. C.: Winston, 1973.(a)
Estes, W. K. Phonemic coding and rehearsal in short-term memory for letter strings. *Journal of Verbal Learning and Verbal Behavior,* 1973, *12,* 360–372.(b)
Estes, W. K. The state of the field: General problems and issues of theory and meta-theory. In W. K. Estes (Ed.), *Handbook of learning and cognitive processes* (Vol. l). Hillsdale, N. J.: Lawrence Erlbaum Associates, 1975.
Garcia, J., & Ervin, F. R. Gustatory-visceral and telereceptor-cutaneous conditioning: Adaptation in internal and external milieus. *Communications in Behavioral Biology,* 1968, *1,* 389–395.
Gormezano, I., & Kehoe, E. J. Classical conditioning: Some methodological-conceptual issues. In W. K. Estes (Ed.), *Handbook of learning and cognitive processes* (Vol. 2). Hillsdale, N. J.: Lawrence Erlbaum Associates, 1975.
Hebb, D. O. *The organization of behavior.* New York: Wiley 1949.
Heth, C. D. Simultaneous and backward fear conditioning as a function of number of CS–UCS pairings. *Journal of Experimental Psychology: Animal Behavior Processes,* 1976, *2(2),* 117–129.
Hilgard, E. R., & Marquis, D. G. *Conditioning and learning.* New York: Appleton-Century-Crofts, 1940.
Hull, C. L. Mind, mechanism and adaptive behavior. *Psychological Review,* 1937, *44,* 1–32.
Hull, C. L. *Principles of behavior.* New York: Appleton-Century-Crofts, 1943.
Kamin, L. J. Selective association and conditioning. In N. J. Mackintosh & W. K. Honig (Eds.), *Fundamental issues in associative learning,* Halifax: Dalhousie University Press, 1969.
Konorski, J. The role of central factors in differentiation. In R. W. Gerard & J. W. Duyff (Eds.), *Information processing in the nervous system.* Amsterdam: Excerpta Medica Foundation, 1962.
Konorski, J. On the mechanism of instrumental conditioning. *Proceedings of the Seventeenth International Congress of Psychology.* Amsterdam: North-Holland Publishing Company, 1964.
Konorski, J. *Integrative activity of the brain.* Chicago: University of Chicago Press, 1967.
Lee, C. L., & Estes, W. K. Order and position in primary memory for letter strings. *Journal of Verbal Learning and Verbal Behavior,* 1977, *16,* 395–418.
Mackintosh, N. J. From classical conditioning to discrimination learning. In W. K. Estes (Ed.), *Handbook of learning and cognitive processes* (Vol. 1). Hillsdale, N. J.: Lawrence Erlbaum Associates, 1975.
Medin, D. L., Roberts, W. A., & Davis, R. T. *Processes of animal memory.* Hillsdale, N. J.: Lawrence Erlbaum Associates, 1976.
Murdock, B. B. *Human memory: Theory and data.* New York: Wiley, 1974.
Rescorla, R. A. & Furrow, D. R. Stimulus similarity as a determinant of Pavlovian conditioning. *Journal of Experimental Psychology: Animal Behavior Processes,* 1977, *3,* 203–215.

Shimp, C. P. Perspectives on the behavioral unit: Choice behavior in animals. In W. K. Estes (Ed.), *Handbook of learning and cognitive processes* (Vol. 2). Hillsdale, N. J.: Lawrence Erlbaum Associates, 1975.

Spence, K. W. The differential response in animals to stimuli varying within a single dimension. *Psychological Review,* 1937, *44,* 430–444.

Spence, K. W. Theoretical interpretations of learning. In S. S. Stevens (Ed.), *Handbook of experimental psychology,* New York: Wiley, 1951.

Spooner, A., & Kellogg, W. N. The backward conditioning curve. *American Journal of Psychology,* 1947, *60,* 321–334.

Warner, L. H. An experimental search for the "conditioned response." *Journal of Genetic Psychology,* 1932, *41,* 91–115.

Wickelgren, W. A. Short-term memory for repeated and non-repeated items. *The Quarterly Journal of Experimental Psychology,* 1965, *17,* 14–25.

Wickelgren, W. A. & Whitman, P. T. Visual very-short-term memory is non-associative. *Journal of Experimental Psychology,* 1970, *84,* 277–281.

Zener, K. The significance of behavior accompanying conditioned salivary secretion for theories of the conditioned response. *American Journal of Psychology,* 1937, *50,* 384–403.

BIBLIOGRAPHY

Bibliography of Konorski's publications compiled by R. Glowacka and previously published in *Acta Neurobiologiae Experimentalis*, 1974, *34*, 681–695. Supplementary references supplied by C. Zielinski.

1928

MILLER, S., & KONORSKI, J. Dzialalność kory mózgowej w świetle teorii Pawlowa. 1. Odruchy warunkowe (The activity of the cerebral cortex in light of Pavlov's theories. 1. Conditioned reflexes). *Warsz. Czas. Lek., 5*, 528–530.

MILLER, S., & KONORSKI, J. Dzialalność kory mózgowej w świetle teorii Pawlowa. 2. Nauka o analizatorach (The activity of the cerebral cortex in light of Pavlov's theories. 2. The science of analysers). *Warsz. Czas. Lek., 5*, 555–557, 575–576, 595–596.

MILLER, S., & KONORSKI, J. Le phénomène de la généralisation motrice. *Compt. Rend. Séanc. Soc. Biol., 99*, 1158.

MILLER, S., & KONORSKI, J. Sur une forme particulière des réflexes conditionnels. *Compt. Rend. Séanc. Soc. Biol., 99*, 1155–1157.

1930

KONORSKI, J., & MILLER, S. L'influence des excitateurs conditionnels et absolus sur les réflexes conditionnels de l'analysateur moteur. *Compt. Rend. Séanc. Soc. Biol., 104*, 911–913.

KONORSKI, J., & MILLER, S. Méthode d'examen de l'analysateur moteur par les réactions salivo-motrices. *Compte Rend. Séanc. Soc. Biol.*, 907–910.

KONORSKI, J., & MILLER, S. Zasada dominanty w dzialalności ukladu nerwowego (The principle of dominance in the activity of the nervous system). *Warsz. Czas. Lek. 7*, 33–35.

1932

KONORSKI, J. Odruchy warunkowe analizatora ruchowego (Les réflexes conditionnels de l'analysateur moteur). *Roczn. Psychiat. 18/19*, 308–312.

KONORSKI, J., & MILLER, S. O fizjologicznym traktowaniu postepowania (Über die physiologische Behandlung des Vorgehens). *Now. Psychiat.* (Gniezno), *9*, 255–257.

KONORSKI, J. and MILLER, S. O warunkach wytwarzania ruchowych odruchów warunkowych (Über die Bedingungen der Ausbildung von bedingten motorischen Reflexen). *Now. Psychiat.* (Gniezno), *9*, 257–258.

1933

KONORSKI, J., & MILLER, S. Nouvelles recherches sur les réflexes conditionnels moteurs. *Compt. Rend. Séanc. Soc. Biol., 115*, 91–96.

KONORSKI, J., & MILLER, S. Podstawy fizjologicznej teorii ruchów nabytych. Ruchowe odruchy warunkowe (Les bases de la théorie physiologique des mouvements acquis. Les réflexes conditionnels moteurs). *Med. Dośw. Spol., 16*, 95–187.

KONORSKI, J., & MILLER, S. Podstawy fizjologicznej teorii ruchów nabytych. Ruchowe odruchy warunkowe. Dokończenie (Les principes fondamentaux de la théorie physiologique des mouvements acquis. Les réflexes conditionnels moteurs. Fin). *Med. Dośw. Spol., 16*, 234–298. (French summary)

KONORSKI, J., & MILLER, S. *Podstawy fizjologicznej teorii ruchów nabytych. Ruchowe odruchy warunkowe (Les principes fondamentaux de la théorie physiologique des mouvements acquis. Les réflexes conditionnels moteurs).* Warszawa: Ksiaznica Atlas TNSW. (French summary)

KONORSKI, J., & MILLER, S. Próba fizjologicznego objaśnienia nabytej dzialalności ruchowej zwierzat (Ein Versuch der physiologischen Aufklarüng der erworbenen motorischen

Tätigkeit der Tiere). *Przegl. Fizjol. Ruchu, 5,* 187-216. (German summary)

KONORSKI, J., & MILLER, S. Zachowania sie zwierzat w świetle fizjologii (Physiological basis of animal behavior). *Wszechświat 1933,* 136-141.

1934

KONORSKI, J. O mechanizmie dzialalności ruchowych (On the mechanism of motor activity). *Wszechświat 1934,* 139-145.

KONORSKI, J., & MILLER, S. Die Grundlagen der physiologischen Theorie erworbener Bewegungen. Zentbl. *Ges. Neurol. Psychiat, 69,* 590.

1936

KONORSKI, J. I. P. Pawlow. *Wszechświat 1936,* 67-72.

KONORSKI, J., & LUBINSKA, L. Próba analizy zjawiska "narkozy magneziwej". III. Mechanizm obwodowego dzialania magnezu i pozorny charakter zmian pobudliwości nerwowej (An attempt to analyze "magnesium narcosis." III. The mechanism of the peripheral action of magnesium and the fallacy of observed changes in nerve excitability). *Acta Biol. Exp., 10,* 251-281. (English summary)

KONORSKI, J., LUBINSKA, L., & MILLER, S. Wytwarzanie sie odruchów warunkowych w zajamowanej indukcyjnie korze mózgowej (Élaboration des reflexes conditionnels dans l'ecorce cérébrale a l'état d'inhibition induite). *Acta Biol. Exp., 10,* 297-330. (French summary)

KONORSKI, J. & MILLER, S. Uslovnye reflesky dvigatel'nogo analizatora (Conditioned reflexes of the motor analyzer). *Trudy Fiziol. Lab. I. P. Pavlova 6:* 119-278. (English summary, pp. 285-288)

1937

KONORSKI, J. Wazniejsze zagadnienia w zakresie lokalizacji w korze mózgowej (Major problems concerning localization in the cerebral cortex). *Wszechświat 1937,* 231-237.

KONORSKI, J., & LUBINSKA, L. Sur le caractère apparent des troubles de l'excitabilité nerveuse pendant la curarisation partielle par l'ion magnésien. *Arch. Int. Physiol., 44,* 249-264.

KONORSKI, J., & MILLER, S. Further remarks on two types of conditioned reflex. *J. Gen. Psychol., 17,* 405-407.

KONORSKI, J., & MILLER, S. On two types of conditioned reflexes. *J. Gen. Psychol., 16,* 264-272.

KONORSKI, J., & MILLER, S. sprawie samoistnego przeksztalcania sie nawyków. Badania nad orientacja przestrzenna u szczurów (Spontaneous transformation of habits. Investigations into the sense of space in rats). *Pol. Arch. Psychol. 9,* 68-80. (English summary)

DOLIN, A. O., & KONORSKI, J. Analiz funktsii golovnogo mozga v protsesse oshibochnykh perebezhek krys v labirinte (The analysis of brain function during the process of erroneous maze-running by rats). *Fiziol. Zh. SSSR, 22,* 187-203.

1938

KONORSKI, J. O zmienności ruchowych reakcji warunkowych (Zasady przelaczania korowego) (Sur la variabilité des réactions conditionnelles motrices [Les principles d'aigullage cortical]). *Przegl. Fizjol. Ruchu, 9* (1938/1939), 191-241. (French summary)

KONORSKI, J. Wazniejsze zagadnienia w zakresie lokalizacji w korze mózgowej (dokończenie) (Major problems concerning localization in the cerebral cortex [continuation]). *Wszechś*wiat 1938, 12-16.

KONORSKI, J., & LUBINSKA, L. A propos de l'action de la strychnine sur la préparation neuro-musculaire. *Acta Biol. Exp., 12,* 13-21.

1939

KONORSKI, J., & LUBINSKA, L. Sur les propriétés d'un centre moteur de l'ecorce cérébrale étudiées sur l'animal normal. *Ann. Physiol. Physiochim, Biol.,* p. 372. XIII Reu-
nion Ass. Physiol. Lang. Francaise.
KONORSKI, J., & LUBINSKA, L. Sur un procédé nouveau d'élaboration et réflexes conditionnels du II type et sur les changements d'excitabilité du centre cortical moteur au
cours de l'apprentissage. *Acta Biol. Exp., 13,* 143–152.

1942

KONORSKI, J. Priobretennaya dvigatel'naya deyatel'nost' zhivotnykh s tochki zrenlya
fizjologii (The acquired motor activity of animals from a physiological point of view).
Usp. Sovrem. Biol., 15, 4–26.

1944

KONORSKI, J., & LUBINSKA, L. Skorost' regeneratsii perifericheskikh nervov u krolikov
(The rate of regeneration of the peripheral nerves in rabbits). *Byull. Eksp. Biol. Med.,
18,* 10–12.

1945

KONORSKI, J. Izuchenie vysshei nervnoi deyatel'nosti u kontuzhennykh (Investigations on
higher nervous activity of humans with brain damage). In *Sbornik posvyashchennyi
35-letniyu nauchnoi deyatel'nosti akad.* Tbilisi: D. N. Uznadze pp. 159–172.
KONORSKI, J., & LUBIŃSKA, L. Skorost' regeneratsii perifericheskikh nervov u
mlekipitayushchikh (The rate of regeneration of peripheral nerves in mammals). *Byull.
Eksp. Biol. Med., 19,* 14.

1946

KONORSKI, J. I. P. Pawlow. W. dziesiata recznice śmierci I. P. Pawlowa (I. P. Pavlov. At
the tenth anniversary of I. P. Pavlov's death). *Myśl Wspólczesna, 1,* 63–71.
KONORSKI, J. & LUBIŃSKA, L. Mechanical excitability of regerating nerve-fibres. *Lancet,
250,* 609–612.

1947

KONORSKI, J. Un the summation of the conditioned reflexes. In XVII Int. Physiol. Congr.
(Oxford), Abstr. Commun., pp. 280-282.

1948

KONORSKI, J. Conditioned reflexes and neuron organization. Cambridge: Cambridge
University Press.
KONORSKI, J. K. voprosu ob vneutrennom tormozhenii (Concerning the problem of
internal inhibition). *In Obedinennaya sessiya posvyaschchennaya 10-yu so dnya smerti I.
P. Pavlova.* Moskva: Izdat. AN SSSR, p. 225.
KONORSKI, J. Podstaway fizjologiczne pamieci (The physiological basis of memory).
Mysl Wspólczesna, 3, 215–232. (English summary)

1949

KONORSKI, J. Nowsze badania nad nerwicam doświadczalnyrni (New researches into
experimental neurosis). *Roczn. psychiat., 37,* 389–393. (English summary)
KONORSKI, J. I. P. Pavlov. *Brit. Med. J.* 1949 (2), 944–950.

KONORSKI, J. Pavlov. *Sci. Amer.*, *181*, 44–47.

KONORSKI, J. Znaczenie badań naukowych Pawlowa dla patologii (The importance of Pavlov's scientific research for pathology). *Pol. Tygod. Lek, 4*, 1275–1279.

1950

KONORSKI, J. Mechanisms of learning. *Symp. Soc. Exp. Biol.*, *4* 409–431.

KONORSKI, J. Pawlow jako badacz i twórca fizjologii i patologii wyzszych czynności nerwowych (Pavlov ans an investigator and creator of the physiology and pathology of higher nervous activity). *Acta Physiol. Pol. 1*, 32–48.

KONORSKI, J., & SZWEJKOWSKA, G. Chronic extinction and restoration of conditioned reflexes. I: Extinction against the excitatory background. *Acta Biol. Exp.*, *15*, 155–170.

KONORSKI, J. & SZWEJKOWSKA, G. Zagadnienie chronicznego wygaszania i wznawiania odruchów warunkowych (The problem of chronic extinction and restoration of conditioned reflexes). *Acta Physiol. Pol. 1* (Suppl.), 61–64.II Zjazd Pol. Tow. Fizjol., Warszawa, 1950.

KONORSKI, J., & WYRWICKA, W. Hamowanie nastepcze ruchowych odruchów warunkowych (Inhibitory after-effect of motor conditioned reflexes). *Acta Physiol. Pol. 1* (Suppl.), 64–70. II Zjazd Pol. Tow. Fizjo., Warszawa, 1950.

KONORSKI, J., & WYRWICKA, W. Researches into conditioned reflexes of the second type. 1. Transformation of conditioned reflexes of the first type into conditioned reflexes of the second type. *Acta Biol. Exp. 15*, 193–204.

1951

KONORSKI, J. Niektóre wezlowe zagadnienia fizjologii wyzszych czynności nerwowych (Certain principal problems of the physiology of higher nervous activities). *Pol. Tygod. Lek. 6*, 289–295.

1952

KONORSKI, J. Podstawy fizjologii wyzszych czynności nerwowych (The physiological basis of higher nervous activity). *In* K. Jusowa (Ed.), *Konferencja naukowa poswiecona nauce Pawlowa (Krynica).* Warszawa: PZWL pp.37–67. Dyskusja (Discussion), p. 123; Odpowiedze referentów i koreferentów (Replies of the lecturers and assistant reporters) pp. 152–155.

KONORSKI, J. Zwiazki czasowe jako podstawowa wlasność czynności mózgowej (Temporal connections as the basic property of cerebral activity). *Problemy, 8*, 162–170.

KONORSKI, J. & STEPIEŃ, L. Wplyw czynności pressoreceptoroów zatoke szyjno tetniczej (Sinus caroticus) na mieśnie somatyczne ustroju (The effect of pressoreceptors of the carotid sinus zone on the muscular activity of the animal). *Neurol. Neurochir. Psychiat. Pol. 2*, 522–540. (English summary)

KONORSKI, J., STEPIEŃ, L., BRUTKOWSKI, S., LAWICKA, W. & STEPIEŃ, I. The effect of the removal of the cerebral cortex on the higher nervous activity of animals. *Bull. Soc. Sci. Lett. Lodz, Cl. IV: Sci. Med. 3*(4), 1–5.

KONORSKI, J., STEPIEŃ, L., BRUTKOWSKI, S., LAWICKA, W., & STEPIEŃ, I. Wplyw cześciowego usuwania platów czolowych i ciemieniowych na odruchy warunkowe (The effect of partial removal of the frontal and the parietal lobes of cerebral cortex upon conditioned reflexes). *Neurol. Neurochir. Psychiat. Pol. 2*, 197–210. (English summary)

KONORSKI, J. & SZWEJKOWSKA, G. Chronic extinction and restoration of conditioned reflexes. III. Defensive motor reflexes. *Acta. Biol. Exp. 16*, 91–94

KONORSKI, J. & SZWEJKOWSKA, G. Chronic extinction and restoration of conditioned reflexes. IV. The dependence of the course of extinction and restoration of conditioned reflexes on the "history" of the conditioned stimulus (The principle of the primacy of first training). *Acta Biol. Exp. 16*, 95–113.

KONORSKI, J. & SZWEJKOWSKA, G. Z badań nad ruchliwościa proceśow korowych (On the dynamics of the cortical processes). *Acta Physiol. Pol. 3,* 25-38.

KONORSKI, J. & WYRWICKA, W. Badania nad warunkowymi orduchami analizatora ruchowego. Nastepcze hamowanie warunkowych odruchów analizatora ruchowego (Inhibitory after-effect in conditioned reflexes of the motor analyser). *Acta Physiol. Pol. 3.,* 63-84 (English summary)

1953

KONORSKI, J. & SZWEJKOWSKA, G. O chronicznym wygaszaniu i wznawianiu odruchów warunkowych (On the chronic extinction and restoration of conditioned reflexes). *Acta Physiol. Pol. 4,* 37-51.(English summary)

1955

KONORSKI, J. Kompensacja czynności ruchowych u zwierzat po uszkodzeniach kory mózgowej (Compensation of motor activities in animals after damage to the cerebral cortex). *Zesz. Probl. Nauki Pol, 5,* 46-58. Zagadnienie zastepczości czynności ruchowych. Mat. sesji nauk. komitetu szerzenia nauki Pawlowa, Warszawa, 1955. Referat końcowy (Concluding report), pp. 203-209.

KONORSKI, J. Prace i osiagniecia Zakladu Neurofizjologii Instytutu im. Nenckiego w zakresie fizjologii i patologii wyzszych czynności nerwowych (Work and achievements of the Department of Neurophysiology of the Nencki Institute in the field of physiology and pathology of the higher nervous activity). *Post. Wiedzy Med. 2,* 15-58.

BRUTKOWSKI, S., KONORSKI, J., LAWICKA, W., STEPIEŃ, I., & STEPIEŃ, L. Wplyw usuwania platów czolowych pólkui mózgowych na odruchy warunkowe u psów. (The effect of the removal of prefrontal areas of the cerebral hemispheres on the conditioned motor reflexes in dogs). *Place Lodz. Tow. Nauk. Wydz. III: Nauki Mat.-Przyr, 37,* 1-60. (English summary)

1956

KONORSKI, J. Analiza nadmiernej ruchliwości zwierzat po uszkodzeniach okolic czolowych kory mózgowej (Analysis of hyperactivity of animals after the removal of prefrontal areas of the cerebral cortex). *Neurol. Neurochir. Psychiat. Pol. 6,* 865-873. (English summary)

KONORSKI, J. Vliyanie udaleniya lobnykh dolei bol'shikh polusharii na vysshuyu nervnuyu deyatel'nost' sobak (On the influence of the frontal lobes of the cerebral hemispheres on higher nervous activity in dogs), *In* S. P. Narikashvili (Ed.), *Problemy sovremennoi fiziologii nervnoi i myshechnoi sistemy.* Tbilisi: Izdat, AN Gruzinskio SSR, pp. 343-356. (English summary)

KONORSKI, J. & SZWEJKOWSKA, G. Reciprocal transformations of heterogeneous conditioned reflexes. *Acta Biol. Exp. 17,* 141-165.

BRUTKOWSKI, S., KONORSKI, J., LAWICKA, W., STEPIEŃ, I., & STEPIEŃ, L. The effect of the removal of frontal poles of the cerebral cortex on motor conditioned reflexes. *Acta Biol. Exp., 17.* 167-188.

BRUTKOWSKI, S., KONORSKI, J., LAWICKA, W., STEPIEŃ, I., & STEPIEŃ L. Impairment of inhibitory conditioned reflexes resulting from prefrontal lesions in dogs. *In* XX Int. Physiol. Congr. (Brussels), Abstr. Commun., pp. 1938-1939.

1957

KONORSKI, J. Kierunki rozwoju fizjologii mózgu (Trends in the development of the physiology of the brain). *Nauka pol. 5,* 33-56.(English summary, p. 213)

KONORSKI, J. O giperaktivnosti zhivotnykh posle udaleniya lobnykh dolei bol'shikh polusharii (On the hyperactivity of animals after the removal of the frontal lobes of the brain hemispheres). *In* A. V. Solov'ev (Ed.), *Problemy fiziologii tsentral'noi nervnoi sistemy.* Moskva: Izdat. AN SSSR, pp. 285-293.

1958

KONORSKI, J. Procesy pobudzenia i hamowania w korze mózgowej (Processes of stimulation and inhibition in the cerebral cortex). *Acta Physiol. Pol. 9,* 17-32.

KONORSKI, J. Trends in the development of the physiology of the brain. *J. Ment. Sci. 104,* 1100-1110.

KONORSKI, J. Zagadnienie struktury i funkcji w odniesieniu do kory mózgowej (The problem of structure and function in relation to the cerebral cortex). *Zesz. Probl. Kosmosu, 9,* 35-51. Z zagadnień stosunku miedzy struktura i funkcja mózgu. Mater. Konf. Problemowej PTP im. Kopernika, Warszaka, 1955. Podsumowanie (Conclusion), pp. 115-117.

1959

KONORSKI, J. A new method of physiological investigation of recent memory in animals. *Bull. Acad. Pol. Sci. Ser. Sci. Biol. 7,* 115-117.

KONORSKI, J., & LAWICKA, W. One trial learning versus delayed response in normal and prefrontal cats. *In* XXI Int. Physiol. Congr. (Buenos Aires), Abstr. Commun.

KONORSKI, J., & LAWICKA, W. Physiological mechanism of delayed reactions. I. The analysis and classification of delayed reactions. *Acta Biol. Exp. 19,* 175-197.

ASRATYAN, E. A., GUTMANN, E., & KONORSKI, J. Konferencja neurofizjologów w Osiecznej 9-16 września 1958 (The conference of neurophysiologists in Osieczna, September 9-16, 1958). *Acta Physiol. Pol 10,* 135-139.

ASRATYAN, E. A., GUTMANN E., & KONORSKI, J. Mekhanizmy dvigatel'noi deyatel'nosti zhivotnykh (Mechanisms of motor activity of animals). *Zh. Vyssh. Nerv. Deyat. I. P. Pavlova, 9,* 301-303.

LAWICKA, W., & KONORSKI, J. Physiological mechanism of delayed reactions. III. The effects of prefrontal ablations on delayed reactions in dogs. *Acta Biol. Exp. 19,* 221-231.

SZWEJKOWSKA, G., & KONORSKI, J. The influence of the primary inhibitory stimulus upon the salivary effect of excitatory conditioned stimuli. *Acta Biol. Exp. 19,* 161-174.

ZERNICKI, B., & KONORSKI, J. Fatigue of acid conditioned reflexes. *Acta Biol. Exp. 19,* 327-337.

1960

KONORSKI, J. Faits nouveaux et hypothéses concernant le mécanisme des réflexes conditionnels du deuxieme type. *Psychol. Franc. 5,* 123-134.

KONORSKI, J. On the functional organisation of the frontal lobes in dogs. *Stud. Cerc. Neurol, 5,* 255-264.

KONORSKI, J. The cortical "representation" of unconditioned reflexes. *Electroenceph. Clin. Neurophysiol. Suppl., 13,* 81-89. The Moscow colloquium on elec-troencephalography of higher nervous activity, Moscow, 1958.

KONORSKI, J. Yavlyatsya li otsrochennye reaktsii sledovymi uslovnymi refleksami? (Delayed response or trace conditioned reflex). *Fiziol, Zh, SSSR, 46,* 244-246.

ASRATYAN, E. A., GUTMANN, E. I., & KONORSKI, J. Predislovie (Introduction). *In* E. A. Asratyan (Ed.), *Tsentral'nye i pericheskie mekhanizmy dvigatel'noi deyatel'nosti zhivotnykh. Sbornik dokladov mezhdunarodnogo Simpozyuma, Polsha, Osieczna, 1958.* Moskva: Izdat, AN SSSR, pp. 3-4.

SANTIBANEZ-H, G., TARNECKI, R. ZERNICKI, B., & KONORSKI, J. Korowa reprezentacja struny bebenkowej u psów (Cortical representation of the chorda tympani nerve in dogs). *Acta Physiol. Pol. 11*, 882–883. (VIII Zjazd Pol. Tow. Fizjol. [Poznan] 1960.

STEPIEŃ, I., STEPIEŃ, L., & KONORSKI, J. Dvigatel'nye uslovnye refleksy (II tipa) posle udaleniya senso-motornoi oblasti kory golovnogo mozga u sobak (The effects of the ablation of various parts of the sensory-motor cortex on conditioned reflexes type II in dogs). *In* E. A. Asratyan (Ed.), *Tsentral'nye i pifericheskie mekhanizmy dvigatel'noi deyatel'nosti zhivotnykh.* (Sbornik dokladov mezhdunarodnogo Simpozyuma, Polsha [Osieczna] 1958). Moskva: Izdat. An SSSR, pp. 267–277.

STEPIEŃ, I., STEPIEŃ, L., & KONORSKI, J. The effects of bilateral lesions in the motor cortex on type II conditioned reflexes in dogs. *Acta Biol. Exp., 20*, 211⤳223.

STEPIEŃ, I., STEPIEŃ, L., & KONORSKI, J. The effects of bilateral lesions in the premotor cortex on type II conditioned reflexes in dogs. *Acta Biol. Exp., 20*, 225–242.

STEPIEŃ, I., STEPIEŃ, L, & KONORSKI, J. Znaczenie funkcjonalne okolicy przedruchowej kory mózgowej u psów (The functional role of the premotor region of the cerebral cortex in dogs). *Acta Physiol. Pol. 11*, 886–887.

1961

KONORSKI, J. Analiza patofizjologiczna róznych rodzajów zaburzeń mowy i próba ich klasyfikacji (Pathophysiological analysis of various forms of speech disorders and an attempt of their classification). *Rozpr. Wydz. Nauk Med. PAN, 6* (Vol. II), 6–32. In J. Konorski, H. Koźniewska, L. Stepień, & J. Subszyński (Eds.), *Z zagadnien patofizjologii wyzszych czynnosci nerwowych po uszkodzeniach mozgu u czlowieka.* Slowo wstepne (Introduction), pp. 5–7. (English summary).

KONORSKI, J. Nowsze osiagniecia w dziedzinie organizacji funkcjonalnej kory mózgowej (More recent developments in the field of the functional organization of the brain cortex). *Acta Physiol. Po., 12*,611–629.

KONORSKI, J. The Physiological approach to the problem of recent memory. *In* J. F. Delafresnaye (Ed.), *Brain mechanisms and learning. A symposium.* Oxford; Blackwell Sci. Publ., pp. 115–132.

LAWICKA, W., & KONORSKI, J. The effects of prefrontal lobectomies on the delayed responses in cats. *Acta Biol. Exp., 21*, 141–156.

STEPIEŃ, I., STEPIEŃ, L., & KONORSKI, J. The effects of unilateral and bilateral ablations of sensorimotor cortex on the instrumental (type II) alimentary conditioned reflexes in dogs. *Acta Biol. Exp. 21*, 121–140.

1962

KONORSKI, J. Changing concepts concerning physiological mechanisms of animal motor behaviour. *Brain, 85*, 277–294.

KONORSKI, J. The role of central factors in differentiation. *In* Proc Int. Union Physiol. Sci. XII Int. Congr. (Leiden), Vol. 3, Information processing in the nervous system. A symposium, pp. 318–329.

KONORSKI, J., & DOBRZECKA, C. The effects of incision in the sensori-motor cortex on instrumental conditioned reflexes in dogs. *In* XXII Int. Congr. Physiol. (Leiden), Abstr. Commun., Exerpta Medica, Int. Congr. Ser. 48, Amsterdam, No. 1180.

CHORAZYNA, H., & KONORSKI, J. Absolute versus relative cues in differentiation of tones in dogs. *Acta Biol. Exp., 22* (2), 11–21.

DOBRZECKA, C., & KONORSKI, J. On the peculiar properties of the instrumental conditioned reflexes to "specific" tactile stimuli." *Acta Biol. Exp., 22* (3), 215–226.

LAWICKA, W., & KONORSKI, J. The properties of delayed responses to double preparatory signals in normal and prefrontal dogs. *Acta Biol. Exp., 22* (2), 47–55.

1963

KONORSKI, J. Neurofizjologia (Neurophysiology). *In* J. Hurwic (Ed.), *Encyklopedia Przyroda i Technika. Zagadnienia wiedzy współczesnej.* Warszawa: Weidza Powszechna, pp. 734–741.

KONORSKI, J. Nowe dane dotyczace funkcji okolic czolowych u zwierzat i czloweika (New findings concerning the functioning of the frontal areas in animals and man). *In* IX Zjazd Pol. Tow. Fizjol. (Toruń), Stresz. Ref. Komunikat., P. 7.

KONORSKI, J. W. A. Rosenblith, (Ed.), *Sensory communication.* Massachusetts and New York: MIT Press; Wiley, 1961. *Acta Biol. Exp., 23,* 151–153.

ASRATYAN, E. A. GUTMANN, E., & KONORSKI, J. Introduction. *In* E. Gutmann & P. Hnik (Ed.), *Central and peripheral mechanisms of motor functions.* Proc. Conf., Liblice, 1961). Prague: Publ. House Czech. Acad. Sci., p. 7.

LAWICKA, W., & KONORSKI, J. Analysis of the impairment of delayed response after prefrontal lesions. *In.* E. Gutmann & P. Hnik (Eds.), *Central and peripheral mechanisms of motor functions.* Proc. Conf., Liblice 1961. Prague: Publ. House Czech. Acad. Sci., pp. 123-132.

TARNECKI, R., & KONORSKI, J. Instrumental conditioned reflexes elaborated by means of direct stimulation of the motor cortex;. *In* E. Gutmann & P. Hnik (Eds.), *Central and peripheral mechanism of motor functions.* (Proc. Conf. [Liblice] 1961. Prague: Publ. House Czech. Acad. Sci., pp 177-182.

1964

KONORSKI, J. On the mechanism of instrumental conditioning *Acta Psychol., 23,* 45–59. XVII Int. Congr. Psychol., Washington, 1963.

KONORSKI, J. Some problems concerning the mechanism of instrumental conditioning. *Acta Biol. Exp., 24,* 59–72.

KONORSKI, J., & LAWICKA, W. Analysis of errors by prefrontal animals on the delayed-response test. *In* J. M. Warren & K. Akert (Eds.), *The frontal granular cortex and behavior.* International Symposium, Pennsylvania State Univ., 1962. New York: McGraw-Hill, 1964, pp. 271–294.

ELLISON, G. D. & KONORSKI, J. Separation of the salivary motor responses in instrumental conditioning. *Science, 146,* 1071–1072.

SZWEJKOWSKA, G., LAWICKA, W., & KONORSKI, J. The properties of alternation of conditioned reflexes in dogs. *Acta Biol. Exp., 24,* 135–144.

1965

DOBRZECKA, C., SYCHOWA, B., & KONORSKI, J. The effects of lesions within the sensori-motor cortex upon instrumental response to the "specific tactile stimulus." *Acta Biol, Exp., 25,* 91–106.

ELLISON, G.D., & KONORSKI, J. An investigation of the relations between salivary and motor responses during instrumental performance. *Acta Biol. Exp., 25,* 297–315.

GAMBARIAN, L.G., TARNECKI, R., KONORSKI, J. Effects of cerebellectomy on cortical action potentials evoked by stimulation of muscular nerves in cats. *Bull. Acad. Pol. Sci. Sér. Sci. Biol.,* 13, 373–376.

1966

KONORSKI, J. Stefan Brutkowski, 1924-1966. Obituary. *Acta Biol. Exp., 26,* 373–374.

KONORSKI, J. & ELLISON, G. Vzaimootnosheniya slyunnoi i dvigatel'noi reaktsii v uslovnykh refleksakh II-go tipa (Interrelationships between salivary and motor reactions in type II conditioned reflexes). In E. A. Asratyan (Ed.), *Nervnye mekhanizmy dvigatel'noi deyatel'nosti.* Tretii mezhdunarodyni Simpozyum, Dilizhan, 1964. Moskva: Izdat. Nauka, pp. 351-356.

ASRATYAN, E. A., GUTMANN, E. I., & KONORSKI, J. Predislovie (Introduction). *In* E. A. Asratyan (Ed.), *Nervnye mekhanizmy dvigatel'noi deyatel'nosti.* Tretii mezhdunarodyni Simpozyum, Dilizhan, 1964. Moskva: Izdat. Nauka, pp. 5-6.

DOBRZECKA, C., SYCHOWA, B., & KONORSKI, J. Vliyanie povrezhdenii sensomotornoi oblasti kory bol'shikh polusharii na uslovnye refleksy II tipa iz raznykh analizatorov (The influence of damage to the sensori-motor cortex on type II conditioned reflexes established to stimuli of various modalities). *In* E. A. Asratyan (Ed), *Nervnye mekhanizmy dvigatel'noi deyatel'nosti.* Tretii mezhdunarodnyi Simpozyum, Dilizhan, 1964. Moskava: Izdat. Nauka, pp. 216-222.

DOBRZECKA, C., SZWEJKOWSKA, G., & KONORSKI, J. Qualitative versus directional cues in two forms of differentiation. *Science, 153,* 87-89.

ELLISON, G. D., & KONORSKI, J. Salivation and instrumental responding to an instrumental CS pretrained using the classical conditioning paradigm. *Acta Biol. Exp., 26,* pp. 159-165.

SOLTYSIK, S., & KONORSKI, J. Relations between classical and instrumental conditioning. *In* XVIII Int. Congr. Psychol. (Moscow), Symp. 4, Classical and instrumental conditioning, pp. 66-73.

1967

KONORSKI, J. *Integrative activity of the brain: An interdisciplinary approach.* Chicago: University of Chicago Press.

KONORSKI, J. Introduction to the symposium "The functional properties of hypothalamus." *Acta Biol. Ex., 27,* 265-267.

KONORSKI, J. Pavlov-uchenyi i chelovek (Pavlov-scientist and man). *In* I. P. Pavlov v vospominaniyakh sovremennikov.

KONORSKI, J. Obecny stan badán nad czynnóscia mózgu i perspektywy ich rozwoju w Polsce (The present stage of the brain research and its perspectives in Poland). *Kosmos, Ser. A, 17,* 225-235.

KONORSKI, J. Some new ideas concerning the physiological mechanisms of perception. *Acta Biol. Exp., 27,* 147-161.

KONORSKI, J. The physiological mechanism of perseveration. *In* J. Choróbski (Ed.), *Neurological problems.* Oxford, Warsawá: Pergamon Press; Pol. Sci. Publ. pp. 83-91.

DOBRZECKA, C., & KONORSKI, J. Qualitative versus directional cues in differential conditioning. I. Left leg-right leg differentiation to cues of a mixed character. *Acta Biol. Ex;. 27,* 163-168.

1968

KONORSKI, J. A review of the brain research carried out in the Department of Neurophysiology of the Nencki Institute of Experimental Biology. *Acta Biol. Exp., 28,* 257-289.

KONORSKI, J. Badania w dziedzini fizjologii mózgu (Investigations in the field of physiology of the brain). In *50 lat dzialalności Instytutu Biologii Doświadczalnej im. M. Nenckiego 1918#1968.* Warszawa: PWN, pp. 25-59. (English summary).

KONORSKI, J. *Conditioned reflexes and neuron organization.* New York: Hafner. (Facsimile reprint of the 1948 edition, with a new foreword and supplementary chapter).

KONORSKI, J. Organizacja funkcjonalna analizatoró́w (Fizjologia percepcji) (The functional organization of analysers [The physiology of perception]). *Zesz. Nauk, Uniw. Jagiellońsk., Prace Psychol.-Pedagog., 13,* 23-27. (English summary).

KONORSKI, J. Zasady neurofizjologicznych mechanizmów percepcji (Neurophysilogical basis of the mechanisms of perceptions). *Studia Psychol., 9,* 5-21. (English summary)

KONORSKI, J. SANTIBANEZ-H. G., & BECK, J. Electrical hippocampal activity and heart rate in clasical and instrumental conditioning. *Acta Biol. Exp., 28,* 169-185.

DOBRZECKA, C., & KONORSKI, J. Qualitiative versus directional cues in differential conditioning. IV. Left leg-right leg differentiation to non-directional cues. *Acta Biol. Exp., 28,* 61-69.

1969

KONORSKI, J. Developmental pathways of research on brain-behavior interrelations in animals. *Acta Biol. Exp., 29,* 239-249.

KONORSKI, J. Integracyjna dzialalność mózgu (Integrative activity of the brain). Warszawa: PWN.

KONORSKI, J. Ogólne uwagi na temat ośrodkowej regulacji funkcji ukladów. Podsumowanie sympozjum (General notes on the central regulations of functions of systems. A summation of the symposium). In *XI Zjazd Pol. Tow. Fizjol.*(Szczecin), Streszcz. Ref. Komunikat. Warsczawa: PZWL.

KONORSKI, J. The sixth Gagra conference, January 13-25, 1969. "On the problem of memory." *Acta Biol. Exp., 29,* 227-228 (Miscellanea)

GLAVCHEVA, L., ROZKOWSKA, E., & KONORSKI, J. Effects of alimentary reflexes on motor gastric activity. *Acta Biol. Exp., 29,* 63-74.

MILLER, S., & KONORSKI, J. On a particular type of conditioned reflex. *J. Exp. Anal. Behav., 12,* 187-189.

TARNECKI, R., & KONORSKI, J. Instrumental conditioning of thalamogenic movements and its dependence on the cerebral cortex. *Acta Biol. Exp., 29,* 17-28.

TARNECKI, R., & KONORSKI, J. Thalamogenic movements in cats: Their characteristics and dependence on the celebral cortex. *Acta Biol. Exp., 29,* 1-15.

1970

KONORSKI, J. Integrative activity of the brain: An interdisciplinary approach (2nd ed.). Chicago: University of Chicago Press.

KONORSKI, J. Integrativnaya deyatel'nost' mozga (Integrative activity of the brain). Izdat. Mir, Moskva.

KONORSKI, J. Pathophysiological mechanisms of speech on the basis of studies on aphasia. *Acta Neurobiol. Exp., 30,* 189-210.

KONORSKI, J. Pavlov and contemporary physiological psychology. *Cond. Reflex, 5,* 241-248.

KONORSKI, J. Rozwój pogladów na mechanizmy fizjologiczne mowy w świetle badań nad afazja (The development of concepts of physiological mechanisms of speech in view of studies on aphasia). In S. Zarski (Ed.), *Zagadnienia patofizjologii wyzszych czynnoś*ci nerwowych po uszkodzeniach mózgu. Vol. 1. Warszawa: PZWL, pp. 5-35. (English summary)

KONORSKI, J. The problem of the peripheral control of skilled movements. *Int. J. neurosci, 1,* 39-50.

KONORSKI, J., & GAWRONSKI, R. An attempt at modelling of the central alimentary system in higher animals. I. Physiological organization of the alimentary system. *Acta Neurobiol. Exp. 30,* 313-337.

KONORSKI, J., & GAWRONSKI, R. An attempt at modelling of the central alimentary system in higher animals. IV. Experiments on classical conditioning. *Acta Neurobiol. Exp., 30,* 371-395.

KONORSKI, J., & GAWRONSKI, R. An attempt at modelling of the central alimentary system in higher animals. V. Instrumental conditioned reflexes. *Acta Neurobiol. Exp., 30,* 397–414.

GAWRONSKI, R., & KONORSKI, J. An attempt at modelling of the central alimentary system in higher animals. II. Technical description of the arrangements involved in modelling. *Acta Neurobiol. Exp., 30,* 333–346.

GAWRONSKI, R., & KONORSKI, J. An attempt at modelling of the central alimentary system in higher animals. III. Some theoretical problems concerning identification and modelling of the simple neural structures. *Acta Neurobiol. Exp., 30,* 347–370.

KONORSKI, J., & TARNECKI, R. Purkinje cells in the cerebellum: Their responses to postural stimuli in cats. *Proc. Nat. Acad. Sci. USA, 65,* 892–897

TARNECKI, R., & KONORSKI, J. Patterns of responses of Purkinje cells in cats to passive displacements of limbs, squeezing and touching. *Acta Neurobiel. Exp., 30,* 95–119.

1971

KONORSKI, J. Higher control of behavioral acts by the prefrontal cortex in animals. In *Proc. Int. Union Physiol. Sci. XXV Int. Congr.* (Munich). Vol. 8. Abstr. Lectures and Symp., pp. 165–168.

KONORSKI, J., ZERNICKI, B., & STAJUDOWA, E. Index and bibliography of papers devoted to brain researches published in *Acta Biologiae Experimentalis* (since 1970, *Acta Neurobiologiae Experimentalis)* in years 1950–1971. *Acta Neurobiol. Exp. Suppl. 1.*

KONORSKI, J. Zasady funkcjonowania ośrodkowego ukladu nerwowego (The principles of functioning of the central nervous system). In L. Woloszynowa (Ed.), *Materialy do nauczania psychologii.* Vol. 4. Ser. I. Psychologia ogólna. Warszawa: PWN, pp. 11–44.

1972

KONORSKI, J. Problema pamyati v fiziologicheskom aspekte (The memory problem in the physiological aspect). *Gagr. Besedy, 6,* 37–56. (English summary)

KONORSKI, J. & TEUBER, H. L. Preface. *In* J. Konorski, H.-L. Teuber, & B. Zernicki (Ed.), *The frontal granular cortex and behavior. International Symposium, Jablonna near Warszawa, 1971. Acta Neurobiol. Exp., 32,* 119–120.

KONORSKI, J. Some hypotheses concerning the functional organization of prefrontal cortex. *Acta Neurobiol. Exp., 32,* 595–613. J. Konorski, H.-L. Teuber, & B. Zernicki (Eds.), *The frontal granular cortex and behavior.* International Symposium, Jablonna near Warszawa, 1971.

KONORSKI, J. Some ideas concerning physiological mechanisms of so-called internal inhibition. In. R. A. Boakes & M. S. Halliday (Eds.), *Inhibition and learning.* London: Academic Press, pp. 341–357.

KONORSKI, J. (The importance of brain physiology for social sciences). *Znaczenie fizjologii mózgu dla nauk spoleczynch* Nauka Pol., 4, 54–60.

CHAMBERS, W. W., KONORSKI, J., LIU, C. N. & ANDERSON, R. The effects of cerebellar lesions upon skilled movements and instrumental conditioned reflexes. *Acta Neurobiol, Exp., 32,* 721–732.

DOBRZECKA, C., KONORSKI, J., STEPIEN, L., & SYCHOWA, B. The effects of the removal of the somatosensory areas I and II on left-right leg differentiation to tactile stimuli in dogs. *Acta Neurobiol. Exp. 32,* 19–33.

1973

KONORSKI, J. In memory of Professor Mieczyslaw Minkowski 1884–1972. *Acta Neurobiol. Exp., 33,* 659–661. (Obituaries)

KONORSKI, J. K. S. Abuladze, 1895-1972. *Acta Neurobiol. Exp., 33,* 661. (Obituaries)

KONORSKI, J. K voprosu neirofiziologicheskikh mekhanizmov pertseptsii (On the neurophysiological mechanisms of perception). *In* A. Prangishvili (Ed.), *Psikhologicheskie issledovaniya posvyashchennye 85-letiyu so dnya rozhdeniya D. H. Uznadze.* Tbilisi: Izdat. Metsniereba, pp. 199-204.

KONORSKI, J. Nekotorye idei, kasayushchiesya fiziologicheskikh.mekhanizmov vnutrennogo tormozheniya. (Some concepts about physiological mechanisms of internal inhibition) In V. S. Rusinov (Ed.), Mekhanizmy formirovaniya i tormozheniya uslovnykh refleksov. Moskva: Izdat. Nauka, pp. 241-256.

KONORSKI, J. On two types of conditional reflex: General laws of association. *Cond. Reflex, 8,* 2-9.

KONORSKI, J. The role of prefrontal control in the programming of motor behavior, *In* J. D. Maser (Ed.), *Efferent organization and the integration of behavior.* New York: Academic Press, pp. 175-201.

KONORSKI, J., BUDOCHOSKA, W. CELINSKI, L., & SZYMANSKI, L. Analysis of perception of complex visual stimulus-patterns. *Acta Neurobiol. Exp., 33,* 497-507.

DOBRZECKA, C., & KONORSKI, J. The effect of dissection of corpus callosum on differentiation of instrumental reflexes to symmetrical tactile stimuli in dogs. *Acta Neurobiol. Exp., 33,* 543-551.

YU, J., TARNECKI, R., CHAMBERS, W. W., LIU, C. N. & KONORSKI, J. Mechanisms mediating ipsilateral limb hyperflexion after cerebellar paravermal cortical ablation or cooling. *Exp. Neurol., 38,* 144-156.

1974

KONORSKI, J. Classical and instrumental conditioning: The general laws of connections between "centers." *Acta Neurobiol. Exp. 34,* 5-13. In R. W. Doty, J. Konorski, & B. Zernicki (Eds.), *Brain and behavior.* International Symposium, Jablonnna near Warsawa, 1972.

KONORSKI, J. Conditioned reflex. In *Encyclopedia of XX Century.* Roma: Instituto dell'Enciclopedia Italiana, in press.

KONORSKI, J. "Jerzy Konorski." In G. Lindzey, (Ed.), A history of psychology in autobiography. Vol. 6, New York: Appleton-Century-Crofts.

KONORSKI, J. Study of behavior: Science or pseudoscience. Unpublished manuscript, Nencki Inst. Exp. Biol., Warsaw.

BUDOHOSKA, W., KONORSKI, J. & CELINSKI, M. Perception of competing visual patterns. *Pol. Psychol. Bull., 5,* 59-65.

1975

ZABLOCKA, T., KONORSKI, J., & ZERNICKI, B. Visual discrimination learning in cats with different early visual experiences. *Acta Neurobiol. Exp. 35,* 387-398.

TARNECKI, R., YU, J., LIU, C. N., KONORSKI, J., & CHAMBERS, W. W. The effect of cerebellar lesions on instrumental responses executed against resistance. *Acta Neurobiol. Exp. 35,* 677-698.

1976

SOLTYSIK, S., KONORSKI, J., HOLOWNIA, A., & RENTOUL, T. The effect of conditioned stimuli signalling food upon the autochthonous instrumental responses in dogs. *Acta Neurobiol. Exp., 36,* 277-310.

1977

KONORSKI, J. Autobiografia. *Kwart. Hist. Nauki i Tech., 22,* 215-250.

Author Index

455

Subject Index

A

Antidrive, 87, 218, 248, 321–323.
  and prefrontal cortex, 323.
Appetitive-aversive interactions, 213, 226–228, 237.
Appetitive behavior, 203.
Associations, *see also* Classical conditioning, Inhibitory conditioning, Stimulus-response (S-R) association.
  expectation method, 58, 350.
  nature of, 162–167, 337–338, 361–362.
  methods for the study of, 338–360.
  silent, 340–342, 360.
  substitution method, 58, 350.
Associative learning *see* Classical conditioning, Inhibitory conditioning, Instrumental conditioning.
Attention, *see also* Latent inhibition and neocortex, 117.
  neuropharmacology of, 126–128.
  and selective association, 377, 391.
  and septum, 122.
Attractive stimulus, 203.
  and defensive responses, 206–207.
Autocontingencies, 340.
Autoshaping, 32–47, *see also* Goal tracking, Observational learning, Omission training, Reinforcement, Sign tracking, Stimulus substitution.

and classical conditioning, 34.
with diffuse CS, 242–244.
direction and form of response, 238–247.
effect of partial reinforcement, 240–241.
procedure in, 32.
and specification of UR, 39–41.
Aversive stimulus, 204
  and appetitive responses, 207.
Avoidance conditioning, 285, *see also* Stimulus-reinforcer interactions and inhibition, 12.
  Konorski's early work on, 11–12.
  passive, *see* Punishment.

B

Backward conditioning, 164, 437–438.
Behavioral contrast, 260–262.
Bidirectional conditioning, 163–165, 178.
Blocking
  with a CS, 29, 61–62, 218–219, 339–340.
  CS relevance and salience, 392–394.
  and hippocampus, 124–126.
  in instrumental conditioning, 156–158.
  with a US, 62–65.
  transreinforcer, 219–220.

C

Classical conditioning, *see also* Autoshaping, Contiguity, Classical-instrumental interactions, Drive, Eyelid conditioning, Inhibitory conditioning, Instrumental conditioning, Reinforcement, Stimulus substitution, US representation.
associations in, 339–346.
associative theories of, 27–30, 424–425.
classical CRs recorded in instrumental experiments, 152.
and consummatory-preparatory distinction, 22–23, 44–46, 210–212, 244–247.
and hippocampus, 121–122, 134–136.
Konorski's theory of, 20–32, 210–213, 424.
and mass action, 119–120.
and neocortex, 116–120.
relation to instrumental conditioning, 9, 46–47.
Classical-instrumental interactions, 235–238, *see also* Conditioned suppression, Transfer
Communication, 258–260.
Compound stimuli, 349–360, *see also* Differential conditioning, Generalisation
Conditioned emotional response (CER), *see* Conditioned suppression
Conditioned inhibition, *see* Inhibitory conditioning
Conditioned suppression
with attractive CS, 12–13, 181, 247–254.
with aversive CS, 236–238, 301–310.
and drugs, 303–310.
and hippocampus, 304.
and septum, 304–310.
Constraints on conditioning, 368, 399–400, *see also* CS-US similarity, Stimulus-reinforcer interactions, Response-reinforcer interactions
Konorski's theory of, 400.
Consummatory responses, 22–23, 44–46, 210, 244–247, *see also* Classical conditioning
Contiguity, 28–30.
Counterconditioning, 213–217.
CS-US similarity, 383–385.

D

Defensive behavior, 204.
Discrimination learning, 258–259, *see also* Differential conditioning.
Differentiation, *see* Differential conditioning.
Differential conditioning, 279–289, *see also* Orientating reactions, Stimulus intensity
with compound stimuli, 351–360.
cue specificity in left-right differentiation, 404–406.
with different drives, 182–184.
with different reinforcer, 184–185.
Konorski's definition of, 279.
Pavlovian differentiation, 281–283.
and prefrontal cortex, 287–288, 317–325.
and septum, 122–123.
with symmetrical and asymmetrical reinforcement, 283–287, 318–319, 325–335.
with tactile stimuli, 403.
Discriminative stimulus, 150–151, *see also* Transfer
Disinhibition, 98.
Drive, 176–177, 179–186, 210–213, 247, *see also* Antidrive, Differential conditioning, Motivation, Preparatory responses
conditioned, 21, 189–190.
inhibition of, 23
role in classical conditioning, 21–23.
shifts, 185–186.
Drugs, *see* Conditioned suppression, Inhibitory conditioning, Punishment.

E

Extinction, 97–108, 270–279, *see also* Latent inhibition, Partial reinforcement extinction effect (PREE), Satiation, Stimulus intensity, Spontaneous recovery
acute and chronic, 273–274.
of conditioned inhibition, 94.
enhancement of, 221–223.
nonassociative factors in, 99–104.
reinstatement of responding after, 99–101, 180, 274.

K

Konorski, *see also* Avoidance conditioning, Classical conditioning, Constraints on conditioning, Differential conditioning, Habituation, Inhibitory conditioning, Instrumental conditioning, Memory, Motivation, Omission training, Perception, Reinforcement, Stimulus-reinforcer interactions.
   influence in West, 2-5, 8.
   life of, 1-2, 10, 15-16.
   and Pavlov, 5, 7, 15-16.

L

Latent inhibition, 65-68, 347-349.
   and cholinergic systems, 127.
   extinction of, 66-67.
   and hippocampus, 123.
   and serotonergic systems, 127.
Law of effect, 147-148, *see also* Instrumental conditioning.
Learned irrelevance, 343.

M

Mass action, *see* Classical conditioning.
Memory, 420.
   Konorski's theory of, 427.
   long-term (associative), 57-58.
   short-term (transient), 58-59, 428-437.
     capacity of, 64.
     and conditioning, 437-439.
     priming of, 60-77.
Misbehavior, 256-257.
Motivation, *see also* Antidrive, Appetitive-aversive interactions, Differential conditioning, Drive, Ideo-motor theory, Satiation, stimulus-response (S-R) associations, US representation.
   and appetitive system, 212, 226-228.
   associative theories of, 171-175.
   and aversive system, 210-212, 226-228.
   Konorski's theory of, 175-179, 210-213.

N

Neocortex, *see* Classical conditioning
Neophobia, 381.

O

Observational learning, 43.
Omission training, 146-147, 286, 296.
   and autoshaping, 41-42, 238-239.
   Konorski's early work on, 10.
Opponent-process, 213.
Orienting (targeting) reaction, 55
   in differential conditioning, 403.
Overshadowing, *see* Instrumental conditioning.

P

Partial reinforcement, *see* Autoshaping, Partial reinforcement extinction effect (PREE), Satiation.
Partial reinforcement extinction effect (PREE), *see also* US representation
   and septum, 310-311.
   in response chains, 193-194.
Passive avoidance, *see* Punishment.
Perception
   and conditioning, 419-422.
   Konorski's theory of, 20-21, 55, 425.
Perceptual judgments, 75-77.
Prefrontal cortex, *see* Differential conditioning
Preparatory responses, 21-22, 44-46, 211, 244-247,
   in selective association, 374-375.
Preservative reflexes, 203.
Protective reflexes, 203.
Punishment, 296, 301-310.
   and drugs, 301.
   and hippocampus, 301-303.
   and septum, 302.

R

Reciprocal inhibition, 213.
Reinforcement, *see also* Instrumental condi-

tioning, Partial reinforcement, Schedules of reinforcement.
in autoshaping, 41–44.
in classical conditioning, 30–32.
Konorski's theory of, 30.
parasitic or superstitious, 31, 41–44, 234.
Releasing stimuli, 179.
Reticular formation, *see* Eyelid conditioning.
Response-reinforcer interactions, 255–256.

S

Satiation
and acquisition, 195–197.
comparisons to extinction, 191–195.
and partial reinforcement, 144–195.
and dissociations between instrumental
    and consummatory responding,
    188–190.
resistance to, 187–194.
and response chains, 190–193.
Schedules of reinforcement, 257–258.
Second-order conditioning, *see* Higher-order
conditioning.
Septum, *see* Attention, Conditioned suppression, Differential conditioning, Inhibitory conditioning, Partial reinforcement
extinction effect (PREE), Punishment.
Sensitization, 373.
Sensory preconditioning, 347.
Sign tracking, 32, 238–239.
Simultaneous conditioning, 343–344, 437–
438.
Spontaneous recovery, 98, 101–102.
Stimulus intensity, 276–278.
in differential conditioning, 288–289.
in extinction, 277.
in inhibition of delay, 277–278.
in stimulus-reinforcer interaction, 371–
372.
Stimulus-reinforcer interactions, *see also*
Stimulus intensity, Taste-aversion conditioning
in avoidance conditioning, 385–390.
Konorski's theory of, 371–376.
methods of study, 368–371, 376–378.
in reward conditioning, 385–391.
Stimulus-response (S–R) associations.
in instrumental conditioning, 149, 163.
role in motivation, 179.

Stimulus substitution, 30, 35–39, 145, 147–
148, 244.
Superconditioning
with appetitive excitor, 209–210.
with aversive inhibitor, 208.

T

Taste-aversion conditioning, 378–382, *see
also* Interstimulus (CS–US) interval
Transfer
between CS and discriminative stimuli,
251
between discriminative stimuli, 180,
402–403.
Type I conditioning, *see* Classical conditioning
Type II conditioning, *see* Instrumental conditioning
Two-factor theory, 235.

U

US representation (center), *see also* Higher-
order conditioning.
in classical conditioning, 186, 212.
in instrumental conditioning, 186–187.
role in motivation, 177–178, 179–187.
in partial reinforcement extinction effect (PREE), 194.